Second Edition

Managing
Turfgrass Pests

Second Edition

Managing Turfgrass Pests

Thomas L. Watschke
Peter H. Dernoeden
David J. Shetlar

CRC Press
Taylor & Francis Group
Boca Raton London New York

CRC Press is an imprint of the
Taylor & Francis Group, an **informa** business

CRC Press
Taylor & Francis Group
6000 Broken Sound Parkway NW, Suite 300
Boca Raton, FL 33487-2742

First issued in paperback 2017

© 2013 by Taylor & Francis Group, LLC
CRC Press is an imprint of Taylor & Francis Group, an Informa business

No claim to original U.S. Government works

ISBN-13: 978-1-4665-5507-5 (hbk)
ISBN-13: 978-1-138-07637-2 (pbk)

Library of Congress Cataloging-in-Publication Data

Watschke, Thomas L.
    Managing turfgrass pests / Thomas L. Watschke, Peter H. Dernoeden, David J. Shetlar. -- Second edition.
        pages cm
        "A CRC title."
        Includes index.
        ISBN 978-1-4665-5507-5 (hardcover)
        1. Turfgrasses--Weed control. 2. Turfgrasses--Diseases and pests--Control. 3. Turf management. I. Dernoeden, Peter H. II. Shetlar, David J. III. Title.

SB608.T87W37 2013
632'.5--dc23   2012050934

Visit the Taylor & Francis Web site at
http://www.taylorandfrancis.com

and the CRC Press Web site at
http://www.crcpress.com

Contents

Preface to the Second Edition

Turfgrass pests have always caused a decline in the playability and aesthetic quality of turfgrasses throughout the world. Turfgrass breeders are in constant pursuit of developing cultivars that are more pest resistant. In addition, the development of pest control products is also a constant pursuit by industry leaders. Consequently, changes in the manner in which pests are managed evolve in response to these developments on the commercial side. As a result, the need to continually update information regarding the way in which pests are managed is critical. This second edition is intended to reinforce those management tactics that are still valuable, and to also introduce new approaches and discuss the new materials that have been introduced to the market since the first edition.

As in the first edition of this book, the concept of integrated pest management, which most turf managers employ regularly, is discussed, and the philosophy of minimizing pests through well-defined and implemented cultural systems is emphasized. Rather than relying on a particular pesticide solution for "control," successful pest management must include the manipulation and fine tuning of cultural practices, so they better address the question of why the pest is present in the first place. Once these finely tuned cultural practices are in place, any pesticide that is still required will be much more effective with respect to controlling the pest.

Inevitably, pests become part of any ecosystem and managing turfgrasses to optimize their competitiveness is the essence of any pest management program.

Preface to the First Edition

Turfgrass pests frequently cause a decline in turfgrass quality (both aesthetically and functionally). The decline in quality can be severe enough that the turfgrass stand becomes sufficiently damaged to require costly renovation. The degree of turf deterioration is a function of the grass species, type of pest, and the environmental conditions before, during, and following the pest involvement.

Often, pests occur in combination and/or as a result of one another. For example, a disease may weaken a turfgrass plant, making it more susceptible to damage from an otherwise subthreshold level of a particular insect. The resulting stand thinning due to the insect activity could allow weed species to germinate and invade the damaged area. Crabgrass levels are frequently proportional to the amount of foliar blighting associated with common springtime diseases. Such a scenario would be aggravated by the existence of adverse environmental conditions, which is often the case.

Inevitably, pests become part of any ecosystem and managing turfgrasses to optimize their competitiveness is the essence of any pest management program. The presence of pests in a turfgrass stand is usually an excellent indicator that management effectiveness has broken down somewhere. An overall examination of the interactive components of the management system is required before an effective pest management strategy can be implemented. Rather than rely on a particular pesticide for control, successful pest management must include the manipulation and fine tuning of cultural practices so they better address the question of why the pest is present in the first place.

Most turfgrass pest control literature relies heavily on appropriate pesticide use for dealing with pests. While pesticides are an important tool in devising sound pest management programs, they should only be used as a supplement to well-coordinated cultural programs. The concept of integrated pest management has been practiced by turfgrass managers for decades, and this book will emphasize the philosophy of minimizing pests through well-defined and organized cultural practices.

These practices may include changes in mowing height and/or frequency, proper irrigation procedures, renovation to convert to resistant species and/or cultivars, adjustments to nitrogen source as well as fertilizer rate and/or timing of application, improved surface and/or subsurface drainage, soil modification, soil pH adjustment, compaction and traffic control, and others. Without well-conceived pest management and cultural programs, the conditions that lead to pest occurrence may not be altered and the pest problem will persist. Without properly adjusting cultural practices, any pesticide solution to a pest problem will not be as efficacious as it could be, and is unlikely to be permanently successful. Unfortunately, when a particular pesticide fails to adequately control a pest, turf managers often are inclined to select an alternative pesticide rather than assessing potential cultural flaws in their management system.

Pesticides are indispensable tools for many pest management situations, particularly when no alternative exists or the pest situation is so bad that a "pesticidal rescue" is required. Most pesticides provide very effective pest control, and their use has evolved in some situations to the point where they are the first line of defense in a pest control program. However, pesticide effectiveness is markedly enhanced when all possible cultural solutions have been imposed. Pesticides often temporarily reduce pest problems, but sometimes the benefits are merely cosmetic. The underlying mismanagement problems that have allowed the pest to become a serious problem in the first place must be resolved. Hence, mismanagement is often the indirect cause of the presence of a pest, and only after corrective changes in management occur will the effectiveness of any pesticide be maximized.

This book has been written in a manner that emphasizes cultural systems. It contains specific recommendations for each pest with respect to integrating all available management tactics. While pesticides are considered a necessary component of managing turf grass pests, they are referenced in this book in the context of being part of the overall strategy. Proper management systems do not always preclude the need for pesticides, but do enhance their effectiveness and reduce the amount of their use.

To increase the global usefulness of this book, references to pest solutions are made using turf grass species (warm or cool season) rather than any specific geographic location in the world. However, some pests unique only to a specific region are discussed. It is hoped that any reader will be able to identify with the management recommendations proposed based upon the turfgrass species being managed. This book is intended to assist turfgrass managers in achieving successful pest control through integrating management strategies.

Authors

Dr. Thomas L. Watschke is presently professor emeritus of turfgrass science at the Pennsylvania State University, where he was on the faculty for 35 years. He obtained a BS from Iowa State University in 1967. He received his MS and PhD from Virginia Polytechnic Institute and State University in 1969 and 1971, respectively. He was an NDEA Fellow at the Virginia Polytechnic Institute and State University between 1967 and 1970. He was an assistant professor at the Pennsylvania State University from 1970 to 1975. Later, he served as an associate professor in the same university between 1975 and 1981. He was upgraded as a professor and continued to work at the Pennsylvania State University from 1981 to 2005. From 2005 to the present, he is professor emeritus in the same university. He taught nine courses in turfgrass management and coordinated the undergraduate degree program in turfgrass science at the Pennsylvania State University. He served as the director of the Center for Turfgrass Science in the College of Agricultural Sciences.

Dr. Watschke's principal areas of research include turfgrass weed control, growth regulation of turfgrasses, effects of fertilizers on turfgrass growth and metabolism, turfgrass physiology and microclimate, roadside vegetation management, and water quality (both runoff and leachate).

Dr. Watschke has been honored nationally by the Golf Course Superintendents Association of America and the American Sod Producers Association, Division C-5 (ASA, CSSA) Grau Award, and has been accorded fellow status by both the Crop Science Society of America and the American Society of Agronomy. He is recognized throughout the world for his teaching and research accomplishments in weed science, plant growth regulation, and water quality. He has made presentations in France, Australia, Spain, England, Scotland, and several locations in Canada and is active in the International Turfgrass Society.

Dr. Peter H. Dernoeden is a professor of turfgrass science in the Department of Plant Science and Landscape Architecture at the University of Maryland, which he joined in 1980. He earned his BS and MS from Colorado State University and his PhD from the University of Rhode Island. He served in the U.S. Army Field Artillery from 1970 to 1973 in West Germany. Dr. Dernoeden's appointment includes research and extension components, and he teaches a course in pest management strategies for turfgrasses. His research and extension programs involve turfgrass pathology, weed science, and the management of turfgrasses for low-mainte-nance sites. He has a very active integrated pest management research program, which has focused on the impact of mowing, irrigation, and fertility on disease severity and weed encroachment. His other research interests include the biology of turfgrass pathogens, etiology of diseases, and chemical control of diseases and weeds. Dr. Dernoeden initiated a laboratory service for turfgrass disease diagnostics in 1980. The laboratory has proved an invaluable teaching tool since he and his graduate students conduct all diagnoses.

Dr. Dernoeden has published more than 100 scientific journal articles, most of which were coauthored with his graduate students. He is the author of *Creeping Bentgrass Management* (CRC Press, 2013) and the coauthor of *Compendium of Turfgrass Diseases* (American Phytopathological Society, 2005). He is a Fellow of the American Society of Agronomy and the Crop Science Society of America (CSSA). He received the Fred V. Grau Turfgrass Science Award from the Turfgrass Science Division of CSSA. He was also the recipient of The Dean Gordon Cairns Award for Distinguished Creative Work in Agriculture from the College of Agriculture and Natural Resources at the University of Maryland. In 2012, he received the Colonel John Morley Distinguished Service Award from the Golf Course Superintendents Association of America.

David J. Shetlar obtained his BS and MS in zoology from the University of Oklahoma in 1969 and 1971, respectively. He obtained his PhD from the Pennsylvania State University in 1977 working on the systematics (taxonomy) of a group of neuropterous insects (lacewings, ant lions, etc.). He then joined the faculty at the Pennsylvania State University to teach beginning entomology and performed research on nursery, green-house, and Christmas tree pests. In 1984, he joined Chemlawn Services R&D in Delaware, Ohio as a research scientist working on turfgrass insect man-agement. For the next six years, he traveled extensively throughout the United States to work on regional turfgrass insect problems. In 1990, he joined the faculty at the Ohio State University and has risen to the rank of professor of urban landscape entomology.

Dr. Shetlar has authored and coauthored numerous trade magazine articles, research journal articles, books and book chapters, and extension factsheets and bulletins. His research emphasis has been on developing alternative materials and methods for management of turfgrass insects, investigating the biology of the bluegrass billbug, hairy chinch bug, and sod webworms, and investigating the biodiversity of the turfgrass ecosystem.

Dr. Shetlar, who goes by the professional nickname of the "BugDoc," pro-duces the popular *P.E.S.T. Newsletter* in association with the Ohio Nursery and Landscape Association. He was one of the recipients of an Annual Leadership Award presented by Lawn and Landscape and Bayer in 2005. He also received the Educator & Public Service Award from the Ohio Nursery and Landscape Association in 2010.

1

Weeds and Their Management

Introduction

Over the years, much has been written about the management of weeds that are pests in various turfgrass situations. Since weeds are defined as plants out of place, some plant species can, simultaneously, be both weeds and desired plants, depending on the specific situation. Much of the early weed management information, which is prevalent to turfgrasses, emphasized cultural controls that combined the best management practices and available mechanical techniques to provide the turf with a competitive advantage over the weeds. As the chemical age developed, turfgrass weed control took on another dimension (herbicide use). Initially, herbicides used in turfgrass weed control were nonselective, mostly acids, such as hydrochloric and euphoric that did little to advance the state of the art of chemical weed control. However, after the introduction of 2,4-D (2,4 di chlorophenoxyacetic acid) in 1944, chemical control quickly became a dominating force in turfgrass weed management programs.

Over the past 60-plus years, many herbicides have been commercialized for use in managing weeds in turfgrasses. Chemical weed control in turfgrasses utilizes compounds that are used as preemergence, postemergence (their combination), differential growth regulation, and/or total vegetation control. There are contact, systemic, selective, nonselective, and various combinations, plus a host of turf growth regulators that are now available for use. At the end of this chapter, a table containing a listing of herbicides registered in the United States for use on turfgrasses can be found. Proper herbicide selection and use is, first, a function of accurate weed identification, and then complete familiarity with the herbicide label.

Managing Turfgrass Weeds

Recommendations for herbicide use can be very specific, particularly with respect to location and grass species. Therefore, it is vitally important that turfgrass managers, regardless of their specific responsibilities, consult all possible sources of expertise in their location before choosing and applying any herbicide.

With regard to biological alternatives to chemical herbicides, such solutions are not as prevalent for weeds as they are for insecticides and fungicides. Some fungi and bacteria have been investigated for their potential herbicidal effects on some weeds in turfgrass situations. However, limited success has been achieved. Interest does remain high and research programs continue to investigate the potential of various substances that possess alternative potential.

Recently, a substance identified as a component of corn meal gluten has been shown to control some summer annual grasses (primarily crabgrass), when applied prior to germination. However, the amount of nitrogen that accompanies the rate of application needed for commercially acceptable control of crabgrass may be considered to result in an excessive amount of nitrogen being applied as part of the application. Researchers in turfgrass weed science will, undoubtedly, continue to identify and develop those weed control substances that have the potential to be alternatives to the herbicides that are used today.

In the final analysis, successful weed control programs begin with cultural practices that favor the competitive nature of the desired turfgrass species over all others. The existence of weeds in a turfgrass stand usually indicates a lack of turfgrass health, vigor, and competitiveness. Consequently, one or more cultural practices and/or environmental influences have probably reduced the competitiveness of the desired turfgrass.

Certain weeds serve as excellent indicators of growing conditions that have favored their development. Examples are as follows.

Weeds as Indicators

- Low pH (sheep sorrel) (cinquefoil)
- Compaction (goosegrass) (knotweed)
- Low N (legumes)
- Poor soil (quackgrass) (tall fescue)
- Poor drainage (sedges) (rushes)
- Excess surface moisture (algae) (moss)
- High pH (plantain)

Therefore, once a weed has been accurately identified, the next course of action is to determine why a void in the turf existed (which allowed the weed to encroach in the first place). The causes for voids can be generally arranged into two groups. The first group could be considered "natural causes" as follows.

Natural Reasons for Voids

- Environmental stresses
- Disruptive use of the area (ballmarks, divots, mechanical wear traf)
- Floods, hail, lightning
- Animals
- Diseases, insects, nematodes

The second group could be considered as management causes for voids that include the following.

Management Reasons for Voids

- Improper mowing height/frequency
- Improper fertilization (rate/timing)
- Improper irrigation (too little or too much)
- Lack of core cultivation (or poor timing)
- Lack of drainage program
- Lack of soil and tissue test information
- Lack of thatch management
- Lack of traffic control

Consequently, the chronological steps for successful weed management are as follows.

Steps in Weed Control Strategy

- Identify the weed
- Determine what breakdown in the management system caused the weed to have an advantage
- Impose the best cultural solution to correct the problem
- Choose an appropriate herbicide, if needed
- Apply herbicide according to the label recommendation using calibrated equipment.

For chemical control, weeds can be divided into four functional categories: annual grasses, broadleaf weeds, sedges, and perennial grasses. A typical herbicide approach to annual grass control is the use of preemergence herbicides. When preemergence control is not successful, postemergence controls can be used. Postemergence herbicides are generally not as consistent or effective as preemergence materials. Although some annual broadleaf weeds can be controlled preemergence, broadleaf weeds most often are controlled postemergence. Postemergence control of broadleaf weeds can be accomplished selectively, but control of perennial grassy weeds cannot be accomplished as readily using herbicides. Selective control of perennial grasses is the result of spot treatment applications of nonselective herbicides. Selectivity is the result of application placement rather than chemical specificity.

Although the positive effects of herbicides on turf–weed competition are well documented, there is some evidence that herbicide use can cause potentially detrimental side effects. Some preemergence herbicides are known to injure some species more than others. Dinitroanaline compounds have been shown to reduce the rerooting of some turfgrasses. However, root injury as a result of preemergence herbicide use, has not been reported to be causal, with respect to increased encroachment by the target weed.

Owing to the effectiveness of existing herbicides and combinations thereof, most weeds in turfgrasses can be successfully managed. However, a limited number of species, which have no known chemical control, are becoming more serious problems particularly on sites that have had the more easily controlled weeds removed. Some speedwell species, for example, are now more prevalent than 50 years ago.

In addition to difficult-to-control weeds becoming more chronic problems, there is some evidence that certain weeds, that is, smooth crabgrass, have developed herbicide resistance.

Some isolated cases of herbicide resistance (primarily to Acclaim Extra [fenoxaprop-ethyl]) have been reported on sites where this herbicide has been used abundantly and over a significant period of time for the control of summer annual grasses.

Future research will be required to develop cultural and chemical strategies for these weeds. Therefore, historical herbicide use on specific sites has, at least, been a partially selecting force in the population dynamics of the existing turfgrass sward.

Managing Summer Annual Grasses

Summer annual grasses are tropical plants that persist and are competitive with turf during frost-free periods. They depend upon seed production as the sole means of propagation and survival. Some examples of summer annuals are smooth crabgrass, large crabgrass, barnyardgrass, fall panicum, green and yellow foxtail, goosegrass, and sandbur. In general, summer annual grasses begin to germinate when the soil temperature in the vicinity of the seed has been 55°F (13°C) for 4 or 5 consecutive days. A soil thermometer is an inexpensive and valuable management tool for monitoring soil temperature. It is important to monitor soil temperature near the surface where most of the previous year's seed production was deposited. The majority of any given summer's annual grass population is mostly the result of the germination of seed produced the previous season. Of the summer annual grasses listed, goosegrass requires higher soil temperatures for germination than smooth crabgrass. Smooth and large crabgrass, goosegrass, sandbur, are more common problems in established turf, while the foxtails, and barnyardgrass are more often a problem when cool-season grasses are seeded in the spring or early summer. Foxtails and barnyardgrass produce their seed on culms that grow taller than most turf, are mowed. Consequently, their reproductive capability, once the turf is established, is quite limited.

Chemical control of summer annual grasses is best achieved by using preemergence herbicides. Proper timing is imperative for successful control. Application of preemergence herbicides should precede weed seed germination by several days. Rainfall and/or irrigation are required to activate and distribute the active ingredient of the herbicide into the upper soil profile. A chemical barrier is created that is toxic to the germinating summer annual grassy weeds. This barrier will last from 6 to 12 weeks or longer, depending upon the chemical. Chemical residual is a function of biodegradability, which is influenced by weather (particularly rainfall), ultraviolet (UV) radiation, and microbial activity. Preemergence herbicide effectiveness can be influenced by physical factors as well. Any soil disturbance, from earthworm activity to core cultivation, will disrupt the chemical barrier. For a high degree of control, the integrity of the chemical barrier should be maintained at the maximum.

Controlling crabgrass in cool-season grasses can often be accomplished with a single application of preemergence herbicide. That being said, in the transition zone, sequential applications of preemergence herbicide are often required for the highest level of control. The second application is typically at half the rate of the initial application. When a herbicide has provided excellent control for a number of years, it has been shown that reducing the rate of the same herbicide can provide commercially acceptable control in subsequent years.

Postemergence control of summer annual grasses that have invaded cool-season grasses can be accomplished using fenoxaprop (Acclaim Extra), dithiopyr (Dimension), and quinclorac (Drive). Postemergence control, however, is usually not as effective as preemergence control. In St. Augustinegrass, asulam is required for postemergence control of crabgrass and goosegrass.

The following descriptions and control strategies are provided for the summer annual grasses.

Crabgrass

Description

Smooth and large crabgrass are warm-season annuals reproducing by seed. They are light green and have a prostrate growth habit. Smooth crabgrass is smaller and less hairy than large crabgrass. Crabgrass leaf blades are short, pointed, hairy to sparsely hairy, and rolled in the bud. The leaf sheaths are

FIGURE 1.1
Crabgrass.

split, compressed, and sparsely hairy. Auricles are present. The collar is broad with hairs along the margin. The ligule is large, membranous, and toothed for large crabgrass and is smooth for smooth crabgrass. The seedhead consists of three to nine spikes atop the main stem. Both species can root at culm nodes (Figure 1.1).

Cultural Strategies

Crabgrass species are relatively easy to manage through sound cultural practices, the use of cool-season grass species that are resistant to early season diseases and insects, and the proper use of preemergence and postemergence herbicides when needed are a must.

Strategy I: Avoid establishing turf at the time crabgrass is germinating (soil temperature near the surface of 55°F [13°C]). Avoid verticutting or core cultivation at the time that crabgrass is germinating or after application of a preemergence herbicide. Irrigate to alleviate moisture stress.

Strategy II: Raise mowing height slightly at the time crabgrass seed is germinating. Lower height of cut slightly and collect clippings during time that crabgrass is setting seed in the late summer and early fall.

Strategy III: Reduce available nitrogen when crabgrass is most competitive (tropical-type weather) via timing and low application rates of soluble nitrogen sources and/or proper selection of slow-release nitrogen sources.

Strategy IV: Use a preemergence herbicide when soil temperature under turf areas has been 55°F (13°C) for 4–5 consecutive days (at one-half inch [13 mm] soil depth). If germination has occurred, use a postemergence herbicide according to label and local recommendations or combinations of pre- and postemergence products.

Goosegrass

Description

Goosegrass is a coarse-textured summer annual, with leaves folded in the bud that germinates later than smooth crabgrass (2–3 weeks). Soil temperatures required for germination are 60–65°F (15–18°C) for 12–15 consecutive days in the upper one-half inch (13 mm) of the soil. Goosegrass has a cluster of tillers arising from the central part of the plant. These tillers typically grow prostrate giving the plant a rosette-like appearance. The leaf sheath is strongly flattened and usually whitish to silvery in color, which has given rise to the common name silver crabgrass in some locations. However, goosegrass as the common name is more frequently used. The most distinguishing feature of goosegrass is the seedhead. Spikes, 3–10 in number, radiate from the end of the seed stalk in a finger-like appearance (some liken it to a herringbone) (Figure 1.2).

Cultural Strategies

Goosegrass is particularly competitive during the summer months and in compacted soil conditions. Proper cultural practices and the use of preemergence herbicides can provide reasonably good management of goosegrass.

FIGURE 1.2
Goosegrass.

Where few plants are competing, the most economical and practical management tactic is to physically remove plants with a knife.

   Strategy I: See Crabgrass.

   Strategy II: See Crabgrass.

   Strategy III: Relieving soil compaction through core cultivation or by reducing or redirecting traffic will increase the competitiveness of the desired turfgrass species. Goosegrass grows well in noncompacted soil than in compacted soil, but competition from desired species is considerably greater when soils are not compacted.

   Strategy IV: Apply preemergence herbicides approximately 10–14 days later than for smooth crabgrass. If lilac is present, petal fall is a fairly reliable indicator for proper timing of application. Check herbicide label for goosegrass control rates, as they frequently differ from those for smooth crabgrass. Do not core cultivate the following application. Postemergence control generally is less effective than for smooth crabgrass, but postemergence treatment prior to basal tillering should result in success.

Foxtail

Description

Foxtails are bunch-type summer annuals that are rolled in the bud and do not persist in established turf. Yellow foxtail is encountered more commonly in turf and tends to be more persistent than green foxtail. The primary distinguishing feature between yellow and green foxtail is the hair on the sheath. Yellow foxtail is smooth to sparsely hairy, while green foxtail is hairy along the sheath margin. Seeds are formed in very dense panicles that resemble "foxtails" in mid- to late summer. Foxtails germinate when soil temperature is above 65°F (18°C) (Figure 1.3).

Cultural Strategies

Foxtails are not severely competitive in established turf, but can seriously compete with desired species when they are being established from seed and planted at the wrong time of the year (spring in northern latitudes).

   Strategy I: See Crabgrass.

   Strategy II: Lower mowing height slightly when foxtails are producing seedheads and collect the clippings. At turf mowing heights of 2 inches (50 mm) or lower, foxtails do not produce large quantities of viable seed.

   Strategy III: See Crabgrass.

FIGURE 1.3
Foxtail. (Photo courtesy of J. Borger, Penn State.)

Strategy IV: Since foxtails are primarily a problem during turf establishment, choice of a preemergence herbicide is restricted to that
does not inhibit seed germination of the desired species. When
foxtails have been a problem during the establishment year, a preemergence herbicide should be used in the spring of the following
year. Postemergence herbicides are effective on foxtails, but desired
species in the seedling stage are often sensitive to applications of
postemergence annual grass herbicides.

Barnyardgrass

Description

Barnyardgrass is an erect, bunch-type coarse-textured summer annual
that typically is shallow rooted and not competitive in established turfs.
Barnyardgrass leaves often appear reddish to purple and the seedheads appear
reddish. Barnyardgrass seeds often have very long awns that are quite conspicuous when observing the seedhead. The upright growth habit and production
of seed at several inches above the soil surface limits the competitiveness of
barnyardgrass to the turf establishment phase. Barnyardgrass germinates in
mid- to late spring once soils have warmed to 60–65°F (15–18°C) and will compete with cool-season grasses that are planted in the spring (Figure 1.4).

Cultural Strategies

Barnyardgrass is not competitive in established turf, but can seriously compete when cool-season grasses are seeded in the spring or early summer.

FIGURE 1.4
Barnyardgrass.

Strategy I: See Crabgrass.

Strategy II: See Foxtail.

Strategy III: See Crabgrass.

Strategy IV: See Foxtail.

Fall Panicum

Description

Fall panicum is a divergently branching, bunch-type summer annual that is shallow rooted and is not competitive in established turfs. It germinates when soils warm to 60°F (15°C) and can be troublesome during turf establishment. The seed stalk is bent abruptly at joints near the base and the seedheads are panicles that are very open. The leaves of fall panicum are rough to touch. Production of viable seed frequently does not occur in turf maintained at 2 inches (50 mm) or lower (Figure 1.5).

Cultural Strategies

Fall panicum does not compete well in established turfs, but can cause problems when cool-season grasses are seeded in spring or early summer.

Strategy I: See Crabgrass.

Strategy II: See Foxtail.

Strategy III: See Crabgrass.

Strategy IV: See Foxtail.

FIGURE 1.5
Fall panicum.

Dallisgrass

Description

Dallisgrass is a coarse-textured, bunch-type grass that spreads primarily by seed, but may have short rhizomes. Lower leaves may be hairy, while all are noticeably shiny. The seed stalks of dallisgrass are distinctive and contrast dramatically in desired turf, particularly at higher heights of cut. The seedheads are loosely ascending with spikelets distinctly egg-shaped and tapering to a point. The seeds are arranged along the spikelet similar to goosegrass (Figure 1.6).

Cultural Strategies

Dallisgrass and hairy crabgrass invade more turf acreage in warmer climates than any other undesirable grasses.

Strategy I: Avoid core cultivation when soils are warming into the range of 60–65°F (15–18°C) in mid- to late spring when dallisgrass is germinating.

Strategy II: Maintain mowing heights as close to the lowest level tolerated by given turfgrass species during the time dallisgrass is producing seedheads.

Strategy III: Use a preemergence herbicide when appropriate. Postemergence control in St. Augustinegrass and centipedegrass cannot be accomplished with currently available herbicides. Postemergence control in bermudagrass and zoysiagrass can be accomplished with disodium and monoammonium arsenates.

FIGURE 1.6
Dallisgrass. (Photo courtesy of A.J. Turgeon, Penn State.)

Managing Winter Annual Grasses

Winter annual grasses are temperate species that germinate predominately when soil temperatures are below 70°F (15°C). Germinated seedlings survive as juvenile plants during winter months, and reproduce via seed set in late spring. The predominate winter annual grassy weed in the United States is *Poa annua*. In accordance with the definition of a weed being a plant out of place, annual bluegrass may or may not fit the definition, depending on whether the management personnel considers it to be "out of place." In many locations where temperatures are moderate and, particular wherein maritime climate exists, annual bluegrass can be competitive throughout the season and is often not considered to be a weed. Indeed, many very high-quality turf sites contain a high percentage of annual bluegrass and it provides exceptional turf quality.

In warmer climatic zones such as the southern United States, annual bluegrass is almost unilaterally considered to be a weed and is, essentially, nonexistent in actively growing warm-season turf species. Control of annual bluegrass is best achieved through cultural and chemical practices intended to slow its competitiveness, while increasing the competitiveness of desired turf species. More research efforts have been targeted on annual bluegrass management than any other winter annual grass species. Successful control has been more readily attained in the southern United States through the use of herbicides that are not labeled for application on cool-season turfgrasses because of their low tolerance. In addition, total vegetation control products can be used to control annual bluegrass in dormant bermudagrass.

The following descriptions and control strategies are provided for annual bluegrass.

Poa annua

Description

Annual bluegrass is mostly bunch-type in growth habit, although the subspecies of a perennial type frequently has stolons. Stoloniferous plants usually are found in irrigated, closely mowed situations. Another subspecies, commonly designated as "greens type" is only found on golf course putting greens, is stoloniferous, very dense, and perennial in nature. Annual bluegrass is folded in the bud and can be distinguished from Kentucky bluegrass very readily by comparing ligules. Annual bluegrass has a very conspicuous, membranous ligule, while Kentucky bluegrass has a ligule that is a very short, blunt membrane. Annual bluegrass can be easily distinguished from creeping bentgrass because bentgrass has a rolled vernation and the veins on the leaf blade are prominent. Most of the time, estimates of the amount of annual bluegrass present in mixed species turf stands are made in mid–late spring when a majority of the annual bluegrass has seedheads. However, this approach often tends to overestimate the actual amount present. When annual bluegrass is in a mixed stand with creeping bentgrass, seasonal fluctuations occur for each species. In the spring and fall, annual bluegrass can out compete creeping bentgrass, while in the summer months the reverse is true. Annual bluegrass can produce viable seed at all cutting heights, even as low as a golf course putting green. Seedheads are a pyramidal-like panicle (Figure 1.7).

FIGURE 1.7
Poa annua.

Cultural Strategies

Annual bluegrass is most competitive under conditions of close mowing, frequent irrigation, moderate nitrogen fertilization during cool weather, and moderate soil compaction. Cultural practices employed to minimize conditions favoring annual bluegrass work best when used in combination with herbicides and/or plant growth regulators.

   Strategy I: Correct drainage problems, raise height of cut when possible, even a millimeter can help, fertilize to favor desired turf species (for creeping bentgrass, nitrogen fertility should be applied during warmer months), and allow turf to wilt during periods of moisture stress. Annual bluegrass does not have a dormancy mechanism for drought, while other cool-season grasses can go dormant, and then recover when irrigated. Annual bluegrass that is allowed to dry down to the permanent wilting point will die, due to this inability to go dormant.

   Strategy II: Renovate turf stand by fumigating the site to kill the annual bluegrass and its seedbank and establishing turf-type perennial ryegrass or creeping bentgrass as the predominant species, and use ethofumesate according to label instructions as a pre-/postemergence herbicidal control once the new turf has become established.

   Strategy III: Avoid turf cultivation during periods when soil temperatures are consistently cool (which will trigger annual bluegrass seed to germinate). Maintain maximum shoot density during this time of the year (early fall), as annual bluegrass seedlings require space and light to become established.

   Strategy IV: Use mowing equipment that causes the least amount of mechanical stress to the turf and minimizes soil compaction (lightweight equipment). Collect clippings, especially during the peak seedhead production period of the year (late spring). Lightly verticut and collect clippings when seedheads are being produced.

   Strategy V: Apply a preemergence herbicide prior to weed seed germination. Two applications may be required during the dormant season in the southern United States. Postemergence control is attained most readily in the southern United States through the use of herbicides that are not tolerated by cool-season species or by using nonselective herbicides when bermudagrass is dormant. Applications of growth regulators labeled for conversion of turf from *P. annua* to creeping bentgrass may be made per label instructions. Successful conversion can be accomplished, but it usually takes 3–5 years and often involves the need to have an overseeding program included in the strategy. However, growth regulator use must be done in conjunction with the other strategies outlined above (except Strategy II). Recently, the herbicide Tenacity has been introduced as a *P. annua* control product; however, research has shown this product to be inconsistent unless used at the highest label rate. When used at the highest label rate, it is important to assess just how much *P. annua* is present in the stand because if excellent control is attained, turf cover will be diminished to an equivalent degree.

Managing Perennial Grasses and Sedges

Perennial grasses are extremely difficult to control because they frequently have morphology and physiology similar to the desired species. Indeed, perennial grassy weeds are also considered as desired species in certain situations. Selective postemergence herbicides are not readily available.

As with annuals, perennial weedy grasses can be either a cool- or warm-season species. Examples of cool-season species are tall fescue, bentgrass, orchardgrass, quackgrass, timothy, and velvetgrass; while examples of warm-season species are bermudagrass, nimblewill, Johnsongrass, and Zoysiagrass. Perennial grasses are considered weeds because of their wide leaf texture, frost sensitivity (sometimes), means of spread (growth habit), mowability, and so on. They typically cause a lack of uniformity in the sward, thus significantly reducing overall quality.

Perennial grassy weed problems often can be avoided during turf establishment, as seed from such species may occur in inexpensive seed sources, mulching materials, noncertified seed, or sod, and unclean topsoil. Seedbeds containing vegetative propagules of unwanted perennial grasses, such as quackgrass, should be fumigated or treated with nonselective total vegetative herbicides prior to the seeding of desired turf species.

The following descriptions and control strategies are provided for perennial grasses.

Creeping Bentgrass

Description

Creeping bentgrass is rolled in the bud, having leaves tapering to a point with prominent veins. It spreads through prolific stolon production that makes it incompatible with bunch-type species or those spreading by rhizomes. Patches of bentgrass become prevalent in the stand. Creeping bentgrass turf quality deteriorates when mowed above one-half inch (13 mm). Therefore, creeping bentgrass use is restricted to monostands that are intensively maintained (Figure 1.8).

Cultural Strategies

Creeping bentgrass should only be seeded with other creeping bentgrasses. When creeping bentgrass begins to encroach into a desired turf, physical

FIGURE 1.8
Creeping bentgrass.

removal, with care taken to remove all stolons, is required. Avoid close mowing of other species, particularly Kentucky bluegrasses.

> Strategy I: Mow desired species as high as feasibly possible. Fertilize
> moderately with complete fertilizer twice a year, with at least two-
> thirds of the yearly nitrogen applied in the fall. Lime to a pH of at
> least 6.7 if needed, as creeping bentgrass is more competitive under
> acidic soil conditions.

> Strategy II: Spot treat with translocatable total vegetation control herbi-
> cide. Treat an area at least twice the size of the visible patch of creep-
> ing bentgrass. Stolons spreading outward from the visible patch are
> underneath the canopy of the desired species and can often escape
> contact with the herbicide. A new herbicide, Tenacity, is labeled in
> most states for some level of selective control of creeping bentgrass
> in Kentucky bluegrass stands.

> Strategy III: Fumigate the site and establish desired turf species from
> seed or vegetative means.

Tall Fescue

Description

Tall fescue is a bunch-type, coarse-textured perennial grass. It is rolled in the
bud with prominently veined, long leaves that are shiny underneath. When
seeded heavily, tall fescue can provide an acceptable turf. New turf-type tall

FIGURE 1.9
Tall fescue.

fescues appear to be reasonably compatible with turf-type perennial ryegrasses, but not with rhizomatous-type species like Kentucky bluegrass (Figure 1.9).

Cultural Strategies

Tall fescue usually can be eliminated as a problem during turf establishment by using certified seed of desired species and by not mixing tall fescue with any other grass. If an established site becomes infested with tall fescue, a different strategy is required.

Strategy I: Eliminate clumps of tall fescue from the seedbed by physical removal or soil screening. Use seed that is free of contamination by tall fescue. No percentage of tall fescue is allowable in seed mixtures containing fine-textured species.

Strategy II: Fumigate site contaminated with tall fescue or use a translocatable nonselective total vegetative control herbicide as a spot treatment. Use topical applications of grass herbicides labeled for tall fescue control; do not apply such herbicides to desired turf.

Orchardgrass

Description

Orchardgrass is a coarse-textured, bunch-grass perennial grass. The leaves are light green, tapering to a boat-shaped tip, and are not prominently veined. The leaves are folded in the bud and the ligule is a prominent membrane.

FIGURE 1.10
Orchardgrass.

Orchardgrass is not compatible with other species because of leaf texture and very clumpy growth habit (Figure 1.10).

Cultural Strategies

Orchardgrass is most easily eliminated as a problem by using certified seed for turf establishment and clean straw as a mulching source.

Strategy I: See Tall Fescue.
Strategy II: See Tall Fescue.

Quackgrass

Description

Quackgrass is a coarse-textured, rhizomatous perennial grass. The leaves are rolled in the bud and the auricles are conspicuously clasping. Quackgrass has been shown to have allelopathic effects on other species that increases its competitiveness. The internodal length on quackgrass rhizomes is quite long compared to many other rhizomatous grasses; therefore, quackgrass does not produce a dense stand when growing alone (Figure 1.11).

Cultural Strategies

Quackgrass can spread by seed or rhizomes from areas adjacent to turf sites. However, quackgrass is stressed by high temperature and can only encroach

FIGURE 1.11
Quackgrass.

areas where the turf is thin. It is often introduced as a contaminant in balled and burlap nursery stock.

Strategy I: See Tall Fescue.

Strategy II: See Tall Fescue.

Strategy III: Mow at the lowest point of the tolerance range for the desired turf species. Increase nitrogen fertility, particularly during the warmest weather, and be sure the nursery stock and topsoil sources are free of quackgrass contamination.

Rough Bluegrass

Description

Rough bluegrass is a creeping fine-textured perennial grass. Spreading by stolons, it is a very opportunistic grass with regard to occupying available space. Owing to growth habit, it forms patches in Kentucky bluegrass turf and other cool-season grasses. It is also characterized by a shiny upper leaf surface and pale green coloration. It is very shade tolerant and will also grow in poorly drained situations. In the south, it is commonly used as an overseeded grass in dormant warm-season grasses. It is a common contaminate in noncertified seed sources. Rough bluegrass does not tolerate wear well, but can be aesthetically pleasing and will compete with mosses in wet, shady areas that are not required to provide a turf for significant use (Figure 1.12).

FIGURE 1.12
Rough bluegrass.

Control Strategies

The best preventative approach is to only use high-quality (certified) seed sources and specify them to be *P. trivialis* free. Do not use seed mixtures that contain any rough bluegrass.

> Strategy I: Physically remove patches when they are small and over-seed with desired species.
>
> Strategy II: Selectively apply a total vegetative killing herbicide approximately one and one-half times the size of the patch and overseed with desired species.
>
> Strategy III: Consult local extension control recommendations for other possible herbicidal solutions.

Nimblewill

Description

Nimblewill is a fine-textured warm-season perennial that spreads by seed, stolons, and rhizomes. Frost will discolor nimblewill, leaving brownish patches throughout the turf stand. Nimblewill is rolled in the bud and the leaves are short, flat, and come to a point. The leaf blades are oriented horizontally, providing the illusion that nimblewill is dense; however, the actual shoot density (plants per unit area) is not particularly high. The growth habit and frost sensitivity make nimblewill undesirable when growing with other species (Figure 1.13).

FIGURE 1.13
Nimblewill.

Cultural Strategies

Nimblewill patches are extremely difficult to manage and competitiveness is severe, particularly when turf is underfertilized with nitrogen. Nimblewill does not invade closely mowed sites (<13 mm).

> Strategy I: Increase overall nitrogen fertility level, particularly when frost has forced nimblewill into dormancy.
>
> Strategy II: Apply a translocatable, nonselective herbicide to nimblewill when it is actively growing. Be sure application is to twice the area of observed nimblewill infestation to ensure the entire infestation has been treated.

Bermudagrass

Description

Bermudagrass is a warm-season perennial that spreads by seeds, stolons, and rhizomes. Like nimblewill, bermudagrass is forced into dormancy by chilling temperatures of <55°F (13°C) and the resulting straw-like color is not compatible when mixed with cool-season turfgrasses. Bermudagrass is folded in the bud and the ligule is a fringe of hairs (Figures 1.14 and 1.15).

Cultural Strategies

Bermudagrass is difficult to manage once it is established and it is strongly competitive during the summer months.

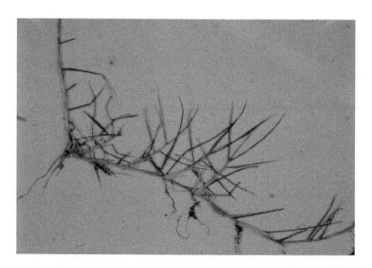

FIGURE 1.14
Bermudagrass.

FIGURE 1.15
Bermudagrass.

Strategy I: See Nimblewill.

Strategy II: See Nimblewill.

Strategy III: Physical barriers (edging, geotextiles, etc.) can impair the spread of bermudagrass into areas where it is not wanted. To reduce bermudagrass encroachment into St. Augustinegrass, maintain a cutting height of 2–3 inches (50–75 mm) so that St. Augustine grass shades out bermudagrass.

FIGURE 1.16
Zoysiagrass.

Zoysiagrass

Description

Zoysiagrass is a warm-season perennial that spreads by stolons and rhizomes. Like nimblewill and bermudagrass, Zoysiagrass is forced into dormancy by near freezing temperatures and the resulting straw-like color is not compatible with cool-season grasses. Zoysiagrass is rolled in the bud and slower growing than bermudagrass. Although slow to establish, once it is established, Zoysiagrass is extremely competitive (Figure 1.16).

Cultural Strategies

Zoysiagrass is difficult to eliminate once it is established and is strongly competitive during the warmest months.

    Strategy I: See Nimblewill.

    Strategy II: See Nimblewill.

    Strategy III: See Bermudagrass.

Nutsedge

Description

Sedges are grass-like plants that are perennial with three-ranked and triangular stems. Yellow nutsedge, sometimes called nutgrass, is more widely encountered in the United States, with purple nutsedge more commonly found in the southern states. Yellow nutsedge is pale green to yellow in color and has a rapid vertical shoot growth rate in the spring and early summer. Sedges are spread by seed, rhizomes, and nutlets, and are good indicators of poor drainage. Purple nutsedge can be distinguished from yellow nutsedge by its brownish to purple seedhead. Yellow nutsedge has a straw-colored seedhead (Figure 1.17).

Cultural Strategies

Sedges are somewhat difficult to eliminate from turf, but are less competitive in dense turf situations. Keeping turf from stress during the times that nutsedge is most vigorous will prevent significant encroachment.

- Strategy I: Improve drainage, both surface and subsurface. Avoid topsoil sources from poorly drained sites (it may contain nutsedge propagules).
- Strategy II: Lower height of cut (on species that will tolerate <20 mm) and increase mowing frequency during periods when nutsedge is rapidly growing.
- Strategy III: When nutsedge has invaded cool-season turf, fertilize with nitrogen in the fall after frosts have become frequent. Nutsedge vigor is reduced by frost more than most cool-season turfgrasses.
- Strategy IV: Use a postemergence herbicide recommended for your area during periods of rapid nutsedge growth.

FIGURE 1.17
Nutsedge.

Managing Summer Annual Broadleaf Weeds

Certain summer annual broadleaf weeds germinate when the soil begins to thaw (i.e., knotweed), while others do not germinate until the soil has warmed considerably (70°F). Many summer annual broadleaf weed species are frost sensitive and therefore germinate only after the danger of frost has past. Regardless of the germination date, summer annual broadleaf weeds are most competitive with cool-season turfgrasses during the warmer months when such weeds are in the vegetative stage of growth. As the day length shortens, flowering and seed set occur and the seeds ripen and shatter usually before the first killing frost. Many species are killed by frost and, in some years, early frosts can reduce the amount of seed set. For cool-season turfgrasses, summer annuals can be avoided by establishing the turf when frosts can occur, such as for fall seedlings in the northern United States. In such locations, late spring and summer turf establishment of cool-season grasses often results in poor stands due to the competition from germinating broadleaf summer annuals, such as purslane, pigweed, and lambsquarters to name a few. On established cool-season turfgrass sites, thinning of the stand can be caused by summer diseases and insect problems which can often result in the invasion of summer annual broadleaf weeds (particularly prostrate spurge). Preemergence herbicides are effective for controlling some broadleaf summer annuals, but timing of application is critical. Since the preemergence herbicides that can control some summer annual broadleaf weeds are also commonly used for crabgrass control, a second application for later germinating summer annuals may not be required. As a result, most summer annual broadleaf weeds are controlled using postemergence herbicides. Summer annuals can also be a problem when establishing warm-season species. Closer spacing of plugs and sprig rows helps reduce invasion, but most summer annuals are still usually controlled by using postemergence herbicides after establishment.

Oxalis

Description

Oxalis is a trifoliate, warm-season annual. The stems branch from the base and are slightly hairy. The leaflets are pale green to yellowish and are somewhat heart shaped. The flower petals are vivid yellow and the seed forms in five-sided pods that are pointed. At maturity the seed pods burst and can disseminate seeds up to several feet (>0.6 m). The leaflets contain calcium oxalate, which gives them a sour or acidic taste. Rooting may occur at nodes when stems are prostrate, but new plants do not form at the rooting site (Figure 1.18).

FIGURE 1.18
Oxalis.

Cultural Strategies

Oxalis is best avoided in cool-season grasses by maintaining a dense stand of turf during warm weather when the seeds most commonly germinate. Hand removal of a few isolated plants when first noticed can impede encroachment.

> Strategy I: Physically remove when sparse in turf. Also, remove from flower beds and the mulch in landscape plantings, as the adjacent turf areas can be readily infested due to exploding seed pods.
>
> Strategy II: Control insects that may weaken the cool-season turf in midsummer. In many cases, turf thinned by grubs and/or chinch bugs is invaded by oxalis.
>
> Strategy III: Use a preemergence herbicide when and where appropriate, but more commonly use a postemergence herbicide following label instructions. Usually only spot treatment is required. Broad area application of an herbicide is usually not necessary.

Knotweed

Description

Knotweed is the earliest germinating summer annual broadleaf weed. Soon after a frozen soil thaws, seed germination can occur. This plant is a low-growing, spreading plant that does not root at nodes, although individual plants may form a thick mat of vegetation up to 3–4 ft (0.9 or 1.2 m) in diameter. The stems are leafy and at each node there is a paper-like sheath. The leaves are bluish-green when plants are growing in compacted soils, but tend to be bright green in noncompacted soil situations. Newly emerged seedlings appear grass-like. Knotweed is capable of persisting in severely compacted soils due to its ability to function under conditions of low soil oxygen diffusion rates (Figures 1.19 and 1.20).

Cultural Strategies

Prostrate knotweed is an excellent indicator of soil compaction and, thus, cultural practices designed to avoid, reduce, minimize, or eliminate soil compaction should be employed.

Strategy I: Core cultivation and the control of traffic to reduce compaction are necessary.

Strategy II: Increase nitrogen fertility in the fall after prostrate knotweed has become reproductive and is less competitive with turf.

Strategy III: Use a postemergence herbicide when knotweed is young, upright, and actively growing. Treat before plants begin to grow prostrate.

FIGURE 1.19
Knotweed.

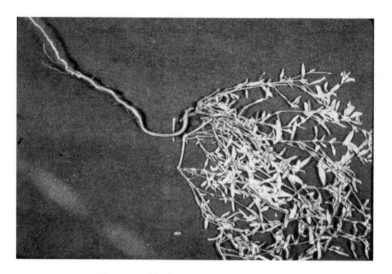

FIGURE 1.20
Knotweed.

Prostrate Spurge

Description

Prostrate spurge is a low-growing, spreading plant that does not root at the nodes, and resembles prostrate knotweed. Like knotweed, prostrate spurge has stems that radiate from a taproot, and the individual plants can form mats 3–4 ft (0.9–1.2 m) across. The stem of prostrate spurge contains a milky sap that is not found in knotweed. Prostrate spurge germinates much later than knotweed as soil temperatures in the 60–65°F (15–18°C) range are required. The underside of the leaves is hairy and a purplish spot may be present on the upper surface (Figure 1.21).

Cultural Strategies

Prostrate spurge is an indicator of soil compaction although it will not tolerate the severity of compaction that knotweed can tolerate. Cool-season turf sites subject to moisture stress that are thinned by disease and insect attacks are typically invaded by prostrate spurge in midsummer. Control of disease and insect pests and relief of soil compaction allows turfs to effectively compete with prostrate spurge. Dense, vigorously growing turf is not readily invaded by prostrate spurge.

  Strategy I: Relieve soil compaction through core cultivation and traffic
    control.
  Strategy II: See Oxalis.

FIGURE 1.21
Prostrate spurge.

Strategy III: Irrigate when moisture stress occurs to avoid wilting or dormancy of desired turf species.

Strategy IV: Use a preemergence herbicide when and where appropriate, but more commonly use a postemergence herbicide.

Purslane

Description

The leaves and stems of this prostrate growing annual are succulent to fleshy and reddish in color. Stems may root, particularly when spreading in very thin turf. The leaves are oblong, clustered, and arranged alternately on the stem (Figure 1.22).

Cultural Strategies

Avoiding late spring and summer seedings of cool-season turf species is the most effective cultural approach to solving a potential purslane problem.

Strategy I: Establish cool-season turfs in the late summer or fall.

Strategy II: See Oxalis.

Strategy III: Use appropriate postemergence herbicide while purslane is in seedling stage or delay herbicide treatment for 3 or 4 mowings and use a postemergence broadleaf herbicide.

FIGURE 1.22
Purslane.

Lambsquarters

Description

This tap-rooted annual has erect-growing stems with ovate alternate leaves. The upper leaves are sometimes sessile. The stems are somewhat square and typically have reddish streaks running along their length. The seedheads are irregular spikes clustered in panicles at the end of branches (Figure 1.23).

FIGURE 1.23
Lambsquarters.

Cultural Strategies

Strategy I: Lower height cut to at least 30 mm after turfgrass seedlings have become established.

Strategy II: Establish cool-season turfs in the late summer or fall.

Strategy III: Use appropriate postemergence herbicide while purslane is in seedling stage or delay herbicide treatment for 3 or 4 mowings and use a postemergence broadleaf herbicide.

Puncturevine

Description

This prostrate, branching annual is slightly hairy. Stems may reach lengths of 5 or 6 ft (1.5 or 1.8 m). Leaves are arranged opposite, pinnately compound, and bright green. Leaflets are narrow, about one-half inch (13 mm) long and hairy. Seed pods have five sections, each with sharp spines resembling horns (Figure 1.24).

Cultural Strategies

Puncturevine is most competitive when soils are compacted and nitrogen fertility is low. Cultural practices to reduce soil compaction should be employed.

Strategies I, II, and III: See Knotweed.

Increase maintenance of turf areas, or where a turf has been thinned by insect attack, or disease incidence.

FIGURE 1.24
Puncturevine. (Photo courtesy of N. Christians, Iowa State.)

FIGURE 1.25
Carpetweed.

Cultural Strategies

See Lambsquarters.

Carpetweed

Description

This late-germinating, but quick-growing annual is prostrate and forms mats
in all directions from the taproot. The stems are green, smooth, and branched
with five to six leaves forming in whorls at each node along the stem. This
growth habit is quite distinctive and most obvious when thin turf has been
invaded. This weed is most often a problem during establishment (Figure 1.25).

Cultural Strategies

See Purslane.

Florida Pusley

Description

Stems and leaves are hairy. Plants grow prostrate in mowed situations, but
tend to be more erect when mowing is infrequent. Leaves are opposite,
somewhat oval, and mostly flat. Plants, when growing prostrate, form dense
patches. Flowers have six parts and appear star-like at the base of upper

FIGURE 1.26
Florida pusley. (Photo courtesy of N. Christians, Iowa State.)

leaves. The plant most commonly is found in locations where warm-season grasses are the predominant turf species (Figure 1.26).

Cultural Strategies

Although an annual plant, management is primarily through use of post-emergence herbicides. Consult local extension recommendations for your area.

> Strategy I: Physically remove any small infestations, and keep turf from wilting through proper irrigation management.
>
> Strategy II: In St. Augustinegrass or dichondra, apply a broad spectrum combination of broadleaf herbicides.
>
> Strategy III: Management of Florida pusley in warm-season grasses that are sensitive to phenoxy herbicides can be attained by using a triazine herbicide according to label recommendations.

Managing Winter Annual Broadleaf Weeds

The preferred time for the establishment of cool-season turfgrasses is late summer through early fall, in part to avoid competition from summer annual weed species. However, several winter annual weed species can become significant weed problems in the seedbed when cool-season turf is established from seed in the fall. In southern latitudes, winter annuals are significant problems because they germinate and become established into warm-season turfgrasses after the onset of winter dormancy. They can also compete with cool-season grasses that have been overseeded into warm-season grasses for winter turf.

Control of broadleaf winter annuals in seedling stands of cool-season turfgrasses can be accomplished by using proper establishment methods to ensure good plant competition, a postemergence herbicide application when the weeds are in the seedling stage, using bromoxynil (if competition is severe), or after the stand has begun to mature after 3 or 4 mowings. In dormant warm-season turf, broadleaf winter annuals can be controlled by treating with conventional broadleaf plus preemergence tank-mixed herbicides or with nonselective plus preemergence tank-mixed herbicides. Preemergence control of some winter annuals can also be achieved; however, such an approach is not feasible in the seedbed. Consult extension recommendations for herbicidal control of broadleaf winter annuals in your area.

Common Chickweed

Description

Common chickweed, found mostly in moist shaded locations, is low growing when mowed, but erect when infrequently mowed. A single row of hairs extends along one side of a square stem. The leaves are teardrop shaped and opposite on the stem (Figure 1.27).

FIGURE 1.27
Common chickweed.

Cultural Strategies

Vigorously growing, dense turf will effectively keep common chickweed from encroaching. Chemical treatment may be required in seedbed situations where common chickweed germination is excessive and the desired turf is unable to compete successfully.

> Strategy I: Reduce shade by eliminating unnecessary trees, removing understory growth and trimming lower limbs of remaining trees up to 8 ft (2.4 m) above ground.
>
> Strategy II: Use a postemergence herbicide in the seedbed that is safe for application to seedling turfgrasses or a preemergence herbicide where appropriate.
>
> Strategy III: Use a combination of postemergence herbicides after seedling cool-season turfgrasses have matured.
>
> Strategy IV: Use a combination of a nonselective herbicide with a preemergence herbicide on dormant warm-season species.

Henbit

Description

This cool-season annual grows upright, but some prostrate stems may sprout from the base of the plant. The stems are square and reasonably weak. The leaves are hairy and the basal ones have a petiole, while those near the top of the plant are sessile. The upper leaf surface is deeply veined with small lobes along the margin. Flowers are blue-purple and trumpet shaped (Figures 1.28 and 1.29).

Cultural Strategies

This winter annual is more competitive than common chickweed, but is not a significant problem in vigorously growing, dense turf. More commonly it is present in seedbeds and dormant warm-season turf.

> Strategy I: See Common Chickweed.
>
> Strategy II: See Common Chickweed.
>
> Strategy III: See Common Chickweed.
>
> Strategy IV: See Common Chickweed.

Shepherd's Purse

Description

This annual germinates when soil temperatures cool, below 60°F (15°C) and forms a rosette of lobed leaves. Considerable variation exists relative to the

FIGURE 1.28
Henbit.

FIGURE 1.29
Henbit.

type of lobes. Some are deeply lobed all the way to the vein, and others are only slightly lobed. The seeds are produced in small heart-shaped capsules that resemble a purse. Small white flowers may be produced in early spring on erect stems (Figure 1.30).

FIGURE 1.30
Shepherd's purse.

Cultural Strategies

Shepherd's purse is not competitive in established turf unless there has been a stand loss. Typically, Shepherd's purse will germinate in voids that are the result of insect or disease damage, which has occurred in mid- to late summer.

Strategy I: See Common Chickweed.

Strategy II: See Common Chickweed.

Strategy III: See Common Chickweed.

Strategy IV: See Common Chickweed.

Corn Speedwell

Description

Corn speedwell germinates in mid-fall and forms a fairly weak seedling that overwinters in the cool-season grass stands but becomes quite conspicuous in dormant warm-season turfgrasses. The entire plant is finely hairy, grows erect, and is most obvious when mowing is infrequent. The upper leaves are smaller than the lower leaves and more pointed (Figure 1.31).

Cultural Strategies

Corn speedwell does well in thin, open, or dormant turf and under these conditions can produce a fairly dense stand. Early fall nitrogen fertilization

FIGURE 1.31
Corn speedwell.

of cool-season grasses, along with control of insects and diseases, help reduce the competitiveness of corn speedwell.

Strategy I: See Common Chickweed.

Strategy II: See Common Chickweed.

Strategy III: See Common Chickweed.

Strategy IV: See Common Chickweed.

Bedstraw

Description

Bedstraw, generally, is considered a winter annual in turf situations and several species exist. Bedstraw is most competitive in the shade. The square stems have serrated structures that often cause the stems to stick together. The leaves are arranged in whorls along the stem and stipales are often present (Figure 1.32).

Cultural Strategies

Elimination or reduction of shade will help competitiveness of desired turf species.

Strategy I: See Common Chickweed.

Strategy II: Apply postemergence herbicide when bedstraw is actively growing. Consult recommendations for your area.

FIGURE 1.32
Bedstraw. (Photo courtesy of A.J. Turgeon, Penn State.)

Managing Biennials

Biennial weeds require 2 years to complete their life cycle. In the first year they grow vegetatively, producing food reserves that are stored in various vegetative plant parts. Consequently, it is during this first year, when they are vegetative, that biennials are most competitive with turfgrasses. They are leafy and aggressively occupy any open space. In the second year, biennials grow primarily in a reproductive mode using the food reserves stored the previous year. In many instances, biennials are less competitive during the second year, as mowing practices continually remove any inflorescence produced by the plant and stored food reserves become depleted. Successful flowering and seed production by biennials usually only occur when mowing is infrequent or the height of cut is higher than normally used in lawn turf. For the most part, biennials are not difficult to manage culturally by maintaining a dense turf, through proper mowing, fertilization practices, and through the use of recommended postemergence herbicides.

Yellow Rocket

Description

This biennial typically germinates when soils cool below 70°F (21°C) and becomes most noticeable in the following spring, as the vigorous early growth exceeds that of most turfgrasses. The leaves are deeply lobed, arranged in a

FIGURE 1.33
Yellow rocket.

rosette, and appear bright green with the terminal lobe large and rounded. In the seedling stage it is quite similar to shepherd's purse. Under infrequent mowing, yellow rocket will produce a bright yellow flower with four petals (Figure 1.33).

Cultural Strategies

Maintain mowing height of <35 mm in early spring to prohibit seed production and consequent reinfestation.

> Strategy I: Maintain proper management of disease and insect pests that cause stand thinning, particularly prior to germination.
>
> Strategy II: Fertilize cool-season species with nitrogen in the fall to encourage increased density.
>
> Strategy III: Apply a selective postemergence herbicide according to recommendations for your area when yellow rocket is in an actively growing vegetative state, with spring being best if it has invaded cool-season grass species.

Wild Carrot

Description

The foliage of this biennial is very similar to the domestic carrot, being multibranched and somewhat lace-like. The root of wild carrot, although edible, is quite tough and fibrous. Like other biennials, wild carrot produces an

FIGURE 1.34
Wild carrot.

aggressive leafy rosette during the first year. Wild carrot is somewhat more adaptable to mowing pressure than yellow rocket. Flowers and viable seed cannot be produced under mowed conditions of less than 45 mm (Figure 1.34).

Cultural Strategies

Wild carrot rarely produces a dense stand of plants in established turf and can be physically removed quite easily.

Strategy I: During physical removal, make sure to remove a significant portion of the taproot, at least two-thirds.

Strategy II: Apply a selective postemergence herbicide according to recommendations for your area.

Black Medic

Description

Black medic is considered an annual, biennial, or perennial depending on the management being employed on any given site. It is a trifoliate species like clover and oxalis, but differs in that the middle leaflet is extended on a

FIGURE 1.35
Black medic.

pedicel or stalk. In most turf situations, black medic persists as a perennial legume through stoloniferous growth, but produces flowers and viable seed under most mowed conditions. The stolons rarely root at nodes. Individual leaflets usually have a short spur at the tip (Figure 1.35).

Cultural Strategies

Black medic is not persistent or aggressive when turf is adequately fertilized with nitrogen. Maintaining proper nitrogen fertility, with a balance between nitrogen and phosphorus, is beneficial. Maintain a ratio of 2:1 for N and P.

   Strategy I: Increase nitrogen fertility level and correct for low phosphorus if soil test results indicate a need.

   Strategy II: Apply a selective postemergence herbicide according to recommendations for your area.

Managing Perennial Broadleaf Weeds

Perennials are weeds that live more than 2 years and are represented by the largest group of species causing problems in turf. Many perennials are very easy to manage, while others are virtually impossible to manage even with herbicides.

Many perennials spread by seed and vegetative plant parts. Often, viable seeds are produced below the cutting height routinely used for a given turf species and/or the seeds are disseminated by the wind.

The presence of abundant plants of a given species is often a good indicator of certain soil conditions or improper management. For example, legumes indicate low nitrogen fertility; sheep sorrel and cinquefoils indicate low pH conditions; yarrow is a good indicator of poor or droughty soil conditions; and creeping buttercup indicates shade and poor drainage. Therefore, modifying soil conditions or changing lime and fertility programs can have a significant positive impact on management programs for perennial weeds.

Wild Garlic

Description

Although both wild garlic and wild onion can be found in turf, wild garlic is significantly more common. Wild garlic leaves are smooth, slender, hollow, and arise from bulbs or bulblets. The leaves have a strong odor that is the result of allyl sulfide contained in the leaf. Wild garlic grows rapidly in early spring and is a significant problem when found in dormant warm-season turfgrasses. Wild garlic reproduces from seed, bulbs, and both aerial and below-ground bulblets (Figure 1.36).

FIGURE 1.36
Wild garlic. (Photo courtesy of J. Borger, Penn State.)

Cultural Strategies

The vigorous, early spring growth requires frequent mowing, perhaps before the turf is in need of mowing or has even broken dormancy in the case of warm-season turfgrasses. Control spring diseases that reduce turfgrass competitiveness.

> Strategy I: Close and frequent mowing, particularly in early spring.
>
> Strategy II: Apply ester-based selective broadleaf herbicides according to recommendations for your area.
>
> Strategy III: Apply a nonselective translocatable herbicide if turf is dormant with a wetting agent according to label directions.

Dandelion

Description

Dandelion has deeply lobed leaves arranged in a rosette. Lobes may be opposite or alternate and the tips may point away from or toward the apex. The large fleshy taproot contains vegetative buds mostly near the soil surface that can develop new plants. The flower is bright yellow turning to white and is wind disseminated (Figure 1.37).

Cultural Strategies

Care must be exercised when physically removing dandelions to insure that the majority of the taproot is removed (at least two-thirds). If not, vegetative

FIGURE 1.37
Dandelion.

buds on the upper portion of the taproot will produce rosettes and two to three plants will grow where one was removed.

Strategy I: Maintain dense turf through adequate nitrogen fertility and disease and insect control to reduce voids in the turf in the spring into which dandelion seed can be disseminated.

Strategy II: Physically remove, being careful to include all reproductive vegetative plant parts.

Strategy III: Apply a selective postemergence broadleaf herbicide when dandelions are actively growing. Fall applications are often most effective. Follow recommendations for your area.

White Clover

Description

This perennial stoloniferous legume has compound leaves, usually palmately divided into three leaflets. Flowers are white, globular, and appear from mid-May through September. White clover can flower and produce viable seed even at very low mowing heights of 6 mm. Low nitrogen nutrition (a third of that typically recommended for a species) and other causes for a thin turf make it particularly susceptible to invasion by white clover (Figure 1.38).

Cultural Strategies

Physical removal of white clover is not a good management strategy as stolons spread further than expected and portions can be left behind. Good

FIGURE 1.38
White clover.

liming and nitrogen fertilization practices that maintain turf density are desired.

Strategy I: Nitrogen fertility level is of greatest importance. When the nitrogen level for a particular turf species is not adequate, white clover or other low-growing legumes commonly invade.

Strategy II: Physical removal is recommended only under conditions of very close mowing of 6–8 mm where pluggers can be used effectively.

Strategy III: Apply a selective postemergence broadleaf herbicide before clover begins to flower in combination with increased nitrogen fertility. Do not apply herbicides when white clover is in flower or under moisture stress, as poor control will result. Follow recommendations for your area.

Common Plantain

Description

This perennial is a commonly occurring weed that has a broad tolerance to soil types. It grows in a rosette with large, broad leaves that have nearly parallel prominent veins. Leaves are often oriented parallel to the ground and the margins are often wavy. Seedstalks grow erect, with seeds forming along more than half the length. Common plantain is green with surface hairs on the leaves, while Rugel's has a purplish cast to the leaves and petioles (Figure 1.39).

FIGURE 1.39
Common plantain.

Cultural Strategies

Common plantain persists in thin turf by competing for space with large, broad, prostrate leaves that shade out new tillers of turfgrass that emerge. Maintain turf density with good management.

Strategy I: Proper pH and adequate fertility levels should be used. Common plantain can compete with turf particularly well where pH is high (>8.0). Areas of lime spills or where irrigation water has a high pH may have more serious common plantain encroachment problems. Acidify soil, 6.5–7.0, if soil testing so indicates.

Strategy II: See Dandelion.

Strategy III: Apply a selective postemergence broadleaf of herbicide in mid-spring or fall. Plantain resumes seasonal growth later in spring than dandelion, and is often poorly controlled with herbicides when applied prior to dandelion bloom.

Buckhorn Plantain

Description

This perennial has long, slender, strap-like leaves growing in a rosette and is frequently found in thinned turf. Perhaps only dandelion is more abundant. The leaves are dark green with nearly parallel venation. The seed stalk is very fibrous and difficult to cut with mowers, particularly reels, and the seeds are formed in a terminal capsule. The modified taproot has many adventitious buds and new plants can form near the bod line (Figure 1.40).

FIGURE 1.40
Buckhorn plantain.

Cultural Strategies

See Common Plantain.

> Strategy I: See Dandelion.
> Strategy II: See Dandelion.
> Strategy III: See Common Plantain.

Mouse-Ear Chickweed

Description

This low-growing perennial spreads by stolons and seed. Mouse-ear chickweed can adapt to mowing heights as low as 6 mm and even produce viable seed under such conditions. The leaves are opposite, oval, and quite hairy, resembling a mouse's ear. The stolons are also hairy with leaves attached directly. Flowers are small, five-petaled, and white (Figure 1.41).

Cultural Strategies

Although mouse-ear chickweed grows quite well in full sun, it can also compete in shaded sites. Use turfgrass species adapted to shade and improve soil drainage. For physical removal, see White Clover.

> Strategy I: Decrease shade and improve drainage.
> Strategy II: See White Clover.
> Strategy III: See White Clover.

FIGURE 1.41
Mouse-ear chickweed.

Ground Ivy

Description

This perennial is a strongly competitive, low-growing weed that spreads by vigorous stolons. Adapted to sun or shade, this species can form thick mats that crowd out existing turf. The leaves are small, roundest, slightly lobed, and have short bristle-like surface hairs somewhat resembling mallow. They are opposite on low-growing squares stems that root at the nodes. Flowers are funnel-shaped, blue to purple in color, and are attractive, particularly when ground ivy is used as a ground cover planting (Figure 1.42).

Cultural Strategies

Ground ivy is difficult to manage without herbicide use. Turf is often invaded by ground ivy that has escaped from mulched landscape plantings. Therefore, physical removal must include the removal of that found in the turf as well as in other plantings.

Strategy I: See Dandelions.

Strategy II: See Dandelions.

Strategy III: Apply a selective postemergence broadleaf herbicide in mid-spring or fall. Repeat herbicide applications every 3 weeks for three applications, if necessary, according to recommendations for your area.

FIGURE 1.42
Ground ivy.

FIGURE 1.43
Sheep sorrel.

Sheep Sorrel

Description

This perennial spreads by a strongly branching rhizome system that can give rise to a very dense mat of leaves that crowd out existing turf. Red sorrel has arrow-shaped leaves 1–3 inches (25–75 mm) long with spreading basal lobes. They are alternate on the stem and have a bitter taste, although they are sometimes eaten in salads. Flowers of red sorrel are borne on separate plants, yellow on male and reddish on female. Flowers seldom occur in mowed turf (G 35 mm) (Figure 1.43).

Cultural Strategies

Maintaining proper pH and provide adequate nitrogen fertility. Inspecting acid-loving nursery stock, such as azalea and rhododendron, prior to planting for red sorrel contamination is a good preventive practice. Do not attempt physical removal, as rhizome fragments are often left behind.

Strategy I: Lime to a pH of 6.7–7.2 for good turf vigor.

Strategy II: Physical removal is not particularly effective.

Strategy III: See Ground Ivy.

Canada Thistle

Description

This thistle is the most troublesome in turf. Spreading by vigorous rhizomes, Canada thistle competes with turf even when mowed. It commonly invades

FIGURE 1.44
Canada thistle.

turf from mulched landscape plantings. Considered a noxious weed in most states, Canada thistle in turf not only reduces aesthetic quality but can also cause physical discomfort for those who come in contact with it. Ridged spines form at the ends of lobes along the leaves. As plants emerge from rhizome buds, they form rosettes that are capable of maintaining a low-growth habit (Figure 1.44).

Cultural Strategies

Managing Canada thistle in turf begins with using certified seed or sod during turf establishment and making sure that plants are not introduced when planting nursery stock. If preventive measures are followed, encroachment by Canada thistle is limited mostly to situations where fields border turf areas.

   Strategy I: Use certified seed and sod during establishment and manage Canada thistle in landscape plantings.

   Strategy II: Use a selective postemergence broadleaf herbicide. Repeat applications at 3-week intervals may be required, according to recommendations for your area.

Chicory

Description

Chicory grows in a rosette and closely resembles dandelion. However, it has blue flowers borne along a tough, fibrous stalk, quite unlike the succulent

FIGURE 1.45
Chicory. (Photo courtesy of J. Borger, Penn State.)

tubular flower stalk of the dandelion. Vegetative buds occur on the chicory flower stalk. Under mowed conditions, these buds can produce aerial rosettes, as chicory does not commonly flower when mowed below 40 mm on a weekly basis. The deep, fibrous taproot can be dried, ground, and used as a supplement in coffee (chicory coffee). The upper surface of chicory leaves is rougher to the touch than dandelion (Figure 1.45).

Cultural Strategies

See Dandelion.

Strategy I: See Dandelion.
Strategy II: See Dandelion.
Strategy III: See Dandelion.

Curly Dock

Description

This perennial has long, strap-like leaves originating from buds at the top of a strong fleshy taproot. The leaves are arranged in a rosette with each leaf having a "wavy" margin; hence the common name curly dock. Under mowed conditions, curly dock rarely produces viable seed, but does persist due to the taproot. Curly dock is most competitive during periods of moisture

FIGURE 1.46
Curly dock.

stress when the turf species wilt, as curly dock rarely exhibits permanent wilt (Figure 1.46).

Cultural Strategies

See Wild Carrot.

Strategy I: See Wild Carrot.
Strategy II: See Wild Carrot.

Bull Thistle

Description

Musk and bull thistles are botanically biennials, but tend to persist as perennials in turf situations. They grow in rosettes, are quite spiny, and commonly develop during establishment of cool-season grasses. Not particularly competitive in established turf, these species are easily identified and can be physically removed (Figure 1.47).

Cultural Strategies

By maintaining the shoot density, using proper seeding rates, and selecting adapted species, bull thistles generally are not serious weed problems.

Strategy I: See Wild Carrot.
Strategy II: See Wild Carrot.

FIGURE 1.47
Bull thistle.

Heal All

Description

Heal all is a widely adapted broadleaf perennial weed that is somewhat difficult to manage. Spreading by fibrous stolons, heal all competes strongly with
weakened turf and in shaded conditions where it tends to be dark green. The
leaves and stems are hairy, the upper leaf surface has a "quilted" appearance,
and the stem is square. Flowers are purplish, borne on compact spikes with
overlapping green bracts (Figure 1.48).

Cultural Strategies

Select turf species that are well adapted to shaded conditions and maintain
maximum shoot density through proper cultural practices.

Ox-Eye Daisy

Description

This perennial spreads by fleshy rhizomes and can compete with low maintenance turf (such as along roadsides). Ox-eye daisy has typical chrysanthemum leaves, deeply lobed, but is not a daisy even though the flower is
daisy-like in shape and color. Leaves are dark green, even when nitrogen
fertility is low (Figure 1.49).

FIGURE 1.48
Heal all.

FIGURE 1.49
Ox-eye daisy.

Cultural Strategies

Ox-eye daisy does compete in turf that is even moderately managed. It is much more prevalent in low-maintenance areas like roadsides and other rights-of-way.

Strategy I: See Dandelion.

Strategy II: See Dandelion.

Strategy III: See Dandelion.

Hawkweed

Description

Hawkweed forms rosettes of very hairy leaves borne on an aggressive rhizome system. The leaves are long and strap-like, with some hairs being quite long. When turf is thin, hawkweed will grow in clusters, while in denser turf, individual rosettes tend to appear. Hawkweed persists through vegetative reproduction in most turf situations, as mowing eliminates the flowering capability (Figure 1.50).

Cultural Strategies

See Dandelion.

Strategy I: See Dandelion.

Strategy II: See Dandelion.

Strategy III: See Dandelion. Include surfactant with postemergence herbicide treatment to improve retention on leaf surface.

FIGURE 1.50
Hawkweed.

Thyme-Leaf Speedwell

Description

This perennial spreads by slow growing but aggressive stolons that compete strongly with turfgrasses. The leaves are dark green, oblong, with a waxy, shiny surface and occur opposite along the stem. Usually thyme-leaf speedwell roots into the soil at each node as the stolon grows. The leaves orient themselves parallel to the soil surface and cannot be removed by close mowing. Thyme-leaf speedwell can be a serious broadleaf problem, even on putting greens and other closely mowed areas (Figure 1.51).

Cultural Strategies

There is no selective chemical management for thyme-leaf speedwell, and physical removal is tedious. When digging it out and pulling it up, be sure to remove all stolon segments.

Creeping Speedwell

Description

This speedwell is the most common one found in cool-season grasses. It spreads aggressively by stolons that frequently grow over the top of a turf without rooting at the nodes. The leaves are pale green, lobed slightly, roundish with upper surface hairs. Rapid stolon development results in dense patches of creeping speedwell even when turf growth is good. Although not restricted to

FIGURE 1.51
Thyme-leaf speedwell.

FIGURE 1.52
Creeping speedwell.

shaded sites, creeping speedwell is more competitive in the shade. In late April and May, bluish to purple flowers form in abundance (Figure 1.52).

Cultural Strategies

Creeping speedwell requires a diligent combination of cultural and chemical programs to effect good management.

Strategy I: Reduce shade where possible and establish turfgrass species best adapted to the prevailing conditions. Improve air and soil drainage to allow turf to become more competitive.

Strategy II: Apply selective postemergence herbicides according to extension recommendations for your area in mid-spring (prior to flowering) and continue every 3–4 weeks. Repeat applications are necessary, as creeping speedwell recovers from herbicide "knockdown" through vegetative bud break on the stolons.

Wild Violet

Description

Numerous species of wild violets occur in turf. Characterized by a vigorous root system, heart-shaped leaves, and purple to blue flowers, this perennial is extremely difficult to manage. It spreads effectively through turf by aggressively growing stolons and forms very dense patches of plants. Although

FIGURE 1.53
Wild violet.

more competitive in shade, wild violets compete in sunny turf sites as well. Violets often invade turf areas from other locations in the landscape where they have been used as a ground cover (Figure 1.53).

Cultural Strategies

Wild violets should not be used as a ground cover unless they can be physically restrained from invading turfed portions of the landscape.

Strategy I: Physically remove all vegetative portions of the wild violet plants. This requires some digging. Remove wild violets from non-turf areas or contain them where they are desired by using physical barriers.

Strategy II: Repeated applications of selective postemergence broad-leaf herbicide combinations will reduce wild violet competition, but management is extremely difficult to achieve. Follow extension recommendations for your area.

Stitchwort

Description

This perennial spreads primarily by rapidly growing stolons that often form dense mats. The leaves are opposite and similar to an elongated teardrop with a somewhat shiny appearance. The square stem forms roots at nodes and often grows over the top of the turf, particularly when mowing

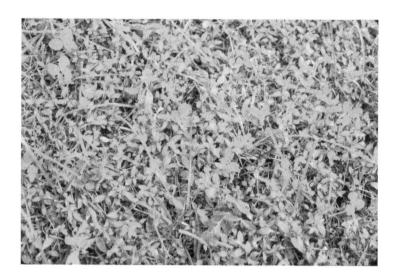

FIGURE 1.54
Stitchwort.

is infrequent. White flowers may form, but usually do not under mowed conditions. Stitchwort resembles common chickweed under certain growing conditions (Figure 1.54).

Cultural Strategies

Management can be achieved by mowing more frequently, reducing height of cut, and proper herbicide use.

Strategy I: Increase mowing frequency to at least weekly and, where possible, lower the height of cut to at least 35 mm.

Strategy II: Apply selective postemergence broadleaf herbicides when stitchwort is actively growing. Addition of nonionic surfactant is often beneficial. Follow herbicide recommendations for your area.

Cinquefoil

Description

Several cinquefoil species occur in turf, but the most common one is the five-leaflet type. Cinquefoil leaflets are palmately divided and arise from fibrous, hairy creeping stems. The leaflet shape is similar to that of a strawberry. In fact, the three-leaflet cinquefoil (rough or sulfur cinquefoil), that grows more upright, looks very similar to the cultivated strawberry. The five-leaflet type can compete effectively in frequently mowed situations, but

FIGURE 1.55
Sulfur/silver cinquefoil.

does not compete strongly when nitrogen fertility is adequate for good turf growth. Cinquefoil is often found when the soil type is gravelly and the pH is below 6.0. The undersurface of the leaflets appears silvery due to pubescence (Figure 1.55).

Cultural Strategies

Maintaining turf shoot density through adequate nitrogen fertilization and liming significantly decreases cinquefoil competition.

Strategy I: Increase nitrogen fertilization and lime to pH range of 6.7–7.2.

Strategy II: See Dandelion.

Strategy III: See Dandelion.

Yarrow

Description

Yarrow, under nonmowed conditions, can grow 2–3 ft (0.6–0.9 m) in height. However, when mowed frequently, it grows more prostrate, spreading by fleshy rhizomes, with individual plants forming modified rosettes. The leaves are compound, finely divided (fern-like), and appear woolly due to grayish fine hairs. When the leaves are crushed, they have an aromatic scent (Figure 1.56).

FIGURE 1.56
Yarrow.

Cultural Strategies

Yarrow is less competitive when frequently mowed below 30 mm and nitrogen fertility is adequate for good turf growth.

Strategy I: Physically remove by digging; being sure to remove all rhizome segments. Improve soil physical conditions by turf cultivation and modification with an organic matter source.

Strategy II: Apply a selective postemergence broadleaf herbicide in late spring or fall. Repeated applications usually are necessary. Addition of a wetting agent to the herbicide treatment improves leaf surface coverage. Follow extension recommendations for your area.

Moss

Description

Moss encompasses a relatively primitive form of plant life that has many species. Forming dense mats of finely divided vegetation that hugs the ground, moss grows well in cool, shaded, and poorly drained areas. Moss does not form true roots, but forms rhizoids, which are filamentous structures that do not provide anchoring for the plants equal to a true root (Figure 1.57).

Cultural Strategies

Moss management is more readily accomplished through cultural practices than through herbicide applications.

FIGURE 1.57
Moss.

Strategy I: Improve drainage by recontouring surface and/or installing subsurface drainage system.

Strategy II: Remove understory growth and trim trees and ornamentals to reduce shade and to allow better air movement and diffuse sunlight to penetrate.

Strategy III: Increase overall fertility level, particularly in fall and early spring when shade conditions are lessened.

Strategy IV: Apply pulverized limestone at 75–100 pounds per thousand square feet (38–50 kg/100 m²) directly to the mossy areas. Rake out moss after it has become dehydrated. Quicksilver, following label instructions, has been shown to provide some level of control.

Betony

Description

This perennial in southern turf often grows more like a winter annual than a perennial. It grows when temperatures are cool and moisture is prevalent. The leaves are opposite on square stems and have serrated margins. Vegetative reproduction occurs from rapidly spreading underground tubers. As hot weather begins, betony is not competitive with warm-season turfgrasses and disappears only to return from tubers when cool weather returns (Figure 1.58).

FIGURE 1.58
Betony. (Photo courtesy of N. Christians, Iowa State.)

Cultural Strategies

Insignificant infestations of betony can be physically removed, using caution to remove the tubers.

Strategy I: Physical removal.

Strategy II: Application of triazine herbicides to existing and established betony in centipedegrass, carpetgrass, zoysiagrass, and St. Augustinegrass. Follow extension recommendations for your area.

English Daisy

Description

Originally used as an ornamental bedding plant, this weed has become a serious turf weed in some parts of the country. Growing in modified rosettes, English daisy forms clusters of vegetation in thin turf. The leaves vary from smooth to hairy, are narrow at the base, and are slightly lobed. Typical daisy flowers are produced on 3–4 inch (7.5–10 mm) flower stalks (Figure 1.59).

Cultural Strategies

Physical removal must include removal of the extensive taproot.

Strategy I: See Dandelion.

Strategy II: See Dandelion.

Strategy III: See Dandelion.

Oxalis

Description

This perennial oxalis is quite similar to the summer annual yellow wood-sorrel, but it spreads by stolons. Creeping oxalis leaves are trifoliate,

FIGURE 1.59
English Daisy. (Photo courtesy of N. Christians, Iowa State.)

heart-shaped, and have a reddish-purple tint. The stolons are slightly hairy and root at the nodes. This species maintains a prostrate growth habit and is competitive in closely mowed situations (20–30 mm). As with other oxalis, creeping oxalis forms yellow flowers that mature into seed pods that explode when dried (Figure 1.60).

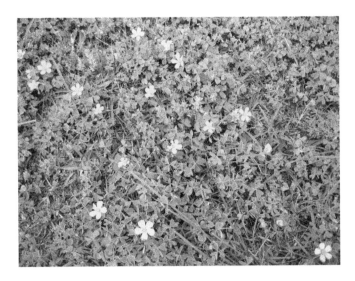

FIGURE 1.60
Oxalis (creeping).

Cultural Strategies

See Oxalis.

Strategy I: See Oxalis.

Strategy II: See Oxalis.

Strategy III: Apply a selective postemergence broadleaf herbicide in late spring and fall. Repeated applications are usually required. Make applications after turf has been mowed to improve coverage. Follow extension recommendations for your area.

Dollar Weed

Description

This perennial has rounded shiny leaves with crennated margins (some resemblance to ground ivy). The leaf is also attached to the petiole in the center of the leaf (again resembling ground ivy in appearance). The flowers are white and appear above the leaf canopy. This weed spreads vegetatively via stolons, which produce new rooting at the nodes and eventual new seedling plants (Figure 1.61).

Dollarweed is a common weed in St. Augustinegrass lawns in the southern states, particularly Florida. Owing to its growth habit, dollarweed can be very competitive in most turfgrass swards.

FIGURE 1.61
Dollar weed. (Photo courtesy of N. Christians, Iowa State.)

Cultural Strategies

Strategy I: Maintain a high level of competitiveness for the desired species.

Weed Management: Integrated Pest Management

As with other pests, integrated pest management (IPM) for weeds is a viable management philosophy that has been practiced for years. Managing weed problems is, perhaps, more complex than managing other pests, in that the pest is also a plant rather than a completely different organism. Therefore, successful weed management requires skillful manipulation of plant competition through proper cultural practices, in addition to exercising outstanding control of insect and disease pests. Lack of control of these pests often leads to turf losses, which ultimately result in weed encroachment.

In this chapter, weeds have been shown to be good indicators of certain soil conditions and that some species often occur as a result of mismanagement of the turf. Such knowledge provides the basis for developing a scouting plan.

The first part of the plan should include a historical account of the property (including records of previous weed problems, cultural attempts to manage, herbicide use and success rate, and changes in species predominance). After the historical record is assembled and understood, the next step is to scout the property. This requires an appropriate transect of the site, accompanied with a detailed record of those weeds identified and their distribution and abundance. Scouting provides the information needed to develop a customized and efficient weed management strategy. Particular attention should be given to those weed species that have escaped previous control procedures. Any unknown weeds should be sampled and identified by any and all appropriate means possible. It is not necessary to identify and record all weed species encountered during scouting. From a scouting perspective, in areas where control has been limited, a simple notation that an array of broadleaf weeds were found is sufficient. In any event, the scouting report form that is developed should provide the opportunity to generalize, but also allow for specific notation of troublesome weeds and those that require particular management attention. For example, the occurrence of several summer annual weeds can be reported as "summer annual weeds." However, a notation that goosegrass is the predominant species of the group signals the important fact that the site is probably compacted. Therefore, core cultivation will undoubtedly improve turf competition and enhance the effectiveness of any herbicide that might be chosen as a chemical control strategy.

Scouting for weeds can be done in any number of ways. Walking or riding the turfgrass site in a random zig-zag fashion, stopping periodically to sample and identify existing weeds, will begin to build a plant inventory of the site. Once this inventory of weeds and their distribution is assembled, underlying reasons for the particular array of weeds might be discerned. Such information should provide the basis for a management strategy that will maximize the competitive nature of the desired turfgrass species and enhance the effectiveness and efficiency of any herbicide that might be chosen as a chemical control agent.

Chemical Control Recommendations

Before this book is available for purchase, the availability of herbicides will have changed. There will be new materials commercialized, older products dropped from production or removed by regulations, and individual states may change their recommendations. Consequently, this chapter does not contain any specific herbicide recommendations for individual weeds. Further, Table 1.1 in the first edition contained trade names, which change rapidly, so this table will not be included. Tables 1.2 through 1.4 will also not be included in this edition. The information found in these tables also changes with regularity and can be found from time to time in most of the trade journals for turfgrass management. It is, however, strongly recommended that the reader access the recommendations provided by the turfgrass extension specialist by individual state (and throughout the world). These recommendations are reliable and current as they are updated routinely on the website for each particular specialist. Further, the Internet offers an abundance of information on herbicides (their manufacturer, MSDS, label, distributor, etc.). Also, the Internet contains valuable information with respect to weed identification, that is, the Weed ID Alert website that is maintained by PBI Gordon. Other weed identification information also exists and can be found on state university websites.

In recent years, plant growth regulators have been used to manipulate plant competition, most specifically with regard to the management of *P. annua*. The definition of a herbicide is that it is a "chemical that when applied to a plant that alters the normal growth of that plant." Therefore, for regulatory purposes, the Environmental Protection Agency considers plant growth regulators to be herbicidal in their action. *P. annua* can be controlled, to an extent, by using selected growth regulators, at very specific rates, and with crucial application timing. Recently, the American Society of Agronomy (Crop Science Division C-5) has offered for sale a CD (*Turf Growth Regulation*, 2012) pertaining to plant growth regulator classification and use on turfgrasses. The CD was developed and prepared by T. L. Watschke, J. M. DiPaola, and D. P. Shepard and is available from the American Society of Agronomy. It is the most comprehensive and current information available with regard to plant growth regulators and their use in turfgrass management.

Further Reading

Carrow, R.N., N.E. Christians, and R.C. Shearman (eds). 1993. *International Turfgrass Society Research Journal Vol. 7*. Intertec Publishing Corp., Overland Park, KS, pp. 238–310.

Lorenzi, H.J. and H.S. Jeffery. 1987. *Weeds of the United States and Their Control*. Van Nostrand Reinhold Company, Inc., New York, NY.

Muenscher, W.C. 1980. *Weeds*. 2nd Edition. Comstock Publishing Associates, Cornell University Press, Ithaca, NY.

Neal, J.C. 1993. *Turfgrass Weed Management—An IPM Approach*. Cornell Weed Management Series No. 8. Cornell University Cooperative Extension Service, Ithaca, NY.

Shurtleff, M.C., T.W. Fermanian, and R. Randell. 1987. *Controlling Turfgrass Pests*. Prentice-Hall, Inc., Englewood Cliffs, NJ.

Turgeon, A.J., L.B. McCarty, and N. Christians. 2009. *Weed Control in Turf and Ornamentals*. Pearson Education Inc., Upper Saddle River, NJ.

Zimdahl, R.L. 1993. *Fundamentals of Weed Science*. Academic Press, Inc., San Diego, CA.

2

Turfgrass Diseases and Their Management

Introduction

Most turfgrass diseases are caused by pathogenic fungi that invade leaves, stems, and roots of plants. Plant parasitic nematodes are known to cause serious damage to turfgrasses in the southeastern United States, but are not as problematic in the northern regions where cool-season grasses are grown. In cool-season grasses, parasitic nematodes are most troublesome in golf greens, especially those that consist mainly of annual bluegrass (*Poa annua* L.). There are very few recorded bacterial or viral diseases of turf.

Plant parasitic fungi and parasitic nematodes use a combination of physical pressure and enzymes to enter plants, invade tissues, and/or disrupt metabolic processes. As a result of the injurious effects of a pathogen (i.e., an organism that causes disease), a plant may exhibit various responses known as symptoms. Examples of symptoms include leaf spots or lesions; blighting (i.e., a sudden and severe death of tissue); rings, spots, or circular patches of unthrifty, discolored or dead turf; root and stem discoloration or rots; water soaking of tissue; bronzing, yellowing, or other changes in leaf color; and death of leaves, tillers, roots, or entire plants in irregular patterns. Symptoms unique to turfgrasses include variously colored circular patches of blighted turf; dead rings with living turf in the center (i.e., "frog-eyes"); small dead or discolored spots and speckles; and arcs, rings, and serpentine patterns of discolored or blighted turf. Symptoms may be unique to a particular pathogen and can be used to easily identify certain diseases. Unfortunately, these symptoms often overlap or share commonality with several different diseases and/or environmental stresses. Fungal pathogens sometimes produce visible structures known as signs. Examples of signs include spores; mushrooms; white powdery mildew; white, fluffy mycelial growth; pink gelatinous mycelial growth; red or black pustules on leaves; fruiting bodies (i.e., spore-bearing structures); and sclerotia (i.e., hard resting bodies). Mycelium (aka hyphae) is the vegetative body of a fungus, and it is composed of a network of fine tubes that often appear cottony or cobweb-like. Sclerotia are compact, hardened masses of mycelium that are produced by some fungi to serve as survival structures. Sclerotia are variously shaped, may have distinctive colors, and can often be seen without the aid of a hand lens. The presence of sclerotia greatly assists in the identification of Typhula blight, red thread, southern blight, *Rhizoctonia* diseases (especially *Rhizoctonia zeae*), and other pathogens. Fruiting bodies (e.g., mushrooms, sporodochia, acervuli, and perithecia) that contain spores are especially helpful in diagnosing several diseases. Fruiting bodies are variously shaped, have different colors, and can be large or microscopic in size. It is through the use of a combination of symptoms and signs that diseases are diagnosed. Many signs, such as spores and some types of fruiting bodies, may only be seen with the help of a microscope.

In addition to symptoms and signs, time of the year, turfgrass species (i.e., the host), and environmental conditions provide very important clues for disease diagnosis. For example, brown patch and *Pythium* blight are seldom a problem in cool-season grasses when night temperatures fall below 65°F (18°C). Pythium blight seldom causes damage to mature Kentucky bluegrass or zoysiagrass lawns, but is common on golf courses where creeping bentgrass, perennial ryegrass, and annual bluegrass are grown. Conversely, just about all turfgrasses are affected by dollar spot and fairy ring. In the transition and northern regions, dollar spot can be active from spring to autumn and can be found at almost any time of the year in some southern regions. It is important to realize that not all grasses are susceptible to all diseases. Summer patch is almost strictly a high summertime temperature disease of Kentucky bluegrass, fine-leaf fescues, and annual bluegrass. Species such as perennial ryegrass, tall fescue, and all warm-season grasses are resistant, if not immune to summer patch. Gray leaf spot is a destructive summer and autumn disease of perennial ryegrass, tall fescue, and St. Augustinegrass, but it is not a problem in creeping bentgrass or other cool-season grasses. To summarize, diseases are diagnosed using a combination of signs, symptoms, weather conditions, and a knowledge of those diseases that are most likely to appear in a given turf species in any particular season of the year. Professional turfgrass managers often turn to diagnostic labs for assistance in dealing with undiagnosed problems. Submitting a proper sample is integral to obtaining a rapid and accurate diagnosis. For more information, see the section "Collecting and Sending Diseased Samples to a Lab."

In this chapter, we focus on disease identification by providing a guide to key factors including field symptoms, distinctive and easily observed signs, predisposing environmental conditions, and grass species most likely to be affected. It should be noted that there are no absolutes in turfgrass pathology and that there can be exceptions to any rule. Approaches to reduce injury through cultural, mechanical, biological, and chemical methods are outlined for each disease.

Monitoring Disease and Establishing Thresholds

Successful disease management is contingent on early detection and a proper diagnosis. Knowledge of the environmental requirements for disease development, symptoms of the disease, pathogen signs, and the diseases likely to affect any particular turfgrass species are the primary factors scouts use to detect and diagnose diseases. Visual monitoring is essential for early detection and selection of a tactic that will effectively address each disease.

Monitoring

Turfgrass managers are responsible for monitoring, establishing acceptable thresholds of disease damage, and selecting the most appropriate management technique(s) (e.g., cultural, biological, chemical, or no action). Monitoring can only be effectively achieved on relatively small, well-defined sites such as a golf course; athletic field; park, campus, church, or public lawn; or selected lawns in a neighborhood. Visually monitoring vast areas or large numbers of lawns is not economically feasible or logistically possible. In the latter situation, chronically affected sites or "hot spots" are preferentially monitored. Hot spots are most often associated with low and frequent mowing; full sun, southern exposure, and other heat sink sites; low-lying areas where water collects; wet shade; high traffic and compacted areas; and heavily thatched sites. Every lawn care organization branch manager, golf course superintendent, landscape management specialist, and sports turf manager should know such sites (i.e., a particular neighborhood lawn, athletic field, low-lying golf green or tree-lined fairway, etc.) that are the first or most likely to develop diseases and their related problems. Some golf course superintendents subscribe to disease forecast websites, while others may install environmental monitoring stations, track growing degree days (i.e., accumulation of heat units), and employ computerized disease forecast models. Disease diagnostic kits are also available for dollar spot, brown patch, and Pythium blight identification from diseased tissues. These kits feature an antigen–antibody reaction to detect one of the aforementioned disease-causing pathogens. Kits are only useful for confirming the presence of the pathogen, rather than accurately predicting disease incidence. Although weather stations, growing degree day, and other prediction models and diagnostic kits have limitations, they provide very useful information that assists in monitoring, diagnosing, and/or helping in identifying conditions favorable for diseases. There is, however, no substitute for vigilant visual scouting and personal familiarity with "hot spot" indicator sites.

The frequency of monitoring depends on the level of management and the economic or aesthetic value of the site. For example, golf greens and stadium athletic fields should be scouted daily during the growing season, and several times daily during periods of high-temperature stress, high humidity, or extended moist and overcast weather. Golf greens covered with winter blankets, particularly when the turf is young, should be checked two or more times weekly for snow molds and/or other damping-off disorders. Similarly, sports fields covered during warm and rainy periods should be scouted every morning and afternoon. High-quality lawns or sports fields with chronic disease problems require frequent scouting during the growing season.

Thresholds

The term "threshold" refers to that level of disease that results in unacceptable injury and thus requires implementation of a management tactic. The threshold is based on the turfgrass species (i.e., its likelihood to be severely damaged); the prevailing environmental conditions (i.e., the likelihood that weather conditions will remain conducive to disease activity); the economic or aesthetic value of the site; and the cost of chemical (fungicides vs. biological agents) treatment versus renovation of damaged sites. In Table 2.1, the major turfgrass species are ranked according to their overall susceptibility to diseases. It should be noted that any attempt to rank species is artificial since there is often great variation in disease susceptibility among cultivars and regions. Intensively managed sites, particularly golf greens, are rendered more susceptible to disease by virtue of high traffic combined with extremely low and frequent mowing and frequent grooming (e.g., vertical cutting, brushing, and sand topdressing). Intensively manicured Kentucky bluegrass or perennial ryegrass lawns or sports turfs are more likely to sustain disease damage than bermudagrass, tall fescue, or zoysiagrass.

Because of their importance to the game of golf and the high maintenance expenses, greens have the lowest disease thresholds (i.e., approaching zero), whereas a utility lawn turf has much higher threshold. Diseases that develop rapidly and are extremely destructive have a lower threshold than less damaging or more slowly developing diseases. For example, Pythium blight can develop extremely rapidly and cause extensive damage in a matter of a day or two under the right set of conditions. Although less potentially destructive than Pythium blight, dollar spot can explode, pit, and disfigure putting surfaces in just a few days. Hence, bentgrass or bermudagrass golf greens have a 0% Pythium blight threshold and a dollar spot threshold of less than 0.5% blighting. In contrast, a Kentucky bluegrass lawn maintained by a lawn care organization may have a dollar spot threshold of less than 10% blighting. The brown patch threshold for the same Kentucky bluegrass lawn may be much greater (say 25%). This is because dollar spot is generally much more destructive than brown patch in Kentucky bluegrass. Conversely, tall fescue is much more likely to be chronically injured by brown patch than Kentucky bluegrass. Tall fescue grown on professionally managed lawns or on green surrounds may have a threshold of only 5–10% injury from brown patch. A tall fescue lawn or green surround affected with net-blotch or red thread would not be a candidate for chemical treatment since tall fescue is

TABLE 2.1

Disease Proneness as Related to the Major Turfgrass Species in the United States[a]

Northern, Western,[b] and Transition Zone Regions	
I. Very high Annual bluegrass (*Poa annua* L.) Perennial ryegrass (*Lolium perenne* L.)	III. Moderate Chewings fescue (*Festuca rubra* L. spp. *commutata* [Thuill.] Nyman) Creeping red fescue (*Festuca rubra* L.) Tall fescue (*Festuca arundinacea* Schreb.) Velvet bentgrass (*Agrostis canina* L.)
II. High Colonial bentgrass (*Agrostis capillaris* L.) Creeping bentgrass (*Agrostis stolonifera* L.) Kentucky bluegrass (*Poa pratensis* L.)	IV. Relatively low[c] Bermudagrass [*Cynodon dactylon* (L.) Pers.] Hard fescue (*Festuca brevipilia* Tracey) Sheep fescue (*Festuca ovina* L.) Buffalograss [*Buchloe dactyloides* (Nutt.) Englem.] Zoysiagrass (*Zoysia japonica* Steud.)

Warm-Season Grasses Grown in Southeastern and/or Southwestern States

I. Very high Kikuyugrass (*Pennisetum clandestinum* Horchst. Ex Chiov.) Seashore paspalum (*Paspalum vaginatum* Swartz)	III. Relatively low[e] Bahiagrass (*Paspalum notatum* Flügge) Carpetgrass (*Axonopus affinis* Chase) Centipedegrass [*Eremochloa ophiuroides* (Munro) Hack.] Common bermudagrass [*Cynodon dactylon* (L.) Pers.]
II. Moderate to high Hybrid bermudagrass [*Cynodon dactylon* (L.) Pers. × *C. transvaalensis* Burtt-Davy] St. Augustinegrass [*Stenotaphrum secundatum* (Walt.) Kuntze] Zoysiagrass[d] (*Zoysia japonica* Steud.) *Zoysia matrella* (L.) Merr. *Zoysia pacifica* (Gaud.) Hotta and Kuroti Hybrid zoysiagrasses	

[a] There can be great variation in susceptibility among cultivars, regions, or levels of turfgrass management.

[b] Turfgrass diseases generally are less severe in arid or semiarid regions of the western United States.

[c] Each species may suffer one or more debilitating diseases and each is generally rendered more disease prone as mowing height is reduced and/or other management inputs are increased.

[d] *Z. japonica* = Japanese lawngrass or Korean common; *Z. matrella* = Manilagrass or matrella; *Z. pacifica* = Mascarenegrass; hybrid zoysiagrasses = *Z. japonica* × *Z. matrella*; *Z. japonica* × *Z. pacifica*, others.

[e] At least one disease can be extremely destructive to each species.

likely to recover rapidly from these diseases. Fungicides generally are not warranted after a turf has been extensively damaged by disease. In that situation, or wherever a disease is a chronic problem, it is wiser to renovate the turf with disease-resistant, regionally adapted species and cultivars to minimize future outbreaks.

Thresholds also are based on the disease "history" of the site. Turf managers should maintain detailed records of all pest problems. Records should be organized by individual management units such as lawns in a particular neighborhood, individual sports fields, or by green, tee, fairway, and so on. The turfgrass species and cultivar(s), if known, should be recorded as well as the following: date of appearance of each disease and duration of disease activity; level of damage tolerated before a management tactic was imposed; and comments regarding the success or failure of the tactic employed. Disease histories for each management unit provide invaluable information for determining future disease management programs and budgets.

Environmental Conditions and Use of Cultural Practices to Manage Diseases

The relationships among environmental conditions, turfgrass plants, cultural practices, and pathogens as they interact with one another are the key factors involved in disease development. Of these factors, the environment is the most important determinant of a disease outbreak. For serious disease problems to occur, the environment simultaneously exerts diverging conditions on both the plant and the pathogen. Hence, in cases of severe disease injury, environmental conditions usually are conducive for the growth of the pathogen, while at the same time they are unfavorable for plant growth. For example, snow mold pathogens only can be destructive when low temperatures slow down or stop growth of the turf. As long as there is sufficient moisture and favorable temperatures (32–60°F; 0–15.6°C), snow mold fungi can grow actively and parasitize grass plants. Similarly, brown patch (*Rhizoctonia solani*) is most damaging to cool-season grasses during summer when high temperatures and humidity reduce turf vigor. Conversely, large patch (also caused by *R. solani*) is most damaging to warm-season grasses in spring and autumn because lower air and soil temperatures retard their growth and vigor at these times of the year.

The intensity and nature of turfgrass management influences the severity, and sometimes types of diseases that occur. Without question, diseases are much more severe on golf courses and fungicide use on these sites is common. This is not only due to heavy traffic, shade, and other poor growing environments, but also due to low and frequent mowing, frequent irrigation, and other intensive management practices routinely carried out on golf courses. Poor air circulation and shade, as well as the aforementioned factors, promote more disease in stadium athletic fields versus the more open and lower-maintenance community fields. Because of greater flexibility in maintenance tactics, most lawn diseases are managed with little use of fungicides. The foundation to any effective disease management program begins with planting regionally adapted, disease-resistant turfgrass species and cultivars. Planting of improved cultivars alone does not eliminate disease problems. Proper mowing, fertilization, irrigation, and cultivation practices (i.e., core aeration, vertical cutting, overseeding with disease-resistant cultivars, etc.) can greatly minimize the severity of many diseases.

A stand of turf normally is composed of plants representing one or two species. Planting a single cultivar of Kentucky bluegrass, the seed of which is mostly genetically identical, or planting vegetatively propagated warm-season grasses is termed monoculture. Where a monoculture exists, little or no genetic variation in disease resistance occurs. As a consequence, if one plant is susceptible to a particular disease, all plants are susceptible. Except where warm-season grasses are grown, monoculture should not be a major factor in disease susceptibility. This is because it is a common practice to blend and mix cool-season turfgrass species and cultivars to provide a broad genetic base, thereby reducing the probability of a disease epidemic. Hence, blending compatible cultivars of cool-season grasses, particularly Kentucky bluegrass and creeping bentgrass, is an extremely important first step in disease management. The best source of information regarding regionally adapted turf species and cultivars is the land-grant university of your state and its cooperative extension websites.

The susceptibility of a particular turf to disease is governed primarily by environmental and cultural factors. Weather conditions obviously cannot be manipulated, but cultural practices can be adapted to reduce disease problems. Management tactics that most influence turf diseases include mowing height and frequency, irrigation frequency and timing, fertilization (especially nitrogen), thatch and soil compaction control, and soil pH.

Mowing favors infection by creating wounds through which a pathogen can easily enter plants. More importantly, low and frequent mowing continuously removes the most photosynthetically active tissues, thereby reducing carbohydrate reserves. The overall effect of low mowing is that it weakens plants, rendering them more susceptible to diseases and environmental stresses. Mowing also spreads spores and fungal mycelium of some pathogens. Hence, when foliar diseases are active, especially Microdochium patch and Pythium blight, it is wise to avoid mowing until leaves are completely dry. Conversely, early morning mowing of creeping bentgrass turf has been shown to reduce dollar spot severity, even when mycelium is present on the foliage. Poling or dragging greens and fairways to speed up leaf drying can help reduce the severity of some diseases such as dollar spot, brown patch, and Pythium blight. Clipping removal often is necessary on golf course fairways and greens because they interfere with playability. Clippings provide nutrients that promote turf vigor, and returning clippings has been linked to reducing dollar spot (a low-nitrogen disease) severity. Recycling of nutrients (i.e., N + P + K) in clippings may help plants recover from injury more rapidly.

Height of the cut is a major factor in disease susceptibility. Close mowing exacerbates most, if not all, turf diseases. This is particularly true for Bipolaris and Drechslera leaf spot and melting-out diseases, summer patch and other root diseases, anthracnose, brown patch, dollar spot, smuts, rusts, and injury caused by plant parasitic nematodes. Cultivars of Kentucky bluegrass that are apparently resistant to summer patch may be predisposed to this disease by low mowing. It is important to note that the term "resistance" means only that a plant is less susceptible to a disease. Resistance is a relative concept that does not imply that a cultivar is immune to a particular disease.

One of the most important management recommendations for most disease problems is to increase mowing height when turf injury is first observed. Continuous removal of the youngest, most photosynthetically productive tissues, below recommended heights, causes depletion of carbohydrate reserves in grass plants. These reserves play a key role in active disease resistance processes and are needed by plants to recover from injury. Increasing mowing height increases leaf area, which results in an increase in carbohydrate synthesis. High mowing also helps moderate soil temperature and stimulate deeper root penetration. For example, studies have shown that summer patch was less damaging to Kentucky bluegrass maintained at 3.0 inches (7.5 cm), when compared to 1.5 inches (3.8 cm) (Davis and Dernoeden, 1991). It was also shown that when Kentucky bluegrass turf was maintained at 1.5 inches and then allowed to grow to 3.0 inches, it recuperated more rapidly from summer patch, when compared to turf maintained continuously at a 1.5 inch mowing height.

Irrigation provides moisture critical to spore germination and fungal development. Timing, duration, and frequency of irrigation may greatly affect disease intensity. Light, frequent irrigations discourage root development, and predispose turf to injury when extended periods of drought occur. Light and frequent irrigations, which keep soils moist during periods of high-temperature stress, greatly increase summer patch in Kentucky and annual bluegrasses. This is because the fungus that causes summer patch (i.e., *Magnaporthe poae*) grows much more rapidly along roots when soils are moist and very warm (>85°F; >29°C in the root zone). Studies have also shown that summer patch in Kentucky bluegrass was much less severe when 3.0 inch (7.5 cm) high turf (4% injury) was subjected to deep and infrequent irrigation when compared to 3.0 inch high turf (16% injury) or 1.5 inch (3.8 cm) high turf (36% injury) subjected to light and frequent irrigation. The light and frequent irrigation treatments in that study kept soil moist for extended periods, which favored root infection by *M. poae* during the warm summer months. Frequent cycles of wetting and drying of thatch enhance spore production by leaf spot pathogens (especially *Pyricularia grisea*, *Drechslera*, and

Bipolaris spp.). Sometimes, however, light and frequent irrigation is recommended in situations where a disease or an insect pest has already caused extensive damage to the roots. For example, studies have shown that turf recovery from necrotic ring spot (NRS), another root disease of Kentucky bluegrass, was improved by light daily watering during dry summer periods (Melvin and Vargas, 1994).

Subjecting turf to extended periods of drought stress can enhance the level of injury in turfgrasses affected by root diseases (e.g., take-all patch and summer patch), anthracnose, leaf spot and melting-out diseases, stripe smut, dollar spot, fairy ring, and localized dry spot. Conversely, excessive irrigation restricts root development and encourages numerous diseases. Turfgrasses grown under wet conditions develop succulent tissues and thinner cell walls, which are more easily penetrated by fungal pathogens. Blue-green algae (aka cyanobacteria), mosses, black-layer, anthracnose, Pythium diseases, brown patch, summer patch, and weed seed germination are encouraged by wet or waterlogged soils, particularly where stand density is poor. Early morning irrigation often is recommended during summer to ensure that plant tissues are dry by nightfall. This practice reduces the number of nighttime hours when leaves are wet and thereby helps minimize the intensity of foliar blighting diseases such as Pythium blight and brown patch.

Proper soil fertility improves plant vigor and the ability of plants to resist or recover from disease. Nitrogen is the most important nutrient used in turfgrass management. Excessive use of nitrogen, however, promotes tissue succulence and thinner cell walls, which as previously mentioned, are more easily penetrated by fungal pathogens. Overfertilization with nitrogen encourages snow molds, leaf spot and melting-out, brown patch, gray leaf spot, Pythium blight, and other diseases. Too much nitrogen applied at inappropriate times diverts carbohydrates and important biochemicals into growth processes devoted to leaf production. Diversion of these plant chemicals into developing tissues, rather than to plant defense, may result in shortages, which adversely affect the capacity of the plant to resist fungal invasion. Conversely, turfgrasses grown under low nitrogen fertility and/or in nutrient-poor soils lack vigor and are more prone to severe damage from anthracnose, dollar spot, red thread, and rust diseases. Application of nitrogen to diseased turf under the aforementioned conditions improves vigor and stimulates new foliar growth at a rate that exceeds the capacity of these fungi to colonize new tissues.

In general, slow-release nitrogen fertilizers (synthetic or organically based) are recommended where chronic disease problems occur. The controlled-release properties of these nitrogen sources result in a moderate rather than rapid or excessive growth rate. Selected composted poultry waste and sewage sludges and other natural organic nitrogen sources can be used effectively for managing foliar diseases such as dollar spot and red thread. While slow-release nitrogen sources are preferred for disease management, there are some exceptions to this generalization. Rapid-release, water-soluble nitrogen sources (e.g., urea) may be recommended during autumn months to help speed up recovery of turf injured by summer diseases or environmental stresses. Furthermore, water-soluble nitrogen from ammonium sulfate when used properly can reduce the severity of several root diseases (e.g., take-all patch, spring dead spot [SDS], dead spot, and summer patch) over time by acidifying soil.

There is little documented evidence that phosphorus (P), potassium (K), or micronutrients alone significantly impact turf diseases. Deficiencies in these nutrients, however, can weaken plants and predispose them to more disease injury than normally would have occurred. For example, take-all patch in creeping bentgrass is more severe where phosphorus is deficient, particularly on sand-based greens and tees. It is likely that Pythium-induced root dysfunction and other root diseases in sand-based golf greens may be more severe in phosphorus-deficient soils. More importantly, complete fertilizers (i.e., N + P + K fertilizers) when applied in a 3N:1P:2K ratio helps reduce the severity of take-all patch, stripe smut, SDS, red thread, and Microdochium patch when compared to nitrogen alone.

Many turfgrass pathogens survive as resting structures or as saprophytes (organisms living on dead organic matter) in thatch. Thatch is a harborage site for pathogens because it provides fungi with nutrients and moisture. As noted earlier, some pathogens such as *Drechslera* and *Bipolaris* spp. (leaf spot and melting-out), *Colletotrichum cereale* (anthracnose), and *P. grisea* (gray leaf spot) produce enormous numbers of spores in thatch. SDS, summer patch, smuts, leaf spot and melting-out, and many other diseases, as well as turf injury due to environmental extremes, are favored by excessive thatch accumulation.

Traffic, like mowing, results in wounds that are easily invaded by some fungal pathogens. Compaction caused by heavy traffic impedes air and water movement into soil and eventually restricts root growth and function. This results in decline in plant vigor and disease resistance. Summer patch, take-all patch, leaf spot and melting-out, and anthracnose tend to be more damaging on highly trafficked or compacted soils. Core aeration helps alleviate compaction, reduce thatch, improves soil aeration and plant vigor, and thus reduces the severity and recovery time from disease. Conversely, plant parasitic nematode populations tend to be higher in light-textured, sandy loam soils than in clays or compacted soils.

Soil pH may also affect disease development in turfgrasses. Most turfgrass pathogens are able to grow at any pH encountered by turf. High soil pH, however, favors dead spot, summer patch, take-all patch, and Microdochium patch. Factors that encourage some diseases under alkaline soil conditions remain imperfectly understood. It is likely, however, that some pathogens achieve a competitive advantage over antagonistic microorganisms in alkaline soils. Acidifying nitrogen fertilizers have been demonstrated to reduce the severity of take-all patch, Microdochium patch, SDS, dead spot, and summer patch. Soil pH, however, does not directly impact many diseases, so it is prudent in most situations to maintain or adjust soil pH to a range 6.0–7.0.

Adherence to sound cultural practices is basic to reducing disease severity. The basic principles of cultural disease management of turfgrasses are outlined in Table 2.2. The turfgrass environment, however, is not static and managers must continually modify cultural practices to encourage vigorous growing conditions. To maintain optimal conditions, turfgrass managers should routinely monitor the nutrient and pH status of soil, adjust mowing heights, judiciously irrigate, overseed with disease-resistant cultivars, manipulate different nitrogen fertilizer sources, and control thatch and soil compaction. Shade is also a key factor that limits root growth and contributes

TABLE 2.2

Basic Principles for Managing Turfgrasses to Minimize Damage Caused by Pathogens

1. Plant disease-resistant species and cultivars. Recommendations for regionally adapted cultivars can be obtained from most land-grant university cooperative extension service websites.
2. Use complete N-P-K fertilizers at the appropriate times. In the absence of a soil test, N-P-K should be applied at least once annually in a ratio of 3N:1P:2K. Most of the total annual nitrogen should be applied to cool-season grasses during autumn months. Conversely, most nitrogen should be applied to warm-season grasses at spring green-up and during summer months.
3. Preferably, half or more of all nitrogen applied per year should be from a slow-release nitrogen source. Some examples of slow-release nitrogen sources include isobutylidene diurea (IBDU), methylene urea, sulfur-coated urea, and composted or activated sludges or poultry waste products (e.g., Milorganite® and Ringer Lawn Restore®).
4. Maintain a high mowing height within a species adapted range. Raise mowing height during periods of environmental stress or disease outbreaks. To reduce disease damage in lawns, it is best to maintain Kentucky bluegrass or perennial ryegrass at 2.5–3.0 inches (6.2–7.5 cm), and all fescue species at 3.0–3.5 inches (7.5–8.8 cm) in height. Bermudagrass and zoysiagrass lawns should be maintained to a height of 1.0–1.5 inches (2.5–3.8 cm); centipedegrass 1.5–2.0 inches (3.8–5.0 cm); St. Augustinegrass 2.5–3.0 inches (6.2–7.5 cm); and bahiagrass 3.0–3.5 inches (7.5–8.8 cm).
5. Reduce shade and improve air circulation and surface water drainage.
6. Irrigate deeply to wet soil to just below root zone depth when turf first exhibits signs of wilt. Avoid frequent and light applications of water, except when root systems are shallow.
7. Test soil every 2 or 3 years for phosphorus and potassium levels as well as soil pH. Adjust soil pH to a range of 6.0–7.0; lower soil pH in the range of 5.5–6.0 is preferred for creeping bentgrass.
8. Routinely monitor soluble salt levels in soils in western regions or wherever high sodium (sodic) soil levels are common or where salt water intrusion of wells is possible.
9. Avoid applying herbicides or plant growth regulators when diseases are active.
10. Alleviate soil compaction and reduce thatch through core aeration, vertical cutting, or a combination of these and other methods.
11. Overseed or renovate chronically damaged sites with disease-resistant species and regionally adapted cultivars.

TABLE 2.3

Cultural and Mechanical Approaches to Alleviating Diseases of Golf Greens

1. Avoid excessive irrigation and water early enough during the day to avoid long leaf wetness duration at night.
2. Remove dew and leaf surface exudates to speed leaf drying by early morning mowing, dragging, poling, or whipping.
3. Avoid mowing when greens are excessively wet and soft.
4. Keep mowers adjusted and blades sharp; use walking greens mowers; and increase height of cut and reduce mowing frequency whenever possible, especially in summer.
5. Core aerate and/or deep tine aerate (e.g., verti-drain and drill and fill) compacted sites.
6. Employ spiking, water injection, and/or small-diameter tine coring routinely during the season.
7. Use soil wetting agents to alleviate localized dry spots or fairy ring damage.
8. Avoid rolling greens when surfaces are wet and soft.
9. Remove trees and brush to improve air movement and sunlight penetration; electric fans are very effective in improving air movement.
10. Prune tree roots around greens to reduce competition for water and nutrients.
11. Spike or solid tine when soils are excessively wet.
12. Keep thatch and/or mat layers to a minimum (i.e., <0.3 inches; <7.5 mm). Control thatch and mat by large-diameter tine core aeration in spring and autumn in cool-season grasses and summer for warm-season grasses and fill holes to the surface with sand. Weather permitting, lightly sand topdress and core with narrow-diameter tines on 2–3-week intervals during the golf season.
13. Avoid aggressive grooming practices when greens are weakened by environmental stress, mechanical damage, or diseases.

to turf decline. Lawns and golf greens shrouded in trees and untrimmed ornamentals are chronically weak turfs. Hence, heavy pruning and sometimes the removal of numerous trees and brush would be sound cultural practices for improving turf vigor and thereby alleviating some diseases. Cultural and mechanical approaches to alleviating golf green diseases are listed in Table 2.3.

Biological Control of Turfgrass Diseases

The development of biological control agents for turfgrass disease management lags behind advances in insect pest control. It is generally easier to identify diseases of insects, isolate the disease agents, grow them in a laboratory, and direct sprays of propagules (infecting units, such as spores or mycelium) to sites infested with these pests. Using bacteria or fungi to control fungal turfgrass pathogens is much more complicated. The microscopic size of the target pest and the complex environmental forces affecting both the biological control agent and the pathogen are significant barriers to progress. Pathologists have demonstrated that several fungi and bacteria provide suppression of some turfgrass diseases in controlled laboratory studies. In field studies, however, these agents often fail to provide acceptable levels of disease suppression. Most field failures are due to the inability of biological control agents to survive, compete, or reproduce in populations large enough to provide disease suppression. The most common biological agents that target turf diseases are bacteria and include *Bacillus licheniformis*, *Bacillus subtilis*, and *Pseudomonas aureofaciens*.

Genetic engineering to improve pest resistance in turfgrasses is underway, but the marketing of transformed cultivars will not be forthcoming for several more years. While genetically transformed grasses may ultimately be the answer to several, common disease problems, other methods and technologies are also being studied for their potential to reduce the use of fungicides on turfgrasses. Composts and natural organic fertilizers are being used in the hope that they can boost the microbial activity in the soil. It is generally believed that enhanced soil microbial activity results in more competition with or antagonism of potential pathogens thereby suppressing disease problems. These materials may be helpful, but the few field studies that have compared natural organic to synthetic organic fertilizers have not provided a great deal of evidence that the natural fertilizers consistently are superior to other slow-release nitrogen sources. Furthermore, when natural and synthetic organics were shown to reduce diseases, such as brown patch or dollar spot, the level of suppression generally was not commercially acceptable after the disease progressed from a low to a moderate level.

The use of biological inoculants is becoming popular and there are numerous products that claim disease suppression. The common soil fungus *Trichoderma harzianum* (Bio-Trek®) was the first biological fungicide registered for turfgrass disease control. The bacteria *P. aureofaciens* strain Tx-1 (Spotless®), *B. licheniformis* (EcoGuard®), and *B. subtilis* strain QST 713 (Rhapsody®) are the primary biological agents currently employed on turfgrasses. All three bacteria target dollar spot and the two *Bacillus* species are also labeled for use against anthracnose and brown patch. Biological inoculants reduce disease by one or more mechanisms, including (1) parasitizing fungi (i.e., mycoparasitism) that cause disease; (2) antagonizing by outcompeting the pathogens for space and infection sites; (3) producing antibiotic substances that suppress the growth of or kill sensitive pathogens; or (4) triggering the plants' own natural defense mechanisms.

Mycoparasitism: Fungi, including some species of *Trichoderma* and *Gliocladium*, presumably work in part by parasitizing the vegetative body (i.e., mycelium) or propagules (spores, sclerotia, etc.) of sensitive pathogens.

Antagonism and competition: Antagonists include bacteria, actinomycetes (i.e., filamentous soil bacteria), and fungi. Antagonists produce antibiotic substances that inhibit growth or spore production, or the antibiotic may be toxic enough to directly kill sensitive microbes. This phenomenon also is known as "antibiosis." Antagonists must be excellent competitors, which grow rapidly to displace potential pathogens as they attempt to colonize roots and other infection sites on plant tissues. Common soil bacteria from the genera *Bacillus*, *Enterobacter*, and *Pseudomonas* are among the best-known antagonists.

Triggering natural defenses: Some microbes can be used to "turn on" the plants' natural defense mechanisms. Plants produce substances in response to infection (i.e., phytoalexins) that are toxic to pathogens. Plants may also produce an array of physical barriers to prevent further ingress of a pathogen following infection. The fungus *Phialophora graminicola* was shown to prevent take-all disease by stimulating plant defense mechanisms in creeping bentgrass roots, but was not commercialized. Today, chemicals are being used in disease management programs to trigger plant defenses (e.g., acibenzolar and phosphonate fungicides).

Most successful examples of biological control were conducted under laboratory, greenhouse, or growth chamber conditions. When field tested, however, they often provided either erratic levels or little or no disease control. There are some significant barriers to success, and perhaps the greatest problem is maintaining high populations of introduced agents in soil and thatch, and/or on leaves, sheaths, and roots. Biological agents are common microbial species found in most soils. Many biotypes within a species evolve in place, but when introduced into new or foreign soils, they are not likely to survive or reproduce in high-enough populations to provide the beneficial effects. For example, the introduced biological agent must be able to grow and reproduce in these new soils or thatch niches where they are likely to encounter unfamiliar types of environmental extremes (i.e., high or low temperatures; wet or dry conditions; soil texture and pH and aeration variations, etc.) They must also be able to compete with a myriad of naturally occurring antagonists, which evolved at the site. Furthermore, scientists do not know much about the survivability of most candidate biological agents or how to handle or formulate these agents for use in turf. Important questions needing answers include: when should applications begin each year for each potential pathogen, and should they be watered into the root zone or are they better placed in the thatch or maintained on leaves and sheaths? We do not know the threshold populations (i.e., how many propagules should be applied per unit area) or how often they need to be applied to sustain beneficial numbers. Furthermore, we know that many (e.g., *Pseudomonas* spp.) are sensitive to UV light, suggesting that daytime application would result in many of the cells dying rapidly.

Also, little is known about the influence of other pesticides, fertilizers, wetting agents, biostimulants, and so on, routinely used on turfgrasses and how they may impact the viability of the introduced agents. Then there are production and storage problems that must be considered. Many of these potential problems were addressed in several university studies involving the Bio-Ject Management System®. The Bio-Ject System ferments large numbers of bacteria and injects them into the irrigation system for nighttime delivery. This system addresses the problem of delivering high populations of disease-suppressing bacteria on a frequent basis during nighttime hours. *P. aureofaciens* strain Tx-1 (Spotless) was registered for use in the Bio-Ject Management System. Research has shown that Tx-1 can provide reductions in dollar spot between 30% and 60%. However, there is little evidence that Tx-1 will suppress other turfgrass diseases in the field.

To summarize, commercially available biological agents generally provide some suppression when disease pressure is low. Once disease becomes severe, they tend to have only minimal effectiveness. It is not unusual to see inconsistencies in pest suppression using biological agents from year to year or even among regions in the same year. This is due in large part to variable environmental conditions. Microbes applied to plants must have favorable moisture and temperature conditions to grow and reproduce. In many situations, the optimal environmental conditions that favor growth of a biological agent may not prevail at the time(s) they are applied. Furthermore, success and failure of biological agents can be influenced by numerous factors such as pathogen biotype at the site (i.e., there are at least four biotypes of *R. solani* that attack turfgrasses); the level of disease pressure (i.e., low vs. high); the relative susceptibility of the turf (e.g., 007 creeping bentgrass has good dollar spot resistance when compared to Penn A-1 creeping bentgrass); the formulation and shelf life of the biological agent; the timing of application (i.e., some organisms are degraded by sunlight); and many other known and unknown factors. Hence, like all new products and technologies, biological agents must be tested over a wide geographic area for several years to learn how to attain the best possible level of disease suppression.

Maintaining turf vigor through sound cultural practices, planting disease-resistant cultivars, and ensuring good soil aeration to promote soil microbial activity are the most effective means of biologically reducing turfgrass diseases. The use of a biological approach is nothing new in turf management. Prior to the advent of synthesized fertilizers and fungicides, golf course managers relied on composts for their nutrients. Composted topdressings were prepared by mixing equal parts of well-rotted cow manure, rotted grass clippings, and soil. The mixture was allowed to compost for a 3-year period prior to being screened, mixed with sand, and spread on golf greens. This mixture probably suppressed some diseases such as dollar spot.

Winter and Early Spring Diseases

Snow protects dormant turfgrass plants from desiccation and frost, but also provides a microenvironment conducive to the development of some low-temperature-tolerant, pathogenic fungi. There is no shortage of fungi capable of damaging turf during cold periods between late autumn and spring. The most common low-temperature fungal diseases are Microdochium patch and Typhula blight. Cottony snow mold (*Coprinus psychromorbidus* Redhead and Traquair), snow scald (*Myriosclerotinia borealis* Bubak and Vleugel), and frost scortch (*Sclerotium rhizodes* Auers) are less common low-temperature diseases, which are normally found in very cold climates. The aforementioned diseases are sometimes collectively referred to as snow molds. Other diseases known to be active under snow cover or at snow melt during winter or early spring months include red thread, Drechslera leaf spot, yellow patch, and Pythium snow blight. Crown hydration, freezing, desiccation, and ice damage also are important problems, especially in the northern-tier states, Rocky Mountain states, and Canada. These winterkill problems may be referred to as abiotic (i.e., nonliving) diseases. Cold-weather-related abiotic disorders can sometimes be confused with snow molds.

Although these fungi are known as snow molds, some can attack turf with or without snow cover. Snow mold fungi are remarkable because they are active at temperatures slightly above freezing. Snow molds are damaging when the turf is dormant or when its growth has been retarded by low temperatures. Under these conditions, turfgrasses cannot actively resist fungal invasion. In general, these diseases develop during overcast periods when temperatures are cool to cold (32–60°F; 0–15.6°C), and there is an abundance of surface moisture.

In the following sections on individual diseases, some fungicide recommendations are provided. All fungicides are not legal for use in all situations and labels are frequently amended or otherwise changed. For example, chlorothalonil and iprodione may not be applied to lawns. Thus, it is important to refer to the most recent label before applying any fungicide.

Microdochium Patch (aka Pink Snow Mold or Fusarium Patch)

Pathogen: Microdochium nivale (Fr.) Samuels and Hallett

Primary hosts: Most turfgrasses, especially annual bluegrass, bentgrasses, fescues, perennial ryegrass, and seashore paspalum

Predisposing conditions: Prolonged periods of snow cover or cold, wet, and overcast weather

Microdochium patch is among the most common and destructive cold weather diseases of turf. This disease may be referred by at least three different names. This is because of two reasons: (1) the disease may occur in the presence or absence of snow, and (2) the taxonomy or Latin binomial for the pathogen has been changed numerous times. Traditionally, the disease was known as Fusarium patch in the United Kingdom and as pink snow mold in the United States. Some pathologists argued that pink snow mold was an inappropriate name because the disease can occur in the absence of snow. Thereafter, the names Fusarium patch and pink snow mold were adopted to refer to the disease when it occurred either in the absence or in the presence of snow, respectively. To confuse matters, mycologists reclassified *Fusarium nivale* as *Gerlachia nivalis* around 1980. In 1983, *G. nivalis* was declared a misnomer and the binomial *M. nivale* was accepted. Hence, the names Fusarium patch, Gerlachia patch, and Microdochium patch appear in the literature, and all refer to the same disease. The latter genera represent the asexual stage of the pathogen. To further confound the issue, each of these asexual binomials was also assigned a sexual-based Latin binomial. For example, the sexual binomial for *M. nivale* is *Monographella nivalis*. Since 1849, the pathogen has been given at least 10 Latin binomials (Smith et al., 1989). Therefore, Fusarium patch, Microdochium patch, and pink snow mold are all commonly used names for the same disease. Microdochium patch is now the accepted name for this disease, regardless of whether or not it is associated with snow cover.

Microdochium patch attacks a wide range of cool-season turfgrass species under snow, at snow melt, or during extended periods of wet, overcast weather from autumn to spring, including perennial ryegrass, annual and Kentucky bluegrasses, bentgrasses, and tall and fine-leaf fescues. Microdochium patch is also a major winter disease in seashore paspalum (a warm-season grass) throughout the southeastern United States to include southern Florida. Among cool-season grasses, this disease generally is most destructive to annual bluegrass and bentgrasses. Microdochium patch is especially destructive to cool-season grass seedlings the first autumn to spring period following seeding. This is due to overstimulation of growth and succulent tissues resulting from high levels of nitrogen fertilizer applied during establishment, as well as the immaturity of plants. The disease can be especially severe when seedlings are covered with blankets. It is, therefore, important to check under blankets frequently for the presence of this and other diseases.

Conditions favoring Microdochium patch include low (32–45°F; 0–7°C) to moderate (46–65°F; 8–18°C) temperatures; high relative humidity; overcast weather, abundant moisture; prolonged, deep snow; snow fallen on unfrozen ground; wet or poorly drained sites; wet shade; lush turf stimulated by late-season applications of excessively high amounts of nitrogen fertilizer; and alkaline soil conditions. Prolonged periods of cool to cold, overcast, and rainy weather are particularly conducive to disease development. The disease may appear anytime between autumn and spring. In alpine regions and some maritime climates, such as in the United Kingdom; Pacific Northwest, United States; and British Columbia, Canada, Microdochium patch (i.e., the snow-free phase) can develop year-round. In some U.S. regions, such as in the mid-Atlantic states, the snow-free phase most often appears in April or May. May outbreaks of Microdochium patch often surprise or confuse turf managers because they think of this disease as a "snow mold" and of the presence of snow.

Symptoms in the Absence of Snow

The snow-free stage of this disease is primarily a problem in close-cut golf course turfs and is not commonly observed in higher-cut cool-season grasses in roughs, lawns, or sports fields. On mature creeping, colonial, and velvet bentgrass or annual bluegrass greens, tees, and fairways, the snow-free Microdochium patch phase initially appears as circular spots or patches 1–3 inches (2.5–7.5 cm) in diameter (Figure 2.1). These patches generally have a pink or reddish-brown color and may increase in size to six or more inches (15 cm) in diameter. During the early stages of the disease in the absence of snow, however, the small spots (1–2 inches diameter; 2.5–5.0 cm) may have a whitish-tan center with a pink or reddish-brown fringe. Gray-colored smoke-rings of *M. nivale* mycelium may also appear at the edge of affected spots or patches in the absence of snow (Figure 2.1). Reddish-brown rings or frog-eyes can occur, particularly in seedlings (Figure 2.2). Mycelium on the leaf blades produces fruiting bodies called sporodochia (i.e., superficial cushions with clusters of spores) on which large numbers of crescent-shaped spores are borne (Figure 2.3). These white- or salmon-pink-colored sporodochia are very tiny and appear as flecks on necrotic (dead) tissue. These flecks may be seen with a hand lens when the disease is active, but generally cannot be seen on dried tissue. The spores are easily spread by water, machinery, and foot traffic. Therefore, blighting can appear in streaks or even straight lines when spores are dispersed by mowers (Figure 2.3). This streaking effect often is confused with a Pythium disease, especially during late spring outbreaks of Microdochium patch.

FIGURE 2.1
In the absence of snow, Microdochium patch appears as reddish-brown spots or small patches in creeping bentgrass.

FIGURE 2.2
Microdochium patch in creeping bentgrass seedlings, which developed under a winter cover.

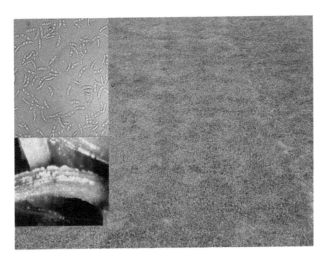

FIGURE 2.3
White cushions of mycelium are fruiting bodies (i.e., sporodochia) on which prodigious numbers of asexual spores (conidia) are produced. Spores can be dispersed by mowers resulting in streaks of blighted turf.

Symptoms in the Presence of Snow

Large, circular patches are most likely to appear when the disease develops underneath a deep snow cover. As snow recedes, the turf at the fringe of the tan-colored patches appears pink (Figure 2.4). The pink appearance is produced by the pink or salmon color of *M. nivale* mycelium. The pink fringe

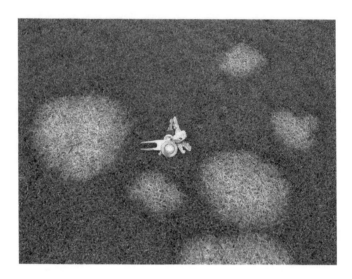

FIGURE 2.4
Microdochium patches forming under snow cover or as snow recedes from golf greens are tan colored and have a pink fringe. (Photo courtesy of C. Bigelow.)

around patches fades in time and may not be evident following the advent of drier conditions. Plants eventually collapse and leaves are matted. Matted leaves have a tan or white color, but on close inspection, they may display a pale, pinkish cast (Figure 2.5). Patches developing under snow range from 3 to 18 inches (7.5–45 cm) in diameter. When damage occurs under snow, the extent of injury usually is more severe than without snow cover.

FIGURE 2.5
Microdochium patch in a tall fescue lawn.

Management

Microdochium patch injury can be reduced by using a complete N-P-K fertilizer in the autumn and by avoiding excessive, late-season applications of water-soluble nitrogen. Ammonium sulfate may be suggested as a nitrogen source where soils are alkaline and Microdochium patch is common. Modest amounts (≤1.0 lb N/1000 ft²; ≤50 kg N/ha) of ammonium sulfate or other water-soluble nitrogen sources applied in late autumn, however, are not likely to enhance Microdochium patch in mature turf. Avoid use of limestone where the soil pH is above 6.5 since soil alkalinity may encourage this disease. Continue to mow into late autumn to ensure that snow will not mat a tall canopy and remove fallen tree leaves. On golf courses, snow fences and windbreaks should be used to prevent snow from drifting onto chronically damaged greens. Divert cross-country skiers and snow mobiles around greens to avoid snow compaction.

Fungicides are best applied preventively in late autumn and just prior to the first major snow storm. It is especially important to protect young seedlings. Subsequent applications to golf greens or other prone locations (e.g., shade and north-facing slopes) should be made during mid-winter thaws and at spring snow melt in areas where the disease is chronic. Azoxystrobin, chlorothalonil, fluazinam, fludioxonil, fluoxystrobin, iprodione, myclobutanil, pyraclostrobin, propiconazole, PCNB (aka pentachloronitrobenzene or quintozene), thiophanate-methyl, triadimefon, trifloxystrobin, and others provide effective Microdochium patch control. Most fungicides in any given year may provide only marginally acceptable Microdochium patch control when applied alone. Therefore, two or more of these fungicides are normally applied in a tank-mix combination. Tank-mix combinations improve the level of control as well as provide a broader spectrum of control for other diseases that may occur at the same time such as Typhula blight and yellow patch. Chlorothalonil, fludioxonil, and iprodione make good tank-mix partners with demethylation or sterol-inhibiting (DMI/SI) and QoI/strobilurin fungicides (see the section "Types of Fungicides" for more information).

As noted previously, turf covered with blankets should be monitored frequently for disease between autumn and spring. During extremely wet or snowy winters, Microdochium patch can cause extensive injury to lawns. It is best to spot apply fungicides to lawns where the disease chronically develops in localized pockets. In most regions of the United States, however, Microdochium patch prevention with fungicides is warranted for golf course turf, bentgrass/annual bluegrass bowling greens and tennis courts, and stadium athletic fields.

Pythium Snow Blight

Pathogens: Pythium graminicola Subramanian, *Pythium iwayamai* Ito, others

Primary hosts: Annual bluegrass and creeping bentgrass golf greens

Predisposing conditions: An abundance of snow, surface water, and saturated soils during cool or cold periods

Creeping bentgrass and annual bluegrass golf greens can be damaged in winter by several *Pythium* species. These low-temperature-tolerant *Pythium* spp. cause both a foliar blight and stem or root rot. Pythium snow blight is also known as Pythium patch and is favored by abundant moisture, particularly when snow falls on unfrozen soils. It can potentially affect many grasses, but currently is recognized as a disease primarily restricted to golf greens and its occurrence is uncommon. This disease often follows surface water drainage patterns and is most injurious in low areas of poorly drained golf greens.

Symptoms

Damage appears as small reddish-brown, tan, gray, or yellow to orange spots or patches about 1.0–4.0 inches (2.5–10.0 cm) in diameter (Figure 2.6). Injury often appears first at the outer peripheral areas of the putting surface, near the interface between native soil and the sandy greens mix. Foliar mycelium may or may not be evident. While the disease occurs on mature greens, it is most common in relatively young bentgrass turf grown on high-sand-content mixes. In late winter or early spring, dead patches may mimic old dollar spot damage, that is, patches are 1.0–3.0 inches (2.5–7.5 cm) in diameter and leaves appear gray or tan. To confirm the diagnosis, pathologists will inspect diseased tissues for the presence of *Pythium* spp. oospores, sporangia, and/or mycelium.

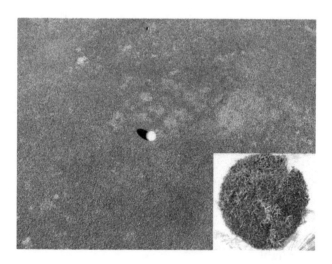

FIGURE 2.6
Symptoms of Pythium snow blight on a golf green appear as small yellow, tan, or bronze-colored spots. (Large photo courtesy of M. Fidanza and inset photo courtesy of D. Settle.)

Management

There are no known cultural measures for managing this disease. Improving surface water drainage would be expected to reduce the potential for severe injury. Check underneath winter blankets frequently for blight development. If there is a persistent snow cover, low areas on pocketed golf greens should be cleared and scouted periodically for disease. Where the disease is chronic, apply a *Pythium*-targeted fungicide (e.g., cyazofamid, fosetyl-aluminum, propamocarb, and mefenoxam) that penetrates tissue in combination with a contact such as chloroneb, ethazole, or mancozeb prior to the first major snow storm of the year. Apply a fungicide again at snow melt or during a mid-winter thaw to greens with a past history of this disease.

Typhula Blight or Gray Snow Mold

Pathogens: Typhula incarnata Fr., *Typhula ishikariensis* Imai

Primary hosts: All cool-season turfgrasses

Predisposing conditions: Prolonged periods of deep snow cover

Typhula blight also is known as gray snow mold. Typhula blight normally occurs under snow cover or in the presence of melting snow; outbreaks in the absence of snow are uncommon. *T. incarnata* is the most common incitant, whereas *T. ishikariensis* is generally found in colder mountainous regions where snow cover is more prolonged. Cool-season grasses are hosts and both species of *Typhula* begin the disease cycle as saprophytes colonizing dead organic matter. Under snow, the fungus attacks leaves and sheaths. Typhula blight is more damaging under prolonged deep snow, particularly during melt as snow recedes. Normally, *Typhula* spp. do not infect and kill stems of mature plants and blighted turf would be expected to recover during spring as soils warm. Conversely, *M. nivale* can invade stem tissues and kill tillers and/or entire plants.

Symptoms

Symptoms initially appear as tan, light-brown, yellow-brown, or gray patches 2.0–4.0 inches (5–10 cm) in diameter (Figure 2.7). Under snow cover, patches may enlarge to 2 ft (60 cm) in diameter and coalesce. A sparse to dense amount of grayish mycelium may be seen on patches as snow recedes. Margins of active patches may have a "halo" of grayish-white mycelium (Figure 2.8). Hyphae bear a bridge-like protrusion called a clamp connection (Figure 2.8). Blighted leaves are grayish-white and leaves are matted. Typhula blight patches are usually larger in size than those of Microdochium patch. Patches may coalesce, resulting in large irregular areas of blighted turf (Figures 2.9 and 2.10).

FIGURE 2.7
Typhula blight (aka gray snow mold) patches in tall fescue.

Typhula spp. survive unfavorable environmental conditions and over-summer as sclerotia. Sclerotia are compact masses of fungal mycelium covered with a dark-colored protective rind. *T. incarnata* sclerotia are initially smooth, round, and pink, but as they mature and dry, they become wrinkled, rounded or flattened, chestnut-brown in color and are often less than 0.125 inch (3.1 mm) in diameter (Figure 2.11). Sclerotia produced by *T. ishikariensis*

FIGURE 2.8
Typhula blight with a halo of mycelium in a bentgrass fairway. Hyphae are made distinguishable by the presence of a prominent bridging projection called a clamp connection (inset).

FIGURE 2.9
Typhula blight patches can coalesce and blight large, irregularly shaped areas.

are black and irregular in shape. Sclerotia may also appear as tiny, rounded specks or as elongated, blackened or reddish-brown crusts embedded in or clinging to necrotic tissue. When cool, moist weather conditions return in late autumn, these sclerotia germinate to produce fungal mycelium or a specialized fruiting body (i.e., basidiocarp) upon which spores are borne. Basidiocarps are club-shaped, 0.125–0.250 inch (3.1–6.2 mm) tall, and are

FIGURE 2.10
Turf in the foreground was blighted under snow by *Typhula incarnata*, but greener turf in the approach and on the green were protected by a preventive fungicide application.

FIGURE 2.11
Sclerotia of *Typhula incarnata* initially are pink and eventually turn red or chestnut brown. Sclerotia may be flattened or rounded and wrinkled when dry. (Large photo courtesy of J. Kaminski.)

white or pinkish-white in color. Spores (i.e., basidiospores) are produced on the surface of basidiocarps. Basidiocarps appear in late autumn or early winter but they are inconspicuous and difficult to find. Both *Typhula* species that attack turf produce similar symptoms. Sclerotial color is the primary characteristic pathologists use to differentiate between *Typhula* spp. Large numbers of sclerotia may give affected patches a speckled appearance. After sclerotia appear, the disease subsides and it is unlikely to reactivate.

Management

As soon as it is possible in spring, it is helpful to brush or rake blighted areas to break up matted leaves and encourage drying. As soon as it is warm enough, apply about 1.0 lb N/1000 ft² (50 kg N/ha) to stimulate growth and thus promote recovery of the turf. Refer to the Microdochium patch section above for information on cultural and chemical approaches for managing both Typhula blight and Microdochium patch. Azoxystrobin, chlorothalonil, fluoxystrobin, iprodione, polyoxin D, pyraclostrobin, PCNB, trifloxystrobin, and most DMI/SI fungicides suppress or control both Microdochium patch and Typhula blight. Chloroneb and flutolanil also control Typhula blight, but not Microdochium patch. Many turf managers mix two or more fungicides to provide a greater spectrum of activity and longer and more consistent control of winter diseases.

Yellow Patch or Cool-Temperature Brown Patch

Pathogen: Rhizoctonia cerealis Van der Hoeven

Primary hosts: Annual bluegrass and creeping bentgrass golf greens and fairways

Predisposing conditions: Cold, wet periods with abundant surface moisture

Yellow patch, sometimes called cool-temperature brown patch, develops during prolonged cool (50–65°F; 10–18°C), overcast, and moist periods from late autumn to early spring. Yellow patch is a disease of bentgrasses, annual bluegrass, and sometimes perennial ryegrass and Kentucky bluegrass turf. The disease may develop in late autumn or winter, but in many regions it is most prevalent in early spring.

Symptoms

Yellow patch is most frequently observed on bentgrass or annual blue-grass golf greens, tees, and fairways where it produces yellow or reddish-brown rings, or circular yellow patches a few inches (8 cm) to one or more feet (>30 cm) in diameter (Figures 2.12 and 2.13). The yellow ring symptom mimics and can easily be confused with brown ring patch (aka Waitea patch). Brown ring patch, however, is a disease favored by mild, rainy weather in spring. A grayish smoke-ring may appear at the periphery of yellow patches; however, they are uncommon and generally only develop during extended rainy periods when golf greens have not been mown for several days. The yellow or reddish-brown rings can loop or intermingle like rings on the Olympic flag, and over 100 rings or yellow patches can develop on a single green (Figure 2.13). When weather conditions shift to

FIGURE 2.12
Yellow ring (aka cool temperature brown patch) symptoms of yellow patch on a golf green. (Large photo courtesy of S. McDonald.)

FIGURE 2.13
Reddish-brown ring symptoms of yellow patch on a creeping bentgrass golf green.

sunny and dry periods, blighted turf at the edge of patches or rings may develop a brown or tan color. These brown rings or arcs are disfiguring and can remain evident for many weeks. Oftentimes, however, the rings and patches fade during sunny and dry periods causing little apparent damage if temperatures are warm enough to promote foliar growth. This disease can be confused with Microdochium patch, take-all patch, brown ring patch, or fairy ring, and diagnosis should be confirmed if in doubt. Damage generally is superficial, but significant thinning of the turf may occur during prolonged wet and overcast weather between late autumn and early spring. Yellow patch tends to be most severe in wet, poorly drained, or shaded sites.

Management

Yellow patch severity is reduced by improving surface water drainage and by squeegeeing-off standing water on golf greens during rainy weather. Turf should be mowed as needed to avoid a tall canopy prior to winter. The last application of nitrogen prior to winter should be in a slow-release form. Preventive applications of azoxystrobin, fluoxastrobin, flutolanil, polyoxin D, and others in autumn suppress yellow patch in the spring. Curative fungicide applications usually involve tank-mixing one of the aforementioned penetrant fungicides with chlorothalonil, fludioxonil, or mancozeb (i.e., contact fungicides) to subdue the disease and prevent thinning. Even following multiple applications of different fungicides, the rings and patches may remain evident for long periods. There are no known fungicides or cultural practices, however, that consistently prevent the formation of these rings and patches. In general, fungicide-treated turf will heal more rapidly than untreated turf with the advent of warmer spring temperatures. An application of a water-soluble nitrogen fertilizer will help speed up recovery once winter dormancy has broken and turf has resumed active growth.

Diseases Initiated in Autumn or Spring That May Persist into Summer

The diseases described in this section normally develop in response to cool, overcast, and moist weather. Many of these diseases, however, can persist or develop anew during the summer. For example, red thread can be active at almost any time of the year when weather conditions are wet and overcast. Red thread is one of the first diseases to appear in spring and can be active in late winter during snow melt. Similarly, anthracnose basal rot can actively attack annual bluegrass at almost any time of the year, but frequently develops in summer and can remain active in autumn and during mild winters. The seasonal appearance of most diseases can never be precisely predicted since environmental conditions are impossible to predict. Furthermore, there always are exceptions to the rule, and the precise conditions that trigger most turfgrass diseases are not well understood.

Anthracnose

> *Pathogen: Colletotrichum cereale* Manns sensu lato Crouch, Clarke, and Hillman
>
> *Primary hosts:* Annual bluegrass and creeping bentgrass
>
> *Predisposing conditions:* Diverse and imperfectly understood conditions ranging from cold and wet to hot and dry environments

C. cereale is a common saprophyte found colonizing thatch or dying plant tissues, but can also behave as a pathogen. Anthracnose is primarily a serious problem of annual bluegrass and creeping bentgrass turf grown on golf courses. Anthracnose has also been reported to attack creeping red fescue, perennial ryegrass, Kentucky bluegrass, and hybrid bermudagrass. The latter species, however, are seldom damaged severely.

Anthracnose is commonly observed in mixed annual bluegrass–creeping bentgrass golf greens in which one or the other species becomes diseased, but seldom both. In most cases, it is the annual bluegrass and not creeping bentgrass that is attacked on golf greens with mixed species. The fungus may cause either a foliar blight or a basal rot. Foliar blighting generally occurs during periods of high temperature and drought stress. Foliar blight is a distinct type or phase of anthracnose, which may or may not progress into basal rot.

Anthracnose basal rot is often extremely destructive to golf greens, but it also develops on tees and fairways. Basal rot anthracnose in annual bluegrass and creeping bentgrass golf greens is exacerbated by very low mowing (i.e., <0.125 inches; 3.1 mm), double cutting, low nitrogen fertility, and drought conditions. The disease also is associated with soil compaction, poor surface water drainage, and intense traffic, particularly on older golf courses with small greens. The disease occurs at varying times of the year and produces different symptoms on different hosts. Basal rot in annual bluegrass occurs both during cool periods in the spring and autumn and during warm to hot periods of summer. Anthracnose basal rot can remain active in annual bluegrass throughout mild winters. Annual bluegrass infected with *C. cereale* in the autumn, and entering winter in a weakened state, may develop severe basal rot the following spring. Severe spring outbreaks of basal rot have also been associated with annual bluegrass that was diseased or otherwise severely stressed the previous summer, thus entering winter in a weakened state.

Basal rot generally does not appear in creeping bentgrass before summer. Spring outbreaks can occur, particularly following cold and dry winters that result in desiccation of the bentgrass. Prolonged periods of overcast and/or rainy weather can trigger or intensify the disease. Shaded and wet sites and severely scalped turfs are particularly vulnerable.

Foliar Blight Symptoms

The foliar blight phase of anthracnose that occurs during high-temperature stress periods in summer causes a yellowing or a reddish-brown discoloration and eventually a loss of shoot density. The distinctive fruiting bodies (i.e., acervuli) with protruding black hairs (i.e., setae) can be observed on green or yellow leaf or sheath tissue. The presence of large numbers of acervuli on dead tissue in thatch does not always indicate that healthy plants are also infected. Therefore, it is important to carefully inspect green or discolored tissue of living plants for the distinctive fruiting bodies. The foliar blighting phase is easier to control, and perhaps is less common than basal rot in most regions.

Basal Rot Symptoms

In late winter or spring, annual bluegrass plants infected by *C. cereale* appear as orange or yellow spots or speckles that can coalesce (Figure 2.14). Some refer to this stage of basal rot as "winter anthracnose," which affects annual bluegrass, but not creeping bentgrass. Individual annual bluegrass plants may have both green, healthy-appearing tillers, and yellow-orange

FIGURE 2.14
Winter anthracnose in annual bluegrass appears as orange- or yellow-colored speckles on golf greens.

infected tillers. The central or youngest leaf is the last to show the yellow-orange color change. Removal of all sheath tissue to expose the infected stem base often reveals a water-soaked and/or black rot of crown tissues where roots and new buds are produced. By mid-spring, the dime-spot or speckle symptoms are less common, and infected annual bluegrass plants coalesce into large, nonuniformly affected areas, which initially appear reddish-brown before turning yellow and dying. Large areas may thin or die out completely. Basal rot can be extremely difficult to suppress where it becomes a chronic, spring problem on annual bluegrass golf greens.

In summer, annual bluegrass and creeping bentgrass plants affected by anthracnose basal rot initially develop a reddish-brown or yellow color and turf thins out in irregularly shaped patterns. In mixed, predominantly annual bluegrass–bentgrass greens, the annual bluegrass is more likely to be infected. Plants often turn a brilliant yellow or orange-yellow color before dying, and this symptom can be confused with summer patch in annual bluegrass (Figure 2.15). During hot and humid weather, annual bluegrass may quickly turn yellow and die leaving large irregular areas of dead turf. Unaffected creeping bentgrass slowly colonizes dead areas by spreading stolons. In creeping bentgrass, infected plants normally appear reddish-brown or bronze during the early stages of disease (Figure 2.16). Irregularly shaped, bentgrass-infected areas thin out and tend to die less rapidly than infected annual bluegrass. On rare occasions, small gray-brown or reddish-brown circular patches can appear on golf greens.

When discoloration or thinning is first observed, managers are advised to carefully inspect stem bases for the infection mats, which during early stages of the disease appear as small (pinhead sized), black "fly specks" (Figure 2.17). This will involve removing leaf sheaths to expose the whitish inner sheath tissues or stem areas. There will be no tell-tale signs of the pathogen on leaf

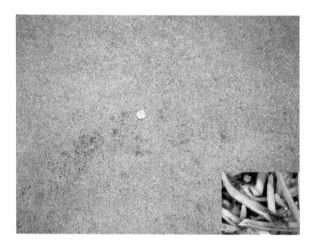

FIGURE 2.15
Anthracnose basal rot causing annual bluegrass to develop a yellow-orange color, while creeping bentgrass on the same golf green remains unaffected.

or sheath tissue during early stages of basal rot. In advanced stages, black aggregates of fungal mycelium and acervuli are often present on infected stolons or stem bases (Figure 2.17). Spore-bearing acervuli with short, black hairs and the black mycelial aggregates can be seen on stem bases and stolons without a hand lens (Figure 2.18). Once acervuli develop on sheath tissue, the basal rot phase is advanced and plants generally die. For mysterious reasons,

FIGURE 2.16
Anthracnose-affected creeping bentgrass turf develops a reddish-brown color and dies in irregular patterns.

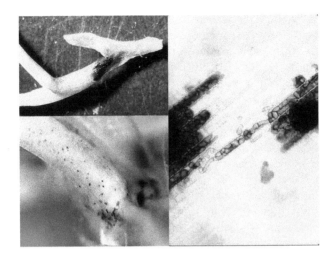

FIGURE 2.17
Infection cushions initially appear as black specks on stem bases (lower left). Infection sites enlarge forming infection mats (right) and eventually stem bases are rotted (top).

FIGURE 2.18
Acervuli developing in the collar region of an annual bluegrass leaf (bottom), and on creeping bentgrass stolons.

the disease seldom attacks both annual bluegrass and creeping bentgrass on the same green or even on the same golf course.

Management

Anthracnose basal rot is very difficult to control once turf shows signs of thinning. This is especially true for annual bluegrass, particularly when it develops in spring. To alleviate basal rot, increase height of cut immediately, reduce mowing frequency and increase nitrogen fertility. Research has shown that regular summer spoon-feeding with water-soluble nitrogen (e.g., ammonium nitrate and urea) at rates of about 0.125 lb N/1000 ft² (6 kg N/ha) every 7–10 days greatly minimizes anthracnose on golf greens. While abrasive cultural practices aimed at improving green speed do not appear to impact anthracnose severity, it seems prudent to avoid sand topdressing, double cutting, brushing, vertical cutting, and other abrasive practices until the disease is brought under control. It is very important to avoid mowing greens when excessively wet (i.e., soft or spongy), which injures turf and intensifies the disease. While irrigation is often withheld in summer to provide for firm and fast playing conditions, it is best to avoid wilt. Research has shown that daily irrigation of annual bluegrass golf greens at 80% evapotranspiration provides sufficient water to maintain growth of annual bluegrass, while helping to minimize disease severity (Roberts et al., 2011). Conversely, the extremes of both wilt and excessive soil wetness can intensify anthracnose. Hence, it is important to avoid excessive irrigation, especially in poorly drained greens. Some preemergence herbicides (e.g., bensulide and dithiopyr) may encourage anthracnose.

Chemical Control

Where foliar blight or basal rot are a chronic problem, a preventive fungicide program needs to be imposed. A modest amount of nitrogen (0.1–0.125 lb N/1000 ft² or 5–6 kg/ha) combined with a contact fungicide (e.g., chlorothalonil or fludioxonil) tank-mixed with QoI/strobilurin fungicide (e.g., azoxystrobin, fluoxastrobin, pyraclostrobin, and trifloxystrobin), DMI/SI fungicide (e.g., metconazole, propiconazole, tebuconazole, triadimefon, and triticonazole), fosetyl-aluminum, polyoxin D, or thiophanate-methyl is effective, but does not always eradicate basal rot. For curative control, include a high label rate of a contact fungicide in the mixture. Several fungicide applications on a 7–10-day interval often are required to arrest basal rot in annual bluegrass golf greens. Basal rot is less common in creeping bentgrass grown on fairways and tees. When treated during the early disease stages, creeping bentgrass fairways and tees recover fairly quickly. Avoid applications of iprodione, vinclozolin, and flutolanil whenever anthracnose is active. Resistance to azoxystrobin and thiophanate-methyl is common.

In extreme cases, golf greens that consist mostly of annual bluegrass that are chronically infected should be renovated using creeping bentgrass. Similarly, for fairways and other large areas that are chronically affected, it is best to eliminate annual bluegrass and renovate or overseed with less susceptible grasses.

Ascochyta and Leptosphaerulina Leaf Blights

Pathogens: Ascochyta spp.

Leptosphaerulina australis McAlpine

Primary hosts: Kentucky bluegrass, fescues, and perennial ryegrass

Predisposing conditions: Hot and dry periods preceded or followed by extended rainy weather

Both diseases are common and can cause extensive foliar blighting. Both agents are regarded as being weak pathogens or senectophytes (i.e., colonizers of senescent or dying tissues). These weak pathogens damage plants that have been severely stressed by environmental conditions, the activity of other primary pathogens, or from mechanical or chemical injury. Both diseases occur from spring to autumn and are often associated with wet weather either preceding or following a period of drought stress. These diseases can be found in association with many turfgrasses. Ascochyta blight is most common in Kentucky bluegrass lawns in spring, whereas Leptosphaeulina blight is most troublesome in spring and summer in perennial ryegrass. Leptosphaerulina blight may also occur in annual bluegrass, creeping bentgrass (particularly the cultivar "SR 1020"), and zoysiagrass grown on golf courses.

Symptoms

Ascochyta and *Leptosphaerulina* spp. produce a myriad of symptoms, which largely appear as nonuniform blighting. Both cause a tip die-back, and their brownish to black and rounded fruiting bodies are invariably found embedded in blighted leaves. These fungi primarily restrict their activity to leaf blighting and are not capable of attacking stem or root tissue.

Ascochyta blight can develop between late spring and throughout the growing season. It is most prevalent in late spring and appears during hot and dry periods that were preceded by cool and rainy weather. Ascochyta blight in Kentucky bluegrass, and to a lesser extent perennial ryegrass and tall fescue, develops rapidly (sometimes overnight). Tips of infected leaves are bleached and turf develops a straw to bleached appearance similar to drought stress (Figure 2.19). Leaves may have white bands or be shriveled. Numerous, small brown fruiting bodies (i.e., pycnidia) can be seen embedded in blighted leaf tissue with the aid of a hand lens. Affected areas survive, but in severe cases, it can take three or more weeks of good growing conditions for the turf to fully recover.

Leptosphaerulina blight appears to be mostly triggered by compacted or droughty soil conditions. The disease can begin in spring and remain active in the summer. *L. australis* is also a common colonizer of leaf and sheath tissues damaged by other diseases (especially Pythium blight and gray leaf spot). Infected leaves often develop a reddish-brown color prior to

FIGURE 2.19
Ascochyta leaf blight in a Kentucky bluegrass lawn. Tip dieback and embedded black fruiting bodies (i.e., pycnidia) are shown on the left leaf in the inset. (Photos courtesy of N. Tisserat.)

turning completely brown. In perennial ryegrass fairways, the disease initially appears as reddish-brown- or orange-colored spots and dead plants eventually develop a tan color (Figure 2.20). Large losses of perennial ryegrass turf may occur in summer in turf subjected to heat and drought stress. Blighting ceases once temperatures moderate and soil moisture is replenished

FIGURE 2.20
Leptosphaerulina blight in perennial ryegrass. Inset shows black fruiting bodies (i.e., perithecia) embedded in a blighted leaf.

by rain or irrigation. The black fruiting bodies (i.e., pseudothecia) are invariably found embedded in blighted tissues and can be seen with a hand lens (Figure 2.20).

Management

Fungicides are seldom recommended for these diseases since significant blighting often occurs rapidly and before the disease can be either properly diagnosed or treated. Furthermore, there are no fungicides known to control these diseases and normally there is a lack of a response from a broad-spectrum fungicide. Cultural practices that may assist in the recovery of turf include increasing mowing height and the employment of a complete N-P-K fertilizer program. Soil moisture should be kept uniform by irrigating turf to avoid drought stress. Thatch control in Kentucky bluegrass lawns is recommended.

Brown Ring Patch or Waitea Patch

Pathogen: Waitea circinata var. *circinata* Warcup & Talbot

Primary hosts: Annual bluegrass grown on golf greens

Predisposing conditions: Mild and rainy spring weather

Brown ring patch is the name given to this disease in creeping bentgrass in Japan, but it does not accurately describe the symptoms in the United States. For this reason, brown ring patch often is referred to as Waitea patch. Little is known about brown ring patch, but the disease has been reported to appear mainly in spring and sometimes in autumn throughout the United States. Annual bluegrass is the most common host, but the disease sometimes is seen in creeping bentgrass golf greens. Brown ring patch is mostly cosmetic and turf normally recovers.

Symptoms

Symptoms of brown ring patch in annual bluegrass golf greens appear as clusters of bright yellow rings 6–18 inches (15–45 cm) or larger in diameter (Figure 2.21). These yellow ring symptoms are similar to yellow patch (aka cool-temperature brown patch). When the disease dissipates, rings may appear brown and in some cases there is thinning of turf. While the yellow rings are distinctive in annual bluegrass, they normally break up into ribbons or crescents when they encounter creeping bentgrass in mixed stands. Turf in the center of the yellow rings sometimes appears much darker green than the surrounding turf. In rare cases, degradation of thatch by the fungus may cause a sunken ring effect, which negatively impacts ball roll.

FIGURE 2.21
Brown ring patch (aka Waitea patch) appears as a cluster of yellow rings in annual bluegrass golf greens. (Photo courtesy of J. Kaminski.)

Management

Normally, turf is temporarily affected by this disfiguring disease. An application of a modest amount of water-soluble nitrogen fertilizer normally speeds up recovery. Fungicides are generally not required, but are often used to hasten recovery. Flutolanil, polyoxin D, and QoI/strobilurin fungicides are highly effective. Chlorothalonil is weak, and thiophanate-methyl is ineffective in controlling brown ring patch.

Dollar Spot

> *Pathogen: Sclerotinia homoeocarpa* F.T. Bennett
>
> *Primary hosts:* Most turfgrass species
>
> *Predisposing conditions:* Warm days, cool nights, and heavy dew formation

Dollar spot is widespread and extremely destructive to turfgrasses. The taxonomy of *S. homoeocarpa* is under study, and at some point in time it will be reclassified and assigned a new Latin binomial. Dollar spot is known to attack most turfgrass species, including both cool-season (i.e., annual bluegrass, bentgrasses, fine-leaf fescue, Kentucky bluegrass, perennial ryegrass, tall fescue) and warm-season (bermudagrass, centipedegrass, seashore paspalum, St. Augustinegrass, zoysiagrass) grasses.

Dollar spot is favored by warm and humid weather, and when night temperatures are cool, and long enough to permit early and heavy dew formation. In cool-season grasses in most U.S. regions, there are two to three annual dollar spot epidemics. The first epidemic usually begins in mid-to-late spring and peaks in early to late summer. There is a period of stasis where there normally is little or no activity and then dollar spot recurs in late summer or early autumn. The second late summer or early autumn epidemic is the most severe. A third epidemic may occur in mid-late autumn in some areas. In southern U.S. regions, however, dollar spot may be active at almost any time of the year. Conversely, in colder regions such as those found in Canada, there is usually only one dollar spot epidemic beginning in late spring or summer.

Some cultural and management factors intensify dollar spot severity such as lower inputs of nitrogen, lower mowing heights, more frequent mowing and the removal of clippings, dry soil conditions, intense play and wear, a lack of good thatch and soil compaction control programs, and the planting of very susceptible cultivars. Conversely, removing leaf surface exudates early in the morning by mowing and rolling golf greens reduces dollar spot severity. For some turf managers, the greatest contributing factor to chronic dollar spot problems has been the planting of cultivars highly susceptible to the disease.

Symptoms

The symptomatic pattern of dollar spot varies with turfgrass species and mowing height. Under close mowing, as with intensively maintained bentgrass, annual bluegrass, bermudagrass, or zoysiagrass on golf courses, the disease appears as small, circular, straw-colored spots of blighted turfgrass about 0.5–2.0 inches in diameter (1.2–5.0 cm) (Figure 2.22). With

FIGURE 2.22
Dollar spot appears as circular, straw- or white-colored spots 0.5–2.0 inches (1.2–5.0 cm) in diameter in close-cut creeping bentgrass. Foliar mycelium often is found on canopies moist with dew in the morning.

FIGURE 2.23
Dollar spot patches are larger (2–4 inches in diameter; 4.0–10 cm) in higher-cut lawns. The grayish structures in the photo are fruiting bodies of a slime mold fungus.

coarser-textured grasses that are suited to higher mowing practices, such as Kentucky bluegrass or perennial ryegrass, the blighted areas are considerably larger and straw-colored patches range from 3.0 to 6.0 inches (7.5–15 cm) in diameter (Figure 2.23). Affected patches frequently coalesce and involve large areas of turf. A fine, white or grayish-white, cobwebby mycelium may cover diseased patches during early morning hours when the fungus is active and leaf surfaces are wet (Figure 2.22).

Grass blades often die back from the tip, and have straw-colored or bleached-white lesions shaped like an hourglass (Figure 2.24). The hourglass banding

FIGURE 2.24
Hourglass-shaped banding symptom of dollar spot in Kentucky bluegrass and tip dieback symptom in tall fescue (lower left).

on leaves is often made more obvious by a narrow brown or purple band, which borders the bleached sections of the lesion from the remaining green portions. Hourglass bands may not appear on warm-season grasses, and are difficult to find in tall fescue and on close-cut bentgrass or annual bluegrass on golf greens, tees, and fairways. On close-cut golf greens and warm-season grasses, lesions are oblong or oval-shaped, and there is a brown-colored band of tissue where the tan or white lesion and green tissue meet. Tip dieback of leaves is common and blighted tips appear tan to white in color, and may also have a brown band bordering tan and green leaf tissue.

Cultural Management in Golf and Sports Turf

Dollar spot tends to be more damaging to turfs insufficiently fertilized with nitrogen. In poorly nourished turf, an application of nitrogen (50% water-soluble plus 50% slow-release) will stimulate growth and mask the disease. Subsequent applications at low rates of water-soluble nitrogen (i.e., 0.125–0.20 lb N/1000 ft^2; 6–10 kg N/ha) in foliar or spoon-feeding programs on a 1- or 2-week application interval also help to suppress dollar spot, but only when disease pressure is low. Potassium and phosphorus have little known impact on dollar spot but may help reduce disease when applied in a complete N-P-K fertility program.

Removal of dew and leaf-surface exudates by poling, dragging, or whipping early in the morning is beneficial. Research indicates that one of the most effective cultural means of reducing dollar spot severity is to mow as early as possible in the morning. Mowing early in the morning will speed surface drying and limits the capacity of the pathogen to infect additional tissue during morning hours. Similarly, use of wetting agents, which reduce leaf wetness periods, may help to reduce dollar spot severity. Rolling golf greens has also been shown to reduce dollar spot, whereas wear from turning mowers can increase dollar spot severity. Minimize wear damage by skipping perimeter mowing 1 or 2 days a week. Avoid other types of mechanical injury (i.e., sand topdressing, vertical cutting, brushing, etc.) during periods when dollar spot is active and turf is not growing vigorously.

Thatch layers and soil compaction have long been recognized to promote dollar spot and other diseases. Hence, core aeration, sand topdressing, vertical cutting, and other practices that alleviate soil compaction and control thatch should assist in improving turf vigor and thus reduce dollar spot incidence and severity. These practices are best performed during disease-free periods when the turf is actively growing.

Turf generally should be kept on the dry side; however, soils kept at or below their wilting point intensifies dollar spot in late summer. Hence, a balanced approach to irrigation is needed to maintain good soil moisture as well as the playability of golf courses or sports fields. Dollar spot is mostly promoted in late summer by drought conditions, which limit growth of turf. Hence, avoiding wilt and not allowing soil moisture to approach the permanent wilting point should help keep the disease less potentially severe. Unlike several other turf diseases, the time of day (i.e., day vs. night) when turf is irrigated does not impact dollar spot severity.

Avoid seeding highly susceptible turfgrass cultivars. For example, the creeping bentgrass cultivars "L-93," "007," and "Declaration" have good dollar spot resistance, whereas "Crenshaw," "Southshore," and "Penn A-1" are highly susceptible. Blending a resistant cultivar (e.g., 007) with a susceptible cultivar (e.g., Southshore) results in less dollar spot compared to a susceptible cultivar grown in monoculture. Land-grant universities and the National Turfgrass Evaluation Program (NTEP; Beltsville, MD) are the best sources of information on disease-resistant cultivars.

Cultural Management in Lawns

Dollar spot is best managed in lawns by increasing mowing height, returning clippings, applying proper amounts of nitrogen at appropriate times, maintaining good soil moisture without overwatering in the summer, controlling thatch, and alleviating soil compaction. An application of modest amounts of nitrogen when dollar spot is active reduces dollar spot severity. Most nitrogen, however, should be applied in the autumn to cool-season grasses and in late spring and summer for warm-season grasses. Attempt to maintain good soil moisture levels by irrigating deeply to root zone depth. Mow early in the morning to speed drying of the canopy. Control thatch and alleviate soil compaction by vertical cutting and core aeration when turf is actively growing (i.e., spring and autumn for cool-season grasses and summer for warm-season grasses).

Chemical Management

Fungicides are routinely used on golf courses and stadium athletic fields, but they are not commonly used on lawns and other less intensively managed turfgrasses. Most fungicides that target dollar spot are from the chemical class known as demethylation or sterol inhibitor (DMI/SI), which include metconazole, myclobutanil, propiconazole, tebuconazole, triadimefon, and triticonazole. Boscalid, iprodione, penthiopyrad, thiophanate-methyl, and vinclozolin represent different classes of fungicides that target dollar spot. The aforementioned fungicides are referred to as penetrants since some of their molecules enter plant tissue. Chlorothalonil, fluazinam and mancozeb are contact fungicides, which remain on the outside of contacted tissue surfaces. These contact fungicides target dollar spot and have no resistance issues. Dollar spot is best managed using preventive applications of fungicides. Curative applications will require higher rates and closer spray intervals. For curative control, it is best to tank-mix a contact with a penetrant for more rapid knockdown of the pathogen and longer residual control. As noted below, there are resistant issues with several fungicides that target dollar spot and tank-mixing helps to delay the potential for resistance to occur. Fungicides should be sprayed through nozzles that atomize the droplets in about 50 gallons water per acre (463 L/ha).

Thiophanate-methyl-resistant (i.e., total failure) dollar spot is common. Reduced sensitivity (i.e., shortened period of residual control), and less commonly, resistance, occurs after many years of applying DMI/SI fungicides. Similarly, reduced sensitivity and resistance has been reported for iprodione; however, these occurrences are sporadic and not very common. Loss of good residual dollar spot control with any particular fungicide, however, does not necessarily mean resistance is developing. Overuse of materials within the same chemical class can result in enhanced microbial degradation. That is, continuous use of fungicides from any class could give rise to a buildup of bacteria and possibly other microorganisms that use chemically related fungicides as an energy source. To delay or avoid reduced sensitivity or resistance, it is important to tank-mix and rotate fungicides from different chemical groups. For example, there is no advantage to rotating myclobutanil, propiconazole, and triadimefon or other DMI/SI fungicides. Because resistance is unlikely, it is especially important to use contact fungicides as tank-mix partners with a penetrant or to just rotate it into the spray program often. Mancozeb is a contact fungicide that has activity on dollar spot but is not nearly as effective as chlorothalonil or fluazinam. More information about the nature of resistance, resistance management programs, and proper application of fungicides can be found in the section "Fungicides Used to Control Turfgrass Diseases."

Paclobutrazol and flurprimidol are plant growth regulators that have fungistatic activity, which significantly reduce the severity of dollar spot. The herbicide bispyribac-sodium also has very good activity against dollar spot. *Bacillus licheniformis* (EcoGuard) and *B. subtilis* (Rhapsody) are bacterial biological agents that suppress dollar spot under low pressure, but seldom provide commercially acceptable control. Some mineral oil products (e.g., Civitas®) and wetting agents also provide significant levels of dollar spot control, but can yellow turf grown on golf greens.

Ultimately, effective dollar spot suppression is going to involve combining those cultural practices that are known to suppress dollar spot into a fungicide program. In sports and golf turfs, nitrogen should be applied on a 7–14-day interval in spoon-feeding programs (i.e., 0.1–0.125 lb N/1000 ft²; 5–6 kg N/ha) throughout the growing season. It is important to mow early in the morning to speed turf drying. Fungicide-treated turf, however, should not be mowed for at least 12 h and preferably 24 h after spraying. Obviously, removal of plant tissues containing fungicides dilutes the total concentration of the active ingredient. This is why using plant growth regulators to reduce mowing frequency can sometimes help to extend the residual effectiveness of fungicides. Returning clippings is helpful, if they do not interfere with play in sports turfs, because they help to recycle nitrogen and other nutrients.

Large Patch

Pathogen: Rhizoctonia solani Kuhn AG 2-2 LP

Primary host: Warm-season grasses, especially kikuyugrass, seashore paspalum, and zoysiagrass

Predisposing conditions: Cool, rainy periods in spring and autumn

Large patch (first known as zoysia patch) primarily occurs in seashore paspalum, kikuyugrass, and zoysiagrass grown on golf courses. Bermudagrass, centipedegrass, and St. Augustinegrass grown in lawns and landscapes are also susceptible. The causal agent is a specific biotype (i.e., AG 2-2 LP) of *R. solani*. Other biotypes of *R. solani* (i.e., AG 1, 1A, AG 2-2 IIIB, and AG 4) are damaging to cool-season grasses during periods of high humidity and high-temperature stress.

Cooler temperatures associated with warm-season grasses either entering or emerging from winter dormancy are required for large patch to develop. In south Florida, seashore paspalum and St. Augustinegrass do not usually go completely dormant, but they do get this disease. Lower temperatures in spring and autumn slow the growth of warm-season grasses, thus giving a competitive edge to the pathogen. This disease appears during extended rainy overcast periods, which further promote growth of the pathogen. Large patch is most severe in shaded and poorly drained sites. The disease can predispose warm-season grasses to winter damage.

Symptoms

Large patch was first recognized as a severe disease of "Meyer" zoysiagrass grown on golf course fairways. Improved zoysiagrass cultivars are also damaged by large patch. The disease is characterized by large rings or circular patches that range from 2 to 10 ft (0.6–3.0 m) or more in diameter (Figure 2.25). Patches are brown or yellow-orange in color. At the periphery of patches, turf may exhibit a brilliant-orange firing. The "orange firing" symptom is most prevalent in autumn or spring during periods of active infection. Affected patches have reduced numbers or living tillers and a reduced leaf growth rate. Living stolons and tillers dispersed throughout patches provide for slow recovery of turf in the spring. Patches and rings tend to develop in the same areas from year to year. In severe cases, symptoms may remain evident in summer due to slow recovery of damaged turf. In shady areas and some other cases, large rings or patches of dead turf persist throughout summer and dead rings can mimic fairy ring.

R. solani attacks basal portions of leaf sheaths of some warm-season grasses (especially zoysiagrass) in the thatch region, producing small reddish-brown or black lesions. Eyespot lesions may appear on basal stems and stolons. Leaves are blighted, and stems may be infected and tillers killed. Turf within affected areas thins out, and 85–90% of the shoots may die. Diseased shoots

FIGURE 2.25
Large patch in a zoysiagrass fairway.

are easily detached from their stolons. Turf within these large, almost dead-appearing rings and patches, eventually recovers but the process is very slow. As temperatures increase in spring, disease subsides and regrowth from stolons results in a slow improvement in turf density.

Symptoms of large patch in other warm-season grasses are similar to those described above. Bermudagrass, centipedegrass, seashore paspalum, kikuyugrass, and St. Augustinegrass also are affected in spring and autumn (Figure 2.26). In bermudagrass, large patch is most commonly observed at spring green-up. Patches can be enormous, ranging from 5 to 10 ft (1.5–3.0 m) or more in diameter (Figure 2.27). Bermudagrass, however, is not seriously

FIGURE 2.26
Large patch in kikuyugrass. (Photo courtesy of L. Stowell, PACE Turf, LLC.)

FIGURE 2.27
Large patch in a bermudagrass sports field.

damaged and turf recovers with the advent of consistently warm soil temperatures. Conversely, centipedegrass, seashore paspalum, kikuyugrass, and St. Augustinegrass can be seriously damaged in both spring and autumn by *R. solani*.

Management

Most of what is known about large patch management was obtained from research conducted in zoysiagrass grown on golf course fairways. These principles are applied here to all warm-season grasses affected, except perhaps for bermudagrass, which normally is not seriously damaged. Increasing mowing height on fairways by 0.25 inch (6.2 mm) or more prior to the onset of wet weather in autumn is perhaps the most effective cultural means of slowing or reducing disease progress and enhancing turf recovery. Thatch reduction is important because most damage to tillers occurs primarily within the thatch layer. Vertical cutting and core aeration of actively growing turf in summer stimulates regrowth and helps to reduce thatch. Coring and vertical cutting, however, should not be employed in the spring or autumn when large patch is active and turf is growing slowly. Redistribution of soil from cores following coring assists in thatch degradation.

Do not apply any nitrogen (N) fertilizer until disease activity has ceased. Most nitrogen fertilizers, including water-soluble urea and slow-release organic fertilizers help to stimulate recovery, but they do not appear to suppress disease development. Spring application of water-soluble nitrogen sources (e.g., ammonium nitrate and urea) can enhance large patch when the disease is still active. Once the disease becomes inactive, nitrogen should be applied at 0.5 lb N/1000 ft² (25 kg N/ha) every 2–3 weeks until recovery has occurred. The total amount of nitrogen applied to mature zoysiagrass, however, should not exceed 2.0 lb N/1000 ft² per year (100 kg N/ha). Seashore paspalum and kikuyugrass seldom require more than 3.0 and 4.0 lb N/1000 ft² annually (150 and 200 kg N/ha), respectively. Large patch also is enhanced by overwatering in the spring and autumn, and tends to be most severe in poorly drained or shaded sites. Hence, irrigate only when turf shows signs of wilt. Improve surface water drainage and prune or remove trees and brush in shaded sites where large patch is a chronic problem.

Fungicides assist in blight reduction but may not prevent large patch from occurring. A fungicide application in late summer or early autumn is recommended where the disease is chronic. It is best to apply fungicide(s) in a high water volume (≥100 gallons water per acre; ≥935 L/ha). Azoxystrobin, fluoxastrobin, flutolanil, iprodione, polyoxin D, metconazole, myclobutanil, propiconazole, pyraclostrobin, tebuconazole, triadimefon, and triticonazole provide satisfactory disease suppression and their use is associated with more rapid recovery. Flutolanil, polyoxin D, or a QoI or strobilurin fungicide (e.g., azoxystrobin, fluoxastrobin, pyraclostrobin, and trifloxystrobin) are preferred for best preventive and curative treatment. Curative control may require several applications of high label rates of fungicide. One or two preventive applications of a fungicide(s) where the disease is a chronic problem is the best approach to large patch management. Normally, it is best to map sites chronically affected and begin treatment in those sites in late summer and/or early autumn followed by a second application in about 4 weeks. Re-treatment as soon as active disease symptoms appear may be necessary, especially if there are prolonged periods of rainy and overcast weather.

Leaf Spot, Melting-Out, and Net-Blotch (Formerly Helminthosporium Diseases)

Pathogen: Drechslera spp. and *Bipolaris* spp.

Primary hosts: Bentgrasses, bermudagrass, bluegrasses, fescues, perennial ryegrass

Predisposing conditions: Varies among turfgrass hosts and fungal species

Many fungi causing leaf spot and melting-out diseases of turfgrasses were once assigned to the taxonomic genus *Helminthosporium*. These fungi are now referred to as species of *Drechslera* or *Bipolaris*, and the common names of the diseases they cause include leaf spot, melting-out, brown blight, net-blotch, and others. A list of the most common of these diseases and pathogens is provided in Table 2.4.

Among the most common spring, autumn, and mild winter diseases of Kentucky bluegrass is leaf spot, caused by *Drechslera poae*. This disease is not as devastating as it once was. This is due to the development and widespread use of resistant Kentucky bluegrass cultivars. In resistant cultivars, leaf spot may develop during rainy weather in spring but the disease seldom progresses into the melting-out stage. Hence, leaf spot-resistant cultivars seldom require fungicide treatment. "Kenblue," "Mermaid," "Park," "South Dakota," and other mostly "common types" of Kentucky bluegrass are very susceptible to leaf spot. Highly susceptible cultivars tend to be less expensive to produce and do find their way into inexpensive grass seed blends and mixtures sold to homeowners.

Symptoms

Drechslera poae and diseases incited by other *Drechslera* and *Bipolaris* spp. may develop in two phases: leaf spot and melting-out phases. In the leaf spot phase, distinct purplish-brown or dark-brown leaf spot lesions with a central

TABLE 2.4

Common Hosts and Names of Diseases Caused by Fungi Previously Known as *Helminthosporium*

Host	Common Name(s) of Disease	Pathogen[a]
Kentucky bluegrass and annual bluegrass	Leaf spot, melting-out	*Drechslera poae* (Baudys) Shoemaker
	Leaf spot, melting-out	*Bipolaris sorokiniana* (Sacc.) Shoemaker
Tall fescue	Net-blotch	*D. dictyoides* (Drechs.) Shoemaker
Creeping red fescue and other fine-leaf fescues	Leaf spot, melting-out Net-blotch	*B. sorokiniana* *D. dictyoides*
Perennial ryegrass	Leaf spot, melting-out	*B. sorokiniana*
	Net-blotch	*D. dictyoides*
	Brown blight	*D. siccans* (Drechs.) Shoemaker
Bentgrasses	Red leaf spot	*D. erythrospila* (Drechs.) Shoemaker
	Leaf spot	*B. sorokiniana*
	Zoneate eyespot	*D. gigantea* (Herald & Wolf) Ito
Bermudagrass	Leaf blotch, melting-out	*B. cynodontis* (Marignoni) Shoemaker
	Zoneate eyespot	*D. gigantea*

[a] Diseases incited by *Drechslera* spp. are most common during cool, rainy, and overcast periods in spring, autumn, and mild winters, whereas diseases incited by *Bipolaris* spp. are usually associated with high-temperature stress in summer.

tan spot are produced on leaves of affected bluegrass plants (Figure 2.28). During favorable disease conditions, numerous lesions coalesce to encompass the entire width of leaves causing tip dieback and a generalized dark-brown blight. Leaf spot lesions initially are associated with older leaves, which die prematurely as a result of the invasion. In a heavily infected stand, leaves and sheaths are blighted and turf appears yellow or red-brown in color when observed from a standing position (Figure 2.29). If environmental conditions favorable for disease continue, particularly overcast, cool, and drizzling weather, successive layers of leaf sheaths are penetrated, and the crown is eventually invaded. Once the crown is invaded, the disease enters the melting-out phase. During this phase, entire tillers are lost, and the turf loses density (Figure 2.29). Hence, it is the melting-out phase that is most damaging.

Net-blotch disease of tall fescue, perennial ryegrass, and creeping red fescue is caused by *Drechslera dictyoides*. In general, net-blotch is more destructive to creeping red fescue and perennial ryegrass and tends to be less injurious in tall fescue. *D. dictyoides* is also a pathogen that attacks turf primarily during cool, moist, and overcast periods of spring and autumn. Symptoms initially appear as minute, brown, or purple-brown specks on fescue leaves. As the disease progresses, a dark-brown, net-like pattern of necrotic lesions develops on leaves of tall fescue and creeping red fescue. These net-blotches may coalesce, and leaves will turn brown or yellow, and

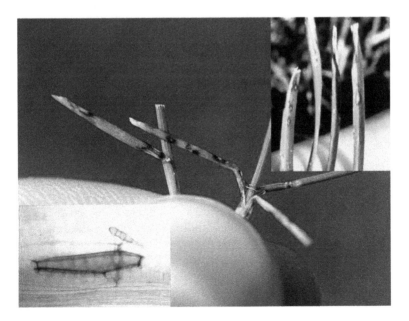

FIGURE 2.28
Purple-brown leaf lesions caused by *Drechslera poae*. Bottom left inset shows *Drechslera dictyoides* spore germinated, penetrating a leaf, and initiating the formation of a lesion.

FIGURE 2.29
During the leaf spot stage of a Drechslera or Bipolaris disease, the turf develops a yellow, brown, or reddish-brown appearance from a standing position (left); during the melting-out stage, the turf loses density when leaves and tillers are killed (right).

die-back from the tip. On leaves of perennial ryegrass, numerous oblong, dark-brown lesions are produced by both *D. dictyoides* and *Drechslera siccans* (i.e., brown blight). Under ideal environmental conditions, these fungi may invade stems, and cause a melting-out of the stand. Both pathogens can be active during relatively warm, rainy periods of winter.

Bipolaris spp. and *Drechslera erythrospila* attack grasses during warm weather from late spring through summer. These pathogens, and the diseases they cause, are briefly discussed below in the "Diseases Initiated During Summer That May Persist Into Autumn" section.

Management

Cultural practices that minimize injury from leaf spot, melting-out, net-blotch, or brown blight are the same regardless of pathogen. Renovation- or overseeding-resistant cultivars are the simplest and best approach to managing these diseases, especially in Kentucky bluegrass. Unfortunately, home-owners continue to buy inexpensive Kentucky bluegrass seed mixes, which invariably contain high percentages of susceptible, common-type cultivars.

These diseases are much more destructive under low mowing. Hence, it is important to increase mowing height when leaf spotting becomes evident. These fungi produce large populations of spores when thatch is subjected to frequent wetting and drying cycles. Irrigation water therefore should be applied deeply and infrequently. Thatch should be controlled by vertical cutting and/or core aeration when layers exceed 0.5 inches (12 mm) in depth. Broadleaf, phenoxy herbicides (e.g., 2,4-D and MCPP or mecoprop) and some plant growth regulators (especially mefluidide) aggravate these diseases and their use should be avoided when they are active. Also, avoid spring application of high rates of water-soluble nitrogen fertilizers that can aggravate leaf spot and melting-out.

Fungicides that effectively control these diseases include chlorothalonil, fludioxonil, iprodione, mancozeb, and vinclozolin. Other fungicides may be labeled for suppression of one phase (i.e., leaf spot vs. melting-out) of the disease. Autumn application of PCNB to Kentucky bluegrass grown in golf course roughs has been shown to control leaf spot during mild winters and early spring periods.

Necrotic Ring Spot

Pathogen: Ophiosphaerella korrae (J. Walker & A.M. Smith) Shoemaker & Babcock

Primary hosts: Kentucky bluegrass and creeping red fescue

Predisposing conditions: Cool, wet weather in spring and autumn, followed by high temperature or drought stress in summer

Necrotic ring spot (NRS) is a disease primarily of Kentucky bluegrass and creeping red fescue. The disease has been documented in annual bluegrass grown on golf greens, but outbreaks have been rare. First described in 1986, NRS is commonly confused with summer patch. *O. korrae* (formerly *Leptosphaeria korrae*) primarily attacks roots during cool, wet weather in spring and autumn. In some regions, disease symptoms may be apparent all year to include mild winter periods. Where NRS is chronic, the disease appears in spring or in autumn. The disease can be found in sunny or shaded sites, but tends to more severe in shaded or wet areas. In contrast, summer patch is a high-temperature summer disease, which normally appears in mid-summer and most injury occurs in sunny rather than shaded sites. Frequently, symptoms of NRS do not become evident until summer, when environmental stresses kill plants whose root systems were damaged by *O. korrae* in the spring. Hence, it can be easy to confuse NRS and summer patch.

Symptoms

The confusion over differentiating between NRS and summer patch in Kentucky bluegrass occurs because both diseases share common symptoms, such as patches and rings of dead grass with living turf in the center (i.e., frog-eye symptom), and both can appear in summer (Figure 2.30). Initially, leaves of NRS-affected plants display a purple or yellow color and plants may be stunted or appear wilted in a circular pattern. During early stages of the disease, symptoms are inconspicuous and often are not detected. Diagnosis is often based on the development of dead rings or thinned-out

FIGURE 2.30
Necrotic ring spot in a Kentucky bluegrass lawn. (Photo courtesy of N. Tisserat.)

turf in circular patches. In chronically affected Kentucky bluegrass areas, NRS patches tend to be large, often greater than 1 ft (>30 cm) in diameter, and the frog-eye (i.e., a dead ring with living grass in the center) symptom is common (Figure 2.31). Conversely, patches associated with summer patch tend to be less than 1 ft (<30 cm) in diameter (a few can exceed 18 inches [45 cm] in width), with perfect doughnuts or frog-eyes being less common. Symptoms of NRS in creeping red fescue are similar to Kentucky bluegrass, but distinct frog-eyes are less numerous.

FIGURE 2.31
Large patches and frog-eye symptoms of necrotic ring spot in Kentucky bluegrass. (Photo courtesy of N. Jackson.)

Sodded lawns in newly cleared woodland sites appear most likely to be affected, but NRS also occurs where lawns are seeded. The disease is particularly severe in wet and/or shaded sites or low-lying areas where water collects. As previously noted, NRS can be severe in both shaded and sunny sites. To date, NRS has been observed primarily in Kentucky bluegrass and creeping red fescue turf grown in the northeast, upper mid-west and Great Lakes region, Rocky Mountain states, and Pacific Northwest regions of the United States and Canada. Its occurrence is believed to be less common than that of summer patch in warmer transition zone regions of the United States.

Management

Because NRS is a root disease, frequent irrigation is required during summer months when heat and drought severely stress injured plants. Daily, light applications of water are recommended to cool turf and assist the dysfunctional root system where NRS is chronic and severe. Use of slow-release and organic fertilizers such as poultry waste products were reported to help reduce disease severity. Acidification with ammonium sulfate as described in the take-all and spring dead spot sections, however, may be more effective in managing NRS. Avoid applying nitrate forms of nitrogen such as calcium, potassium, or sodium nitrate where NRS is a problem since they can gradually increase soil pH and intensify the disease. Improve rooting by coring when the disease is inactive and control thatch.

An early spring application of an appropriate fungicide (e.g., azoxystrobin, fenarimol, fluoxastrobin, myclobutanil, propiconazole, tebuconazole, or thiophanate-methyl) can suppress NRS severity, but fungicide use may not completely control this disease. Curative control would involve higher rates and multiple applications, and only partial control should be expected. It is recommended to water-in fungicides prior to drying on leaves into the upper 1 inch (2.5 cm) of soil.

Powdery Mildew

Pathogen: Blumeria graminis (DC) E.O. Speer

Primary hosts: Kentucky bluegrass, creeping red fescue

Predisposing conditions: Shade

Powdery mildew is a disease mainly confined to shaded sites. The fungus is an obligate parasite (i.e., grow and completes life cycle on living tissues only) and was formerly known as *Erysiphe graminis*. Powdery mildew can be found at almost any time of year, but peak activity normally occurs when days are warm and nights are cool. The white coating of mycelium and spores is most prevalent on leaves during cool, humid, and cloudy periods of late summer and early autumn. Spores are produced in abundance on leaf surfaces and they are disseminated by air currents and equipment to adjacent, healthy leaves. Spores germinate rapidly, even in the absence of dew or water.

Symptoms

The presence of grayish-white mycelium on upper leaf surfaces is a conspicuous, diagnostic sign of this disease (Figure 2.32). The lower, older leaves are more heavily infected than upper, younger leaves. In heavy infestations, leaves appear to have been dusted with ground limestone or flour. The abundant surface mycelium absorbs nutrients from the epidermal cells and leaves turn yellow. Eventually, leaves and tillers may die and the turf exhibits poor density. The fungus seldom kills; however, it weakens plants and can predispose them to injury from environmental stresses or other diseases.

Management

Because shade is the primary predisposing factor for powdery mildew, reducing shade and improving air circulation is a sound but often impractical approach to reduce damage. Planting or overseeding with shade-tolerant cultivars, increasing mowing height, avoiding drought stress, and using a complete N-P-K fertilizer program will promote turfgrass growth and help

FIGURE 2.32
Grayish-white mycelium on Kentucky bluegrass leaves affected with powdery mildew.

to minimize injury from powdery mildew. Fungicides may be applied in situations where the disease is yellowing plants and thinning the stand. The DMI/SI fungicides (e.g., propiconazole, triadimefon, others) and QoI fungicides (e.g., azoxystrobin, pyraclostrobin others) are effective in controlling powdery mildew.

Pythium-Induced Root Dysfunction

Pathogens: Pythium aphanidermatum (Edson) Fitzp.; *Pythium aristosporum* Vanterpool; *Pythium arrhenomanes* Drechs.; *Pythium graminicola* Subramanian; *Pythium torulosum* Coker & F. Patterson; *Pythium vanterpoolii* V. Kouyeas & H. Kouyeas; *Pythium volutum* Vanterpool & Truscott; others

Primary host: Creeping bentgrass and annual bluegrass golf greens

Predisposing conditions: Wet soil conditions from spring to autumn

Pythium species are associated with most turfgrasses and cause seed decay, damping-off of seedlings, foliar blight, root and crown rot, and snow blight. Foliar blight is known as *Pythium* blight, cottony blight, grease spot, and spot blight, which is most destructive during hot and humid periods.

Unlike foliar blight, Pythium-Induced root dysfunction (PRD) primarily is a disease of creeping bentgrass and annual bluegrass grown in sand-based root zones. This disease may be referred to as Pythium root dysfunction or Pythium root rot. PRD develops during cool (54–75°F; 12–24°C) and wet periods of spring or autumn, causing a general decline of the stand. Large areas, however, may not collapse until the advent of warmer and drier conditions in summer. Symptoms of this decline are diverse, involve variously discolored tissues and are difficult to describe.

A positive diagnosis requires a laboratory analysis that involves searching roots or stem bases for the presence of distinctive thick-walled, resistant oospores produced by *Pythium* spp. (Figure 2.33). Sometimes, trapping or isolation of the pathogen becomes necessary. A definitive lab diagnosis is also made difficult since *Pythium* species are common inhabitants of turfgrass roots. Furthermore, the thick-walled oospores are not always present. Even when spores are present, infected roots can appear white and healthy. Several studies have shown that *P. torulosum* is among the most common species associated with PRD; however, the same studies showed the fungus to be weakly pathogenic. It is believed that colonization of roots by several species of *Pythium* causes root systems to decline in size and function (i.e., become dysfunctional rather than destroyed).

PRD primarily is a problem in new creeping bentgrass golf greens or older greens renovated using a fumigant. Annual bluegrass grown on golf greens is also susceptible to PRD, especially in colder climates where winter injury is common. The disease is promoted by cool temperatures and excessive soil

FIGURE 2.33

Symptoms of Pythium-induced root dysfunction in creeping bentgrass golf greens are normally nonspecific and usually involve color changes in the canopy (e.g., yellow, bronze, and purple), and the turf usually dies in irregular patterns. The inset shows distinctive thick-walled oospores of the pathogen in root tissue. (Photos courtesy of S. McDonald.)

wetness. Destructive PRD cases have appeared just after seeding in mid-to-late autumn or in the spring following an autumn seeding. Symptoms of PRD may not appear until late spring or summer when temperatures rise and stress the root systems of infected plants. PRD in creeping bentgrass is most likely to develop in greens less than 3 years old. The disease declines in severity as golf greens age. The disease, however, may be more chronic in annual bluegrass greens in colder climates where turf frequently is weakened by winter stresses.

In 2004, young golf greens in North Carolina were found with circular patches in summer. *P. volutum* was isolated from infected roots and was shown to be highly pathogenic to creeping bentgrass (Kerns and Tredway, 2008, 2010). Symptoms of PRD incited by *P. volutum* (PRD-Pv) are distinctive and appear to differ from other incitants.

Symptoms of PRD in Creeping Bentgrass

Symptoms of PRD are nonspecific but the turf may display color changes (e.g., yellow, bronze, or purple canopies) and gradual or rapid collapse of turf in irregular patterns (Figure 2.33). In immature bentgrass golf greens, leaves of infected plants are yellow and stunted, and tend to be narrower than leaves of healthy plants. In most situations, there are dark-green, perfectly healthy-looking plants dispersed throughout pockets of chlorotic and diminutive plants. Mixed within chlorotic areas, small patches turn brown and die. Leaves of infected plants may develop a reddish-brown, bronze, or even a purplish color. Proneness to wilt is another key symptom of PRD. Loss of density usually occurs in a nonuniform pattern similar to those symptoms associated with black-layer.

Yellow, wilted, or dying plants often first appear at the outer periphery or clean-up pass of golf greens (Figure 2.34). This is due to two likely reasons. First, *Pythium* spp. ingress at the point where the native soil and greens' mix meet. Death of plants, however, is due to a combination of mechanical injury from mowing stressed plants with dysfunctional root systems. More mechanical injury occurs at the point where the mower turns into a curved area. The turning action of mowers twists and wears leaves and stems placing a lethal stress on infected plants. Large areas throughout the entire putting surface may die.

While some researchers report rotting of roots, this symptom is uncommon. In fact, washing soil from infected bentgrass plants and observing roots for symptoms often provides no useful clues. Infected bentgrass roots may have lesions or a generalized light tan to brownish discoloration and water-soaked appearance, which can only be seen with a microscope. It should be noted that in summer, sloughing of root cortexes and necrotic (i.e., brown and rotted) roots are commonplace and are caused mostly by high soil temperatures and not always by pathogens. Diagnosticians base a positive PRD diagnosis on the presence of moderate to large numbers of oospores (i.e., thick-walled resting spores) in a large frequency of roots (Figure 2.33). This method is flawed since oospores are not always present in infected roots. If oospores are present and abundant, a positive diagnosis can be made. Furthermore, there is often not enough time to attempt to isolate a pathogen. Even if time is

FIGURE 2.34
Pythium-induced root dysfunction frequently begins on peripheral, clean-up pass areas of golf greens and infected areas may have a wilted appearance. (Large photo courtesy of D. Bevard.)

not an issue, damage may be been inflicted weeks prior to symptom expression and the pathogen may no longer be detectable. In many cases, turf has been treated with fungicides complicating the diagnosis. Hence, diagnostics for this disease are based mostly on season of occurrence, foliar symptoms, age of creeping bentgrass, and whether or not there is a response from an appropriate fungicide.

Symptoms of PRD in Annual Bluegrass

In annual bluegrass, PRD appears in early spring. Leaves of infected plants appear yellow-brown or reddish-brown in color and stunted before plants die. Symptoms in annual bluegrass mimic those induced by growth regulators like paclobutrazol or flurprimidol. Plants generally die in a nonuniform pattern, but circular patches are sometimes associated with the disease. Disease develops primarily in pockets and is often most pronounced in low areas or follows the surface water drainage pattern. Some injury, however, may be observed throughout the putting surface even in higher, well-drained areas. PRD in annual bluegrass is more common in colder climates, where winter stresses weaken plants prior to spring, rendering them more susceptible to infection. Roots typically appear shortened and brown in color.

Symptoms of PRD-Pv in Creeping Bentgrass

Symptoms of PRD-Pv are different from those described above. Symptoms induced by *P. volutum* appear as distinct, reddish-brown- or bronze-colored circular patches. *P. volutum* is most pathogenic when soil temperatures are cool, but symptoms do not appear until periods of heat and drought stress (Kerns and Tredway, 2008). Initially, symptoms appear as circular areas of wilt followed by chlorosis or drought stress and range from 1.5 to 6.5 inches (4–16 cm) in diameter. Patches can increase in size from 6 to 20 inches (16–50 cm) in diameter and may become irregular in shape. Affected patches may appear yellow or orange and mimic take-all patch. PRD-Pv is associated with a shortened or truncated root system. *P. volutum* is not actively growing in summer and this is why oospores and hyphae of *P. volutum* usually are not found in affected root tissues. Symptoms of PRD-Pv have not been widely reported outside the southeastern United States. It is possible that other *Pythium* species are also capable of producing a patch symptom. Thus, making a distinction between PRD and PRD-Pv may one day prove to be unnecessary.

Management

PRD must be aggressively managed by a combination of cultural and chemical means. *Pythium* spp. grow and reproduce rapidly in wet soils; therefore, drying soil is essential. Irrigating from sprinkler heads should be replaced by frequent hand syringing or hand watering on an as-needed basis. Once soils are dry, turf should not be allowed to wilt for long periods; therefore, greens need to be syringed as needed to alleviate wilt and heat stress. Affected greens should only be mowed when dry and with a lightweight, walk-behind mower. Never mow affected greens on rainy days or when the surface is excessively (i.e., soft or spongy) wet. Mowing height should be increased (i.e., ≥0.156 inches; ≥4 mm) and mowing frequency reduced to four or five times weekly. Remove grooved rollers from mowers and replace them with solid rollers. Avoid other potential mechanical stresses by delaying sand topdressing, brushing, vertical cutting, and core and water injection aeration until the disease has been controlled.

For new constructions, fungicides should be applied preventively in the autumn of seedling emergence and the following spring every 21–28 days or more frequently if unusually long periods of rain should occur. There are few published studies that have investigated the performance of fungicides targeting PRD. An exception is PRD-Pv, where it has been shown that cyazofamid and pyraclostrobin are more effective than other *Pythium*-targeted fungicides. Conventional fungicides that target *Pythium* spp. include chloroneb, cyazofamid, ethazole, fosetyl-aluminum, mefenoxam, and propamocarb. Since there are numerous *Pythium* spp. that can cause PRD, it is unlikely that all species would be sensitive to the same fungicide. Hence, it is prudent to tank-mix fungicides when targeting PRD. If this approach fails to provide satisfactory control, a more comprehensive battery of fungicide treatments may be recommended. This could involve an application of fosetyl-aluminum tank-mixed with mancozeb, followed in 5–7 days with an ethazole or chloroneb drench. Thereafter, alternating or tank-mixing cyazofamid, mefenoxam, and propamocarb on a 10–14-day interval may be warranted. In extreme cases, a curative fungicide application may be required on a 7–10-day interval to suppress the disease. Thereafter, fungicides may be needed periodically on a 14–21-day interval. Of the aforementioned, only fosetyl-aluminum tank-mixed with mancozeb should not be watered-in. The aforementioned tank-mix is a good option should greens be saturated with water at the time the disease appears. Other fungicides should be watered-in to a depth of 0.5–1.0" (1.2–2.5 cm). A syringe cycle from the irrigation system usually provides sufficient water to move fungicides off the canopy and into the effective zone immediately following a fungicide application. Spoon-feeding and applications of biostimulants and micronutrients are suggested along with fungicides, but their overall impact on the condition is unknown.

Rapid Blight

Pathogen: Labyrinthula terrestris

Primary hosts: Annual bluegrass, creeping bentgrass, perennial ryegrass, and roughstalk bluegrass

Predisposing conditions: Sodium-affected soils

Rapid blight is caused by an organism that is neither a fungus nor a bacterium, but belongs to a classification group known as the net slime molds. These single-celled organisms develop in aggregates in culture and have football (i.e., spindle)-shaped cells that glide within a net-like matrix that they produce. These football-shaped cells can be found in infected tissues. Rapid blight has become a problem is various regions where cool-season grasses are irrigated with water containing moderate to high levels of sodium. Indeed, *Labyrinthula* spp. are normally associated with marine environments where at least one species (*Labyrinthula zosterae*) has been shown to be pathogenic on a marine grass. Such water is also found in saline and sodic (i.e., high in sodium) soils in arid regions of the western United States as well as coastal regions in the southeastern United States and Southern California, where there is seawater intrusion in freshwater wells. While cool-season grasses are most susceptible, bermudagrass and seashore paspalum are also hosts for this pathogen, but are not usually damaged. Rapid blight most commonly attacks roughstalk bluegrass that has been overseeded into bermudagrass as well as perennial ryegrass grown on fairways and tees. Rapid blight also occurs in annual bluegrass grown on golf greens. The disease is less commonly found on creeping bentgrass, but has been reported to attack colonial bentgrass golf greens in the United Kingdom. Rapid blight develops during periods of drought when there is inadequate rain water to flush sodium from soils.

Symptoms

In less severe cases, rapid blight may appear as a chlorosis and stunting of plants. Under high pressure, disease develops rapidly and kills turf in just a few days. In severe cases, rapid blight appears as yellow- or brownish-colored turf in circular or irregularly shaped patches (Figure 2.35). At the periphery of patches, there may be a dark-brown and water-soaked border. Yellow leaves on close inspection appear mottled and water-soaked. Plants die once the pathogen attacks meristematic tissues. The football-shaped cells of the pathogen can be seen with a microscope in blighted leaves and roots.

Management

Displacing sodium from soil using freshwater flushes and/or gypsum is recommended. When the disease is active, avoid abrasive cultural practices like sand topdressing and mow when surfaces are dry. The fungicides mancozeb, pyraclostrobin, and trifloxystrobin are effective in controlling rapid blight.

Red Thread and Pink Patch

Pathogen: Red thread = *Laetisaria fuciformis* (McAlpine) Burdsall

Pink patch = *Limonomyces roseipellis* Stalpers & Loerakker

FIGURE 2.35
Brownish patches with dark-brown water-soaked border of rapid blight in a golf green over-seeded with roughstalk bluegrass. (Photo courtesy of B. Martin.)

Primary hosts: Bluegrasses, fescues, and perennial ryegrass

Predisposing conditions: Extended overcast and rainy weather

Red thread and pink patch can be active at almost any time of the year in many regions of the United States. Development of these diseases is favored most by cool (65–75°F; 18–24°C), wet, and extended periods of overcast weather. Both diseases may occur during cool weather in the presence of abundant surface moisture in spring and autumn, during warm and drizzling weather in summer, or at snow melt in winter. Both diseases may become widespread among turfgrass species during mild winters. Spring, however, is the dominant period of red thread and pink patch activity in most regions.

Pink patch once was believed to be a form of red thread, but actually is a distinct disease with the same general hosts and symptoms as red thread. Pink patch is most prevalent in fine-leaf fescues in shaded locations. Pink patch, however, seldom develops in the absence of red thread. *L. roseipellis* frequently complexes with *L. fuciformis* on the same plant and sometimes on the same leaf.

Red thread and pink patch often occur together and are most damaging to perennial ryegrass, common-type Kentucky bluegrasses, and tall and fine-leaf fescues. Improved cultivars of Kentucky bluegrass and bentgrasses may also be affected, but these grasses do not usually sustain a significant level of injury if sufficiently fertilized with nitrogen. Red thread, however, is common in tall fescue serviced by lawn care companies and in properly fertilized roughs. Red thread therefore should not always be thought of as a disease of poorly nourished turf.

Symptoms

Symptoms of red thread and pink patch are concentrated in pink- to red-colored circular patches 2 inches (5 cm) to 3 ft (90 cm) in diameter (Figure 2.36). Patches may coalesce to involve large, irregular-shaped areas (Figure 2.37). On close inspection, one can easily see the gelatinous, pinkish-red mycelium on leaves and sheaths. Infected green leaves soon become water-soaked in appearance. When leaves dry, the fungal mycelium becomes pale pink in color and on close inspection is easily seen on the straw-brown or tan tissues of dead leaves and sheaths. From a distance, affected turf has a straw-brown, tan, or pinkish color.

Signs of red thread are distinctive and unmistakable. In the presence of morning dew or rain, a coral pink or reddish layer of gelatinous fungal growth can be easily seen on leaves and sheaths. Sclerotia or "red threads" are invariably present and can be observed extending from leaf surfaces, particularly cut leaf tips (Figure 2.38). They are initially pink and gelatinous, while mature sclerotia are bright red, hard, and brittle strands of fungal mycelium. White-pink "cotton candy-like" mycelia flocks may appear in the foliage (Figure 2.38). These flocks break up into spores (i.e., arthrospores) when placed in water. These red threads fall into thatch and serve as resting structures for the fungus by surviving long periods that are unfavorable for growth of the pathogen.

The visible sign of pink patch appears as a gelatinous mass of pinkish mycelium associated with a water-soaked appearance of leaves. The pink patch fungus does not produce "red threads"; however, it may produce a whitish-pink foliar mycelium or "cotton candy-like flocks" of pinkish mycelium. Both pathogens can be easily distinguished with a microscope. The

FIGURE 2.36
Red thread appears as patches of pink- or red-colored turf. Inset shows immature, pink, and gelatinous sclerotia. (Inset photo courtesy of J. Kaminski.)

FIGURE 2.37
Red thread in a tall fescue lawn.

FIGURE 2.38
Red, brittle, and antler-like sclerotia and mycelial flock of *Laetisaria fuciformis*.

pink patch fungus produces distinctive clamp connections (i.e., a bridge-like protrusion between cell walls) on hyphae, whereas clamp connections are absent on hyphae of the red thread fungus. The presence or absence of red, antler-like sclerotia, however, is the primary means of distinguishing between red thread and pink patch in the field.

Management

Red thread and pink patch generally are most injurious to poorly nourished turfs, but as noted previously, properly fertilized turf can be blighted to objectionable levels. Both pathogens are foliar blighters that do not infect crowns or kill plants. Frequently, both diseases are best managed by an application of about 1.0 lb N/1000 ft² (50 kg N/ha). Quick-release, water-soluble nitrogen (e.g., urea) is more effective in reducing red thread injury than slow-release nitrogen. The most effective suppression occurs when nitrogen is applied as soon as the disease appears and is less effective after significant blighting has occurred. Application of nitrogen during periods too cool for turf growth will not aid in reducing disease severity. This is because nitrogen alleviates red thread or pink patch injury by stimulating the production of new leaves and tillers to replace blighted tissue. Phosphorus (P) and potassium (K) applied alone have little or no effect on red thread, especially in soils with moderate or high P and K levels. Nitrogen plus potassium has been shown to suppress red thread symptoms more effectively than nitrogen alone. Both red thread and pink patch can develop in well-fertilized turf, and additional nitrogen may not be warranted for general agronomic reasons.

Loss of turf density due to foliar blighting often favors weed encroachment. Spring blighting in particular favors crabgrass (*Digitaria* spp.) and white clover (*Trifolium repens*) invasion. Use of fungicides to control red thread in spring reduces weed levels significantly by preventing deterioration of stand density. Hence, fungicide use may be warranted in some situations. Fungicides that control both red thread and pink patch include azoxystrobin, chlorothalonil, fluoxastrobin, iprodione, myclobutanil, propiconazole, pyraclostrobin, tebuconazole, triadimefon, trifloxystrobin, triticonazole, vinclozolin, and others. Flutolanil is effective against red thread, but research indicates that it is ineffective against pink patch.

Spring Dead Spot

Pathogen(s): Ophiosphaerella korrae (J. Walker & A.M. Smith) Shoemaker & Babcock; *Ophiosphaerella herpotricha* (Fr.: Fr.) J. Walker; and *Ophiosphaerella narmari* (J. Walker & A. M. Smith) Wetzel, Hulbert & Tisserat.

Primary host: Common and hybrid bermudagrass, and less commonly zoysiagrass and buffalograss

Predisposing conditions: Cool to cold and wet weather in autumn and spring

Spring dead spot (SDS) is the most damaging fungal disease of bermudagrass turf. This disease is incited by three root pathogens, including *O. korrae* (formerly *Leptosphaeria korrae*), *O. herpotricha*, and *O. namari* (formerly

Leptosphaeria namari). In the western United States, *O. herpotricha* is the primary pathogen, whereas *O. korrae* is the primary casual agent of SDS in the eastern United States. *O. narmari* is also prevalent in California. *O. narmari* and *O. korrae* are casual agents of SDS in Australia. Both *O. herpotricha* and *O. narmari* were reported to attack roots of the same bermudagrass plants in the southwestern United States. *O. korrae* causes SDS in zoysiagrass (i.e., *Zoysia japonica* and *Zoysia matrella*) in the southeastern United States. SDS in buffalograss has been reported to be caused by *O. herpotricha* in the western United States.

The intensity of SDS varies greatly from year to year, and it is not possible to predict those years when it will be severe. Indeed, turf pathologists are baffled by the unpredictable nature of SDS outbreaks. The disease generally is more common in northern and colder climatic regions of bermudagrass adaptation and during springs following very cold winters. Severe outbreaks of SDS are due to imperfectly understood environmental and soil conditions as well as the fact that three different species of *Ophiosphaerella* can be involved in the SDS complex in some regions. Soil fertility and pH, and some herbicides may potentially impact SDS severity.

SDS can be extremely destructive to bermudagrass under both low and high management. Recovery, however, is very slow in turf maintained with low levels of nitrogen. The disease is most commonly associated with bermudagrass turf older than 3 years, but it may appear the spring following sprigging with stolons from sites previously affected with SDS. SDS injury is most likely to occur where thick thatch layers exist, where soils are compacted, and where high rates of nitrogen fertilizers were applied in the autumn.

Since zoysiagrass is more winter hardy than bermudagrass, it is more resistant to the disease and generally only appears following very cold winters. The patch symptoms are similar to those described below for bermudagrass, but tend to be smaller (4–12 inches in diameter; 10–30 cm) and more widely distributed. While SDS is less severe, it takes longer for zoysiagrass to recover since this grass species grows much more slowly than bermudagrass.

Symptoms

As the name implies, SDS injury becomes apparent in spring. Root infection begins in early autumn, but root injury by the pathogen(s) becomes rapid just prior to spring green-up. As bermudagrass breaks dormancy, circular patches of tan or brown, sunken turf a few inches (5 cm) to 3 ft (90 cm) or greater in diameter become conspicuous (Figures 2.39 and 2.40). Sometimes, patches coalesce and the damage is nonuniform and appears similar to winter-kill or winter desiccation. After a few years, bermudagrass may recolonize patches, and affected areas may take on a ring or "frog-eye" appearance. Rhizomes and stolons from nearby, healthy plants eventually spread into and cover dead patches. This filling-in process is slow, a period that may

FIGURE 2.39
Spring dead spot in bermudagrass.

FIGURE 2.40
Spring dead spot in a bermudagrass sports field.

last 4–8 weeks or longer following spring green-up of bermudagrass. The slowness of the filling-in process is believed to be due to toxic substances generated by the pathogen(s) in the thatch or soil below dead patches. Weeds commonly invade dead patches.

Management

Factors that delay autumn dormancy or reduce winter hardiness tend to promote SDS. This normally is caused by high nitrogen fertility and/or late-season applications of nitrogen to prolong greenness prior to dormancy. Older common types (e.g., "Tufcote" and "Vamont" [*Cynodon dactylon*]) and hybrid

"Tifway" bermudagrass (*C. dactylon* × *C. transvaalensis*) are very susceptible to this disease. Ultradwarf hybrid bermudagrass cultivars (e.g., Champion, TifEagle, and Miniverde), which are grown on golf greens, are also highly susceptible to SDS. Cultivars of bermudagrass with greater winter hardiness such as "Latitude 36," "Midlawn," "Patriot," "Riviera," "T-1," and "Yukon" tend to be less susceptible and generally recover more rapidly from SDS. While the aforementioned cultivars have improved resistance, they do get the disease, which can be severe in some years. Generally, this occurs 3 or more years after establishment and when a significant thatch layer has developed.

It is very important to eliminate weeds from diseased sites as their presence will slow down and in some cases prevent complete recovery of bermudagrass. Eliminating weeds reduces competition, which helps to speed up turf recovery. Some preemergence herbicides used in spring to control annual grass weeds, however, can slow down recovery by inhibiting rooting from bermudagrass stolons. This appears to be mostly a problem with dinitroaniline herbicides (e.g., pendimethalin and prodiamine), whereas oxadiazon is safe. Bermudagrass grown in dense thatch, compacted, and poorly drained soils tends to be more prone to SDS damage. Hence, core aeration and other practices that reduce thatch and improve surface water drainage and soil aeration in the summer when bermudagrass is actively growing are recommended.

Ammonium sulfate (applied at 1.0 lb N/1000 ft^2; 50 kg N/ha) and potassium chloride (applied at 1.0 lb K/1000 ft^2; 50 kg K/ha) applied on 21–28-day intervals beginning at spring green-up speed recovery and help to alleviate SDS severity over time. On golf greens, lower amounts of nitrogen are applied on a 7–14-day interval to speed recovery. The suppression effect provided by ammonium sulfate may take 3 years to develop. Acidification alone does not reduce the number of patches but reduces their size. It is important, however, to cease nitrogen application about 6 weeks prior to the anticipated dormancy of bermudagrass. This is because the application of significant amounts of nitrogen in early autumn has been linked to an increase in SDS severity the following spring. Nitrate forms of nitrogen, such as sodium nitrate, potassium nitrate, and calcium nitrate should be avoided. This is because nitrate forms of nitrogen tend to increase the soil pH in the root zone, which could intensify SDS.

Sulfur reduces SDS severity, but is not recommended since it causes slower spring green-up and thinning of bermudagrass. Potassium can improve low-temperature stress hardiness in bermudagrass and research suggests that potassium chloride used in conjunction with ammonium sulfate may help to reduce SDS. Irrigate SDS-affected turf to prevent wilt during dry periods and maintain nitrogen applications throughout summer to encourage regrowth by stolons and rhizomes.

Higher-cut bermudagrass generally is not as severely damaged by low temperatures and ice cover. Hence, it is helpful to increase mowing height as much as possible for the site (i.e., green, fairway, athletic field, etc.) about 30 days prior to dormancy. Research, however, showed that seeded bermudagrass cultivars were damaged less by SDS at a lower (0.5 in.; 1.3 cm) versus a higher (1.5 in.; 3.8 cm) mowing height (Martin et al., 2001). The impact of mowing height on SDS in vegetatively propagated versus seeded cultivars needs further study. For golf greens, use of blankets in winter to protect bermudagrass from low-temperature injury would be expected to help reduce SDS severity.

Azoxystrobin, fenarimol, fluoxastrobin, myclobutanil, propiconazole, and tebuconazole suppress SDS. Fenarimol is perhaps the most consistently effective fungicide, but its removal from the market is anticipated. A fungicide should be applied once or twice in mid-to-late September or about 30 days prior to anticipated winter dormancy. Fungicides, however, do not provide complete SDS control, and normally spring applications provide little or no benefit since most damage is inflicted prior to spring green-up. Furthermore, most fungicide trials conducted in the eastern United States involving *O. korrae* have indicated that one autumn application of a fungicide can be expected to be just as effective as two applications. However, research conducted in Oklahoma has shown that two fungicide applications in the autumn followed by one application in spring are most effective for controlling SDS caused by *O. herpotricha* (Walker, 2009). Control typically is erratic with any fungicide in any given year, with levels of SDS suppression often ranging from 0% to 75%. As noted previously, complete control with fungicides is seldom achieved, but use of these chemicals can help improve survival and promote more rapid recovery.

Fungicides should be applied in ≥100 gallons (935 L/ha) of water per acre. High water dilutions help move fungicide down to stolons and between leaf sheaths to make contact with vital growing points. Currently, there are no data to support the premise that watering-in of a fungicide to the root zone will improve SDS control. Indeed, bermudagrass generally loses most of its existing root system at spring green-up. Hence, it would appear that protecting stem bases and stolons, which can live for one or more years, is the correct target for a fungicide.

Stripe Smut and Flag Smut

Pathogen: Stripe smut = *Ustilago striiformis* (Westend.) Niessl

Flag smut = *Urocystis agropyri* (G. Preuss) Schrot

Primary hosts: Kentucky bluegrass

Predisposing conditions: Overcast, cool, and wet weather in spring and autumn

Stripe smut (*U. striiformis*) and flag smut (*U. agropyri*) are diseases that occur primarily in mature Kentucky bluegrass stands. Stripe smut occasionally is a problem in tall fescue, perennial ryegrass, and some older cultivars of creeping bentgrass.

Symptoms

Symptoms of stripe and flag smut are identical and are most conspicuous during cool, moist seasons of spring and autumn. Infected plants are often stunted and pale green or yellow in color. Narrow, silvery, or gray-black streaks appear in parallel lines along the leaf surface (Figure 2.41). These streaks are fruiting structures (i.e., sori) in which masses of black spores (i.e., teliospores) are produced. When sori mature, the cuticle and epidermis rupture and leaves shred and curl, which releases the teliospores. Both pathogens are obligate parasites (i.e., grow and complete life cycle in living plants only) and once infected, plants will remain colonized by these fungi until they die or they are treated with an effective fungicide.

Although both pathogens are systemic and persistent in surviving plants, during summer months, infected plants may appear healthy if properly maintained. In spring or autumn, however, badly infected stands may appear chlorotic (yellow) and in need of fertilizer. Turf often losses density and generally has an unthrifty appearance. Large losses of turf may occur, particularly if infected plants are attacked by other pathogens (especially *Puccinia* spp.). From early winter until spring green-up, leaves that had been shredded by matured sori in the autumn develop a gray-brown, desiccated appearance (Figure 2.41). Affected stands may appear thin and winter dormant many weeks prior to bitterly cold weather.

FIGURE 2.41
Silvery or black parallel lines produced by stripe smut fruiting bodies (i.e., sori) in Kentucky bluegrass leaves. Lower right inset shows the gray-brown and unthrifty appearance of shredded leaves. (Upper right photo courtesy of R. Smiley.)

Management

Most Kentucky bluegrass cultivars available today are resistant to these diseases, which has led to a dramatic reduction in the occurrence of stripe and flag smuts. Although leaf shredding symptoms may not be evident, stripe and flag smuts can be very damaging to infected plants during periods of heat and drought stress. If properly irrigated and fertilized, however, smutted stands survive and exhibit only a decrease in turf quality and some thinning during stressful summer months. These smut diseases most commonly occur in mature (2–4 years and older) stands that have been managed with high levels of nitrogen fertilizer. Using a complete N-P-K autumn fertilizer program, increasing the mowing height in summer and deep irrigation at the first sign of drought stress are effective management practices that minimize smut injury. Severely infected stands can be effectively revitalized with a single, spring or autumn application of a DMI/SI fungicide (e.g., propiconazole and triadimefon).

Take-All Patch

> *Pathogen: Gaeumannomyces graminis* (Sacc.) Arx & D. Oliver var. *avenae* (E. M. Turner) Dennis
>
> *Primary host:* Bentgrasses
>
> *Predisposing conditions:* Cool, wet weather in autumn and spring, followed by hot and/or dry summer conditions

Take-all patch (formerly Ophiobolus patch) is primarily a disease of creeping, colonial, and velvet bentgrass (*Agrostis* spp.) turfs. This disease is most common on newly constructed golf courses, particularly those carved out of woodlands, peat bogs, or other areas that have not supported crops or grasses for decades. This disease can be especially damaging to rebuilt greens or tees on old golf courses or where a fumigant was used for renovation. Take-all can also be imported on infected bentgrass sod. The disease tends to spread more rapidly and occurs with greater severity in sandy soils and in soils that have been heavily limed. Take-all may appear immediately following installation of infected sod, but in seeded stands, it generally does not appear until the second year following seeding.

Symptoms

The pathogen actively attacks roots during cool and wet periods, but symptoms may not appear until the advent of warmer and drier conditions. Symptoms of the disease are most conspicuous from late spring and throughout summer and early autumn. Take-all can appear in early spring in response to drought, low mowing, aggressive grooming, and/or use of plant growth regulators. Bentgrass affected by take-all in spring may recover by summer; however, if irrigation is withheld, those areas affected in spring are first to wilt and possibly die from drought stress.

There are numerous color symptoms and circular patch sizes associated with take-all in creeping bentgrass. On golf greens in the morning, there may be a lack of dew on circular-shaped areas (i.e., circular patches). Later in the day, turf in affected patches develops the blue-gray color associated with wilt. This is due to the impaired ability of infected roots to take up sufficient amounts of water. Initially, circular patches are reddish-brown or orange-bronze in color and may only be 3.0–6.0 inches (7.5–15 cm) in diameter. Patches increase to 2 ft (60 cm) or more in diameter, particularly on chronically affected bentgrass sites. Most patches range from 3 to 18 inches (7.5–45 cm) in diameter, but some may exceed 3 ft (90 cm) in diameter (Figures 2.42 and 2.43). Patches may develop in tight clusters that give the appearance of a larger patch. Patches may also coalesce, resulting in large, irregular areas of discolored or dead turf. When the disease is active, the perimeter of the patch usually assumes a bronzed or reddish-brown appearance, and turf eventually turns a bleached or tan color. Frog-eyes, or rings of dead turf with living plants in the center, may also develop. The small, circular patches increase in size over a number of years and dead bentgrass in the center of the patch may be colonized by annual bluegrass, dandelion, and other weeds. Sometimes, small horseshoe-shaped crescents are associated with take-all. Because the pathogen attacks the root system, turf in affected areas is easily detached and is reminiscent of the type of damage caused by root feeding insect pests (e.g., white grubs).

FIGURE 2.42
Classic bronze-colored take-all patches with dandelions. Smaller white spots are dollar spot infection centers.

FIGURE 2.43
Take-all can persist for long periods and patches may grow to 2 ft (60 cm) or more in diameter.

Signs of the take-all fungus are distinctive, but most can only be seen under high magnification (Figure 2.44). Dark-colored runner hyphae are found on roots and sometimes mats of infection hyphae (aka mycelial aggregates or plate mycelia). A prominent diagnostic root symptom is browning of the central portion of the root (i.e., the stele or vascular cylinder), while cortical

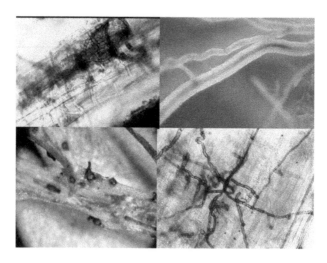

FIGURE 2.44
Mycelial infection mat (top left); brown discoloration of the central cylinder (i.e., stele) of an infected root (top right); black flask-shaped fruiting bodies (i.e., perithecia; bottom left); and darkly pigmented runner hyphae of *Gaeumannomyces graminis* var. *avenae*.

tissue remains white. The black flask-shaped fruiting bodies (i.e., perithecia) are seldom found in nature.

In most cases, take-all naturally declines over time, presumably due to a buildup of antibiotic-producing bacteria that antagonize or in some other way prevent *G. graminis* var. *avenae* from damaging roots. This explains why take-all is normally not a problem on older golf courses. There are, however, exceptions. Golf courses with a past problem with take-all may see the disease redevelop following a heavy application of lime or other alkaline reacting materials. For mysterious reasons, take-all has redeveloped on some very old golf courses in the United Kingdom, usually during excessively wet years. The decline phenomenon usually occurs within 2–5 years from the time that the first disease symptoms were observed. During the decline phase, plants in affected patches appear reddish-brown, tan, or chlorotic (yellow) and turf may or may not thin out. Take-all patch, however, can persist indefinitely in highly buffered or alkaline soils or where irrigation water has a high pH.

Management

Take-all patch management on golf greens will involve alleviating stress. It is important to increase mowing height, reduce mowing frequency, replace grooved rollers with solid rollers, and mow with lightweight mowers. Do not mow greens when they are excessively (spongy or presence of casual water) wet due to the likelihood of mechanical damage. Avoid other practices that promote green speed such as sand topdressing, vertical cutting, brushing, and use of plant growth regulators. Furthermore, core aeration and high-pressure water injection aeration should not be attempted when take-all is active.

Take-all can occur over a wide range of soil pHs, including acidic soils, but it is most severe and persistent where the soil pH is above 6.5 in the top 1.0–2.0 inches (2.5–5.0 cm) of soil. Acidification of soil with ammonium sulfate is the primary cultural approach to managing take-all. Acidification either directly reduces growth of the take-all fungus or favors growth of other microorganisms, which effectively compete with or in some other way antagonize the pathogen. Even in situations where soil pH is as low as 5.5, ammonium sulfate may still be recommended. Nitrate forms of nitrogen, such as potassium nitrate, calcium nitrate, and sodium nitrate, however, may intensify take-all and other patch diseases caused by root-attacking fungi by raising soil pH.

Ammonium sulfate applied at a rate of at least 3.0 lb N/1000 ft² per year (150 kg N/ha/year) should be used as the primary N source as long as take-all remains persistent. Manganese sulfate should be applied in conjunction with ammonium sulfate. Acidification reduces the ability of microbes to oxidize manganese and the resulting increase in manganese availability for root uptake assists plants in their defense against the take-all fungus.

Manganese sulfate is best applied on a monthly interval at 0.02 to 0.04 lb elemental Mn/1000 ft² (1–2 kg Mn/ha) between early spring to early winter for a total of 1.8 to 3.6 lb Mn/1000 ft²/year (9–18 kg Mn/ha/year). Both ammonium sulfate and manganese sulfate can burn turf and should be watered-in immediately. Use of ammonium sulfate provides very good winter color, but also encourages foliar growth. This causes slower green speeds and stimulates the need for frequent mowing into early winter. When using acidifying fertilizers, check soil pH often and avoid acidification if soil pH falls below 5.2.

Applications of phosphorus (P) and potassium (K) have also been linked to reducing take-all severity. For best results, ammonium sulfate should be applied with P and K in a 3N:1P:2K ratio. Use of lime or sand topdressing with a pH above 6.5 should be avoided. If limestone is required, use coarse (i.e., slower reacting) rather than fine grades in areas with a past history of take-all.

Thatch and soil compaction should be controlled through coring and other techniques. Coring, however, should be performed only when symptoms of take-all patch are not evident and should be delayed if it causes lifting of bentgrass sod. Check the pH of your irrigation water. In extreme cases, acid injection of irrigation water may be recommended. Frequent syringing and hand watering of affected areas is often required in summer to prevent death of plants whose root system has been significantly damaged by the pathogen. Finally, use of preemergence herbicides and plant growth regulators should be avoided where take-all is a problem.

The fungicides azoxystrobin, fenarimol, fluoxastrobin, pyraclostrobin, propiconazole, tebuconazole, triadimefon, and triticonazole may be used to target take-all patch. High rates of fenarimol, propiconazole, triadimefon have growth-regulating properties and can cause bentgrass to develop a blue-green color and wider textured leaves. Fungicides should be applied preventively in late autumn and early spring. For additional suppression, fungicides may have to be applied 2–3 times on a 14–21-day interval at the onset of symptoms in spring or summer. Azoxystrobin, myclobutanil, and propiconazole are preferred for curative treatment. Fungicides should be applied in ≥100 gallons of water per acre (≥935 L/ha) using flat fan nozzles to ensure more uniform coverage and better penetration into the canopy. Watering-in fungicides to a 1.0 inch (2.5 cm) soil depth may boost the performance of fungicides where the use of high water volumes is impractical.

Yellow Tuft or Downy Mildew

Pathogen: Sclerophthora macrospora (Sacc.) Thirum., Shaw & Naras.

Primary hosts: Most cool-season turfgrasses, St. Augustinegrass, and zoysiagrass

Predisposing conditions: Extended periods of cool and wet weather

S. macrospora attacks nearly all turfgrasses as well as several major grass crops, including rice, sorghum, and corn. Kentucky bluegrass sod can be rendered temporarily unsalable, and a severe infection mars the appearance and playability of creeping bentgrass and/or annual bluegrass golf greens. Infected plants are not directly killed by this obligate parasite (i.e., grows and completes life cycle in living tissue only). Yellow tuft is primarily a problem on golf greens, where it can disrupt the smoothness of putting surfaces and mar their appearance.

Plants infected with *S. macrospora* can persist in excess of 2 years. The pathogen is a very sophisticated obligate parasite. As a result, most obligate parasites (including rusts, smuts, and powdery mildew) have evolved to the point where they generally do not directly kill plants. Instead, they debilitate or weaken plants, predisposing them to possible injury or death from other stress factors.

S. macrospora produces lemon-shaped asexual fruiting bodies called sporangia on the surfaces of infected leaves (Figure 2.45). Sporangia develop below leaf surfaces and protrude through stomates in the morning. During early morning hours, when leaves are wet, the pearly white sporangia can be seen with a hand lens on the upper and to a lesser extent the lower leaf surfaces of infected plants. Swimming spores called zoospores develop within sporangia and when released they gain entry through meristematic tissue (i.e., stems, buds, or the mesocotyl region of germinating seeds) (Figure 2.45). Seedlings are most vulnerable to infection by *S. macrospora*, which accounts

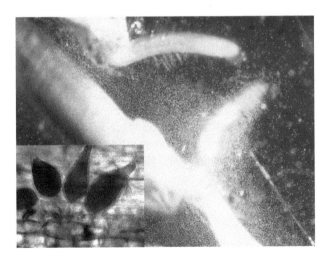

FIGURE 2.45
Sclerophthora macrospora zoospores are borne in lemon-shaped sporangia (stained red) that protrude through stomates on leaf surfaces (inset). Once released, swimming zoospores can be attracted to substances liberated from germinating seeds. Zoospores encyst (i.e., resting stage), germinate, and passively penetrate meristematic cells in the mesocotyl region of developing seedlings.

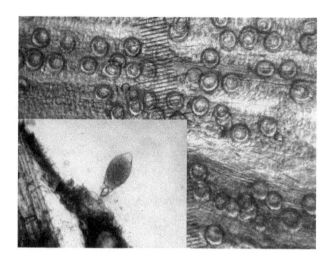

FIGURE 2.46
Sclerophthora macrospora sporangium (i.e., oosporangium) developing from an oospore (inset). Oospores (sexual spores) serve as thick-walled survival structures.

for why the disease is most commonly observed in the spring following autumn seeding. Yellow tuft recurs in older turfs following excessively wet weather in spring and autumn. During most summer months, infected plants appear green and healthy.

Once infected, mycelia grow upwards between cells (i.e., intercellular spaces) from infected buds or stem bases into sheaths and then leaves. If plants are allowed to produce seedheads, the fungus can grow into the culm, invade the inflorescence, and infect seed (hence the disease can be seed borne). When leaves die, mycelium can differentiate into large spores called oospores. Oospores are thick-walled survival structures, which can persist indefinitely in dead leaf tissue (Figure 2.46). While oospores help to disperse the pathogen, the production of zoospores in asexual sporangia is the primary mechanism by which large numbers of plants become infected. Crabgrass (*Digitaria* spp.) is very susceptible to *S. macrospora*, and this weed serves as a major harborage site for the production of zoospores and oospores.

Symptoms

On golf greens, tees, and fairways, the disease appears as yellow spots, 0.25–0.5 inch (6.2–12.5 mm) in diameter (Figure 2.47). In Kentucky bluegrass, and other wider bladed grasses, the yellow spots are 1.0–3.0 inches (25–75 mm) in diameter. In low areas where water collects and puddles, infected stands exhibit a generalized chlorosis (Figure 2.48). Each spot consists of one or two plants having numerous tillers, giving plants a tufted appearance. The tufting, or abnormal tiller production, is induced by

FIGURE 2.47
On creeping bentgrass golf greens, yellow tuft appears as small yellow spots of tufted plants with short bunchy roots.

FIGURE 2.48
Yellow tuft in swale of a Kentucky bluegrass sod field. (Photo courtesy of N. Jackson.) The inset shows dried white *S. macrospora* sporangia on crabgrass leaves.

S. macrospora, which causes a shift in the production of a hormone that regulates tillering. Roots of infected plants are short and bunchy, and tufts are easily detached from the turf. During cool and moist periods in late spring and autumn, plants develop a yellow color at which time the infected plants are "yellow tufted." The yellowing is the indirect result of heavy fruiting body (sporangia) and subsequent spore production by the fungus.

Once leaves dry, sporangia desiccate and when abundant appear as a white residue on leaves (Figure 2.48). These spores (i.e., zoospores) swim, and this accounts for why yellow tuft is more severe in low-lying areas where water puddles.

In St. Augustinegrass, the disease is called "downy mildew" instead of "yellow tuft," and the symptoms are different. The disease appears as white, linear streaks that run parallel to leaf veins. Leaves turn yellow and there may be some browning of leaf tips. Excessive tillering does not occur. The disease is disfiguring and St. Augustinegrass growth may be stunted. In zoysiagrass, however, the "yellow tuft" symptom is common.

Management

Vertical cutting golf greens in spring will physically detach many tufted plants. Improving surface water drainage helps to alleviate yellow tuft since the disease is most severe in low areas where water collects. As noted previously, it is in wet environments that swimming zoospores are able to move easily to uninfected plants.

Yellow tuft is best controlled with mefenoxam. Mefenoxam performs better when it is tank-mixed with fludioxonil. Two or three mefenoxam applications may be required to eradicate the fungus. After fungicide application(s), however, plants can retain their tufted appearance for several weeks. It is only until new tillers replace the older infected shoots that plants regain their normal appearance and growth habit. Mefenoxam works best when applied preventively and prior to rainy weather, particularly where there are seedlings or immature plants. There may be little response from a curative mefenoxam application if it is not tank-mixed with fludioxonil.

Diseases Initiated During Summer That May Persist into Autumn

Many of the diseases in this section are initiated in response to high temperature and/or high humidity. Some are most severe during wet and overcast periods, while others are more damaging during periods of heat and drought. A root pathogen may damage turf during cool and moist conditions, but symptoms may not develop until periods of heat or drought stress. Hence, in some regions, take-all patch, necrotic ring spot and Pythium-induced root dysfunction may be more apparent in summer despite the fact that most damage was inflicted to the root systems earlier in the year. Several of these diseases can occur in some turfgrasses at different times of year in different regions.

Brown Patch and Leaf and Sheath Spot

Pathogen(s): Rhizoctonia solani Kuhn and *Rhizoctonia zeae* Voorhees

Primary hosts: Cool-season grasses, especially bentgrasses, perennial ryegrass, and tall fescue

Predisposing conditions: High humidity, long periods of leaf surface wetness, and high night temperatures

Brown patch, also known as Rhizoctonia blight, is caused by *R. solani* and is a common, summertime disease of cool-season turfgrasses. The most susceptible species include colonial and creeping bentgrasses, perennial ryegrass, and tall fescue. In southern and transition regions, *R. solani* damages bermudagrass, centipedegrass, seashore paspalum, St. Augustinegrass, and zoysiagrass in the spring or autumn. This disease in warm-season grasses is referred to as large patch and was previously described.

In very warm environments, *R. zeae* may also produce symptoms that are similar to brown patch in cool-season grasses, especially in perennial ryegrass. The disease caused by *R. zeae* was named Rhizoctonia leaf and sheath spot by Japanese pathologists, which is not descriptive of the symptoms in turfgrasses in the United States.

Environmental conditions that favor disease development are day temperatures above 85°F (29°C) and high relative humidity. A night temperature above 68°F (20°C), relative humidity ≥90% at night, and periods of leaf surface wetness exceeding 10 h are the most critical environmental requirements for brown patch development in cool-season grasses. This disease becomes extremely severe in cool-season grasses during prolonged, overcast wet periods in summer as long as average daily temperatures remain above 63°F (17°C). *R. solani*, however, can be quite active at lower temperatures. This explains its ability to damage warm-season grasses in the autumn and spring.

Symptoms of Brown Patch

Symptoms of brown patch vary according to host species. On closely mown turf, affected patches are roughly circular and range from 3 inches to 3 ft (7.5–90 cm) or greater in diameter (Figure 2.49). Patches may exhibit significant thinning, especially in immature golf greens overstimulated in spring with nitrogen (Figure 2.50). The outer edge of patches may develop a 1–2 inch (2.5–5.0 cm)-wide smoke-ring. The smoke-ring is blue-gray or black in color and is caused by mycelium in the active process of infecting leaves (Figures 2.49 and 2.51). The mycelia of *R. solani* have distinctive right-angle branching, which is a useful diagnostic sign for all *Rhizoctonia* spp. (Figure 2.51).

On high-cut turfs, smoke-rings may not be present or are less distinct. Patches may coalesce and have an irregular rather than circular shape. Close

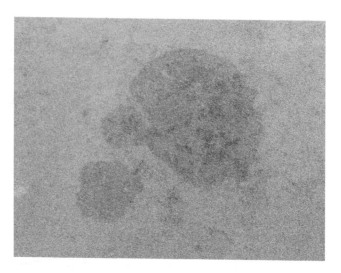

FIGURE 2.49

Circular patches of brown patch with smoke rings in a creeping bentgrass golf green. Blackening within the larger patch is due to colonization by blue-green algae. Smaller white spots are dollar spot infection centers.

inspection of leaf blades reveals that the fungus primarily causes tip dieback or a general blight as a result of numerous lesions. This gives diseased turf its brown color, especially in tall fescue whose leaves are often severely blighted (Figure 2.52). *R. solani* produces distinctive and often elongated lesions on tall fescue leaves (Figure 2.53). The lesions are a light, chocolate-brown color,

FIGURE 2.50

Brown patch can be particularly destructive during establishment when high rates of nitrogen are applied in spring to speed grow-in of bentgrass golf greens.

FIGURE 2.51
Active brown patch with smoke rings in a tall fescue lawn. Inset shows characteristic right-angle branching of *R. solani* hyphae.

and are bordered by narrow, dark-brown bands. On perennial ryegrass and Kentucky bluegrass, smaller leaf lesions are produced and tip dieback commonly occurs. On bentgrasses, distinct lesions may not be evident because leaf blades are too fine textured and usually appear blighted or shriveled. During early morning hours, when the disease is active, a cobweb-like mycelium or tufts of hyphae can develop in sparse to copious amounts on leaves laden with dew (Figure 2.53). Late in the season, distinctive circular patches

FIGURE 2.52
Rhizoctonia solani can severely blight tall fescue leaves, but the turf normally recovers.

FIGURE 2.53
Rhizoctonia solani foliar mycelium and distinctive light brown lesions with dark brown borders on tall fescue leaves.

may not appear. For example, in perennial ryegrass, the turf may simply exhibit a nonuniform thinning out and there may be little or no foliar mycelium evident in the morning.

Symptoms of Leaf and Sheath Spot

R. zeae is more commonly found in warm and humid southeastern and transition regions, and appears to require high night temperatures (≥70°F; ≥21°C) to become destructive. In transition zone regions, *R. zeae* can cause yellow circular patches or small yellow blotches in perennial ryegrass in the summer. Perennial ryegrass fairways can be more severely damaged by *R. zeae* than *R. solani*. In creeping bentgrass, *R. zeae* produces yellow or orange-colored rings and sometimes brown patches similar to *R. solani*, which may or may not have a smoke-ring (Figure 2.54). High-density creeping bentgrass cultivars (e.g., Declaration, Penn A and G-series, and Tyee) appear to be more susceptible than traditional cultivars (e.g., Penncross and Providence) to *R. zeae*. In mature stands of high-density cultivars, the disease is more disfiguring than destructive. Immature stands, however, can be severely damage during the first summer following grow-in when high rates of nitrogen are used to speed establishment.

Management

Nitrogen (N) source and timing of fertilizer application can impact brown patch severity. In particular, autumn applications of a slow-release N source to cool-season grasses would be expected to result in less brown patch the

FIGURE 2.54
Orange- or bronze-colored rings and smoke rings produced by *Rhizoctonia zeae* in high-density creeping bentgrass golf greens.

following summer, when compared to mostly spring applications of water-soluble N. Furthermore, autumn-applied slow-release N plus phosphorus (P) and potassium (K) can lower brown patch severity the following summer, when compared to autumn-applied water-soluble N plus P and K. Applications of high rates of N in the spring or summer intensify brown patch. Spoon-feeding golf greens with low N rates (0.1–0.125 lb N/1000 ft^2; 5–6 kg N/ha) intermittently throughout summer can enhance brown patch; however, this practice has more positive than negative attributes and should be continued if a suitable fungicide program is in place.

Irrigating at dusk intensifies brown patch, whereas irrigation during early morning hours reduces disease potential. Evening irrigation intensifies brown patch by providing for longer leaf wetness durations. Conversely, early morning irrigation does not extend the leaf wetness period and physically knocks *R. solani* foliar mycelium off leaves. Use of wetting agents as well as dragging or poling speeds up leaf drying and may help to reduce disease activity. Frequent irrigation that results in saturated soil conditions favors brown patch, particularly in shaded sites with poor air circulation. It is helpful to improve surface water drainage, improve air circulation, reduce shade, and alleviate soil compaction.

Brown patch is more intense in dense, high-cut turf when compared to lower mowing in more open stands. However, under high disease pressure conditions, mowing height appears to have little impact on brown patch severity. Generally, mowing high within the recommended range for the species helps turf to better tolerate summer stresses, diseases, and insect pests, and helps to reduce weed invasion. Hence, for numerous agronomic reasons, it is generally best to maintain the highest possible mowing height in the summer.

Brown patch is more effectively controlled when fungicides are applied prior (i.e., preventively) to the onset of blighting. Flutolanil, polyoxin D, and QoI/strobilurins (e.g., azoxystrobin, fluoxastrobin, pyraclostrobin, and trifloxystrobin) provide the longest residual effectiveness. Iprodione, penthiopyrad, and thiophanate-methyl provide good levels of residual control. Contact fungicides such as chlorothalonil, fluazinam fludioxonil, and mancozeb also provide good control, but have less longevity compared to the aforementioned penetrants. DMI/SI fungicides such as myclobutanil and triticonazole are also effective when applied preventively. The aforementioned and other DMI/SI fungicides (e.g., metconazole, propiconazole, and triadimefon) perform best when tank-mixed with a contact fungicide. For curative control, it is best to tank-mix a contact fungicide (e.g., chlorothalonil, fluazinam, fludioxonil, or mancozeb) with one of the aforementioned penetrants. Thiophanate-methyl is ineffective against *R. zeae*.

Copper Spot

Pathogen: Gloeocercospora sorghi Bain & Edgerton ex Deighton

Primary host: Velvet and creeping bentgrass

Predisposing conditions: Warm to hot and humid periods in summer

Copper spot is primarily a disease of velvet bentgrass, but it may also develop in creeping bentgrass golf greens, especially in warmer transition and southeastern U.S. locations. High-density creeping bentgrass cultivars grown on golf greens in cooler climates can also contract copper spot.

Symptoms

As the name implies, the disease appears as reddish-brown- or copper-colored spots 1.0–3.0 inches (2.5–7.5 cm) in diameter (Figure 2.55). This disease can mimic Pythium blight, dollar spot, and dead spot. During wet periods, leaves are covered with small, slimy, salmon-pink- to orange-colored masses of spores, which are borne on structures called sporodochia. When dry, sporodochia are bright-orange and resinous. Very small, black sclerotia (i.e., resistant structures of the pathogen) are produced in large numbers in dead leaf tissue. Sporodochia and sclerotia can be seen with the aid of a hand lens. The copper spot pathogen only blights leaves and plants seldom die. Turf slowly recovers from injury following the advent of cool and dry weather.

Management

Speed up leaf drying by mowing early in the morning or by dragging or poling greens. Manage velvet bentgrass with low to moderate amounts of

FIGURE 2.55
Copper spot in velvet bentgrass. (Photo courtesy of N. Mitkowski.)

nitrogen and apply limestone if the soil pH falls below 5.5. Summer spoon-feeding (0.1–0.125 lb N/1000 ft²; 5–6 kg N/ha) with a water-soluble nitrogen source like urea promotes recovery. Broad-spectrum fungicides such as chlorothalonil and iprodione as well as QoI/strobilurin and DMI/SI fungicides effectively control copper spot.

Dead Spot

Pathogen: Ophiosphaerella agrostis Dernoeden, Camara, O'Neill, van Berkum, et. Palm

Primary hosts: Bentgrasses and hybrid bermudagrass golf greens

Predisposing conditions: Young golf greens, heat stress, and sand-based growing media

Dead spot primarily occurs on new golf greens or older greens that were fumigated prior to regrassing. The disease is less common on higher mowed tees and collars. Dead spot is associated with high sand mixes and thus far has not been found on fairways or wherever turf is grown on native soil. Most golf greens affected by dead spot are 1–3 years old, but greens as old as 6 years have developed the disease. Most injury is associated with greens in open or exposed locations, rather than shaded sites. While the disease is most common in creeping bentgrass, colonial and velvet bentgrass are also susceptible. On hybrid ultradwarf bermudagrass golf greens in Florida and Texas, *O. agrostis* may initially attack overseeded roughstalk bluegrass (*Poa trivialis*) during the early spring transition period. Dead spot symptoms on bermudagrass greens can develop as early as spring green-up. Bermudagrass recovers with the advent of higher temperatures, which favor rapid growth of stolons and rhizomes.

Symptoms

Dead spot initially appears as small, reddish-brown spots 0.25–0.5 inch (6.2–12.5 mm) in diameter (Figure 2.56). The initial, small spots mimic ball mark injury. Spots enlarge to only about 3.0 inches (7.5 cm) in diameter, and have tan tissues in the center and reddish-brown leaves on the outer periphery of larger, active patches. The symptoms at times are similar to those associated with copper spot, dollar spot, Microdochium patch, black cutworm (*Agrotis ipsilon*) damage, and ball-mark injury. No foliar mycelium is evident on turf in the field; however, pale-pinkish-white foliar mycelium develops in a laboratory humidity chamber after a few days of incubation.

Dead spot in creeping bentgrass appears during warm to hot and dry weather from late spring to early autumn. The disease is most active and severe during summer months. Dead spot may remain active until hard frosts occur in autumn. Except when the disease is severe, the spots or patches seldom coalesce. Sometimes depressed spots or "crater pits" develop. Crater pits may take from 4–6 weeks to recover. Dead rings with living turf in the center, or the frog-eye symptom, are uncommon. In the absence of using an effective fungicide, new stolons growing into the dead pits are rapidly attacked and killed. The pathogen colonizes leaves, stems, and roots. Darkly pigmented hyphae, typical of the other *Ophiosphaerella* species that attack turf, may be found on stems and nodes of stolons. Infected roots are brown and may appear as shortened, brown stubs. Stolon growth into the dead spots or

FIGURE 2.56
Active (inset) and winter inactive dead spot in creeping bentgrass golf greens.

FIGURE 2.57
Black flask-shaped fruiting bodies (i.e., pseudothecia) of *Ophiosphaerella agrostis*. Inset shows a crushed pseudothecium and brown color of spores (i.e., ascospores in asci) en masse.

patches is restrained or inhibited. Some recovery occurs as a result of tillering of adjacent healthy plants, but many dead spots do not fully recover prior to winter. During winter, inactive spots or patches appear whitish-tan and mimic old dollar spot damage (Figure 2.56). Diseased spots are often void of living tissue, and the underlying bare, sandy soil remains evident in the center of dead spots during winter and the following spring.

Numerous black, flask-shaped fruiting bodies called pseudothecia are often found embedded in necrotic leaf and sheath tissues, and on dead stolons (Figure 2.57). The fruiting bodies contain large numbers of needle-shaped spores (i.e., ascospores). When mature, ascospores are ejected or exude through a pore in the top of the neck of the pseudothecia. Hence, spores are dispersed by wind and water. Fruiting bodies can develop before symptoms, and the pathogen is rapidly disseminated by ascospores. The fungus overwinters in infected stolons, other tissues, and as pseudothecia.

Management

Symptoms are arrested by boscalid, chlorothalonil, fludioxonil, iprodione, propiconazole, pyraclostrobin, and thiophanate-methyl. Control may last for only 7–10 days, after which time active symptoms can recur. A water-soluble nitrogen fertilizer such as urea or ammonium sulfate should be applied every 1–2 weeks at 0.1–0.2 lb N/1000 ft^2 (5–10 kg N/ha) to stimulate growth of surrounding, healthy creeping bentgrass plants. Routine applications of ammonium sulfate have been shown to suppress dead spot in both bermudagrass and creeping bentgrass. Use caution when applying ammonium sulfate since it has a high burn potential and should be watered-in immediately. Nitrate forms of nitrogen (i.e., calcium, sodium, and potassium nitrate) can increase dead spot incidence and severity.

Fairy Ring

Pathogen(s): Agaricus spp., *Lycoperdon* spp., *Marasmius oreades* (Bolt. ex Fr.) Fr, many others

Primary hosts: All turfgrasses

Predisposing conditions: Warm and moist weather with intermittent periods of drought stress in summer

Fairy rings are commonly found in turf and pastures and may be caused by any one of the 50 or more species of fungi. Fairy ring fungi belong to a group known as the basidiomycetes or "mushroom fungi." The most common fungi associated with damaging fairy rings in general turf areas include *M. oreades,* and various species of *Agaricus, Agrocybe, Chlorophyllum, Coprinus, Lepiota,* and *Lycoperdon* (known as puffballs). Members of the puffball family, which include *Arachnion album* Schwein, *Bovista dermoxantha* (Vittad.) de Toni, and *Vascellum curtisii* (Berk.) Kreisel, are common causes of fairy ring in golf greens (Miller et al., 2011).

The fruiting body (i.e., basidiocarp) of *M. oreades* and *Agaricus* spp. is a typical mushroom with a cap and stem. The underside of the mushroom cap is composed of gills, upon which spores are produced. Puffballs do not produce caps and spores are borne within white, fleshy, and egg-shaped fruiting bodies. Puffballs often have fleshy warts or spines on their surfaces. Puffballs turn brown as they age and when they crack or are crushed, they release large numbers of brown or purplish spores. The importance of spores in the spread of fairy ring fungi is unknown.

Fairy rings occur under any soil condition that will support turfgrass growth. Nearly all commonly cultivated turfgrasses are known to be affected by fairy ring fungi. Growth of a fairy ring begins with the transport of fragments of fungal mycelium and possibly spores. The fungus initiates growth at a central point and continues outward in all directions (i.e., radial) at an equal rate. Fairy rings are more numerous in wet years, but are most destructive when weather conditions become hot and dry.

Fairy ring fungi cause the formation of rings or arcs of dead or unthrifty turf, or rings or arcs of dark-green, luxuriantly growing grass. Fairy ring fungi primarily colonize thatch and organic matter in soil and generally do not directly attack turfgrass plants; however, some are weakly parasitic.

Fairy rings are classified into three types according to their effects on turf:

Type 1: Those that kill or badly damage plants in rings or arcs

Type 2: Those that stimulate grass, causing the formation of rings or arcs of dark-green turf

Type 3: Those that do not stimulate grass and cause no damage, but produce fruiting bodies (i.e., mushrooms or puffballs) in rings or arcs

Fairy rings vary in size from 1 ft (30 cm) to 10 ft (3.0 m) or more in diameter, and become larger each year. Rings greater than 200 ft (60 m) across have been reported. The rate of outward movement, as well as overall ring diameter, is determined by soil and weather conditions. In general, rings grow more rapidly in light-textured and moist soils than in heavy clay and dry soils. Rings fade in the autumn or winter, but bare zones remain visible until turf recovers via stolons, rhizomes, and/or tillering. Fairy rings are most conspicuous during hot and dry summer periods, but can fade rapidly with the advent of rainy weather. When soil moisture is abundant, dark-green arcs or rings may be evident, while the dead zone is absent. With the advent of warm to hot and dry weather, however, the dead zone appears. Hence, Type 2 rings can develop into Type 1 rings, and dead zones are most likely to appear during dry summer periods.

Symptoms

The most destructive rings are of the Type 1 variety. Type 1 rings are very common in lawns and on golf courses. In lawns, roughs, and fairways, Type 1 rings initially appear as circles or arcs of dark-green grass, but the dead zone does not appear until summer (Figure 2.58). The most common fungi known to cause Type 1 fairy rings include *M. oreades*, *Agaricus* spp., and *Lycoperdon* spp. Classic Type 1 rings are distinguished by three distinct zones: an inner lush zone where the grass is darker green and grows luxuriantly; a middle zone where the grass may be wilted or dead; and an outer

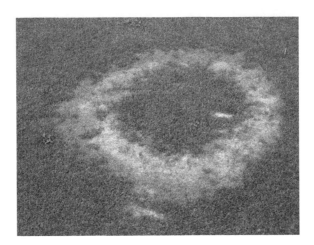

FIGURE 2.58
Type 1 fairy ring in a perennial ryegrass fairway.

zone in which the grass is stimulated and/or darker green (Figure 2.59). The distance from inside of the inner zone to outside of the outer zone may range from a few inches (3–6 cm) to a foot (30 m) or more wide. Darker green, stimulated zones of grass are the result of the breakdown of organic matter, which releases nitrogen and results in more vigorous leaf growth. The outer green zone is caused by the breakdown of thatch and organic matter by the fairy ring fungus, which liberates nitrogen. The inner green zone develops in response to the release of nitrogen as bacteria degrade aging or dead mycelium of the fairy ring fungus produced in previous years. Mushrooms or puffballs are produced in the bare zone or at the junction of the bare and outer green zones. Rings, however, may produce few or no mushrooms, especially on closely cut golf greens. The presence of the three distinct zones may not always be noticeable, but Type 1 rings can remain visible from summer to winter.

The three distinct zones of Type 1 fairy ring previously described may not develop on golf greens. On golf greens, fairy rings appear initially as dark-green Type 2 rings that eventually develop a bluish color associated with wilt. Fairy rings on greens may also appear as solid circular patches of darker green and often wilted patches 6 inches to 2 ft (15–60 cm) or greater in diameter. These circular and wilted patches could be confused with take-all patch. If not properly irrigated or syringed, plants enter a brown, dormant state. Patches may become sunken, which is referred to as "thatch collapse" (Figure 2.60). Fairy rings may appear and disappear on young and old greens.

Type 2 fairy rings are commonplace and can appear in early spring and remain evident until winter (Figure 2.61). While a Type 2 fairy ring can develop

FIGURE 2.59
Arc of a Type 1 fairy ring in Kentucky bluegrass showing inner and outer green stimulated zones. Inset shows mycelium of a fairy ring fungus in thatch and soil.

FIGURE 2.60
Wilted and sunken "thatch collapse" patches in a fine-leaf fescue fairway are attributed to the activity of unknown mushroom fungi.

FIGURE 2.61
Type 2 fairy ring in a fine-leaf fescue fairway.

into a Type 1 ring, Type 2 rings generally only affect the aesthetic quality of turf. Type 2 rings tend to come and go, but are most prevalent during wet summer periods. Fruiting bodies develop within the darker-green zone following rainy weather in summer (Figure 2.62). Type 3 rings of mushrooms or puffballs growing in a ring or arc are mostly found in infrequently mowed fields and lawns and usually appear rapidly after a rainy period (Figure 2.63).

FIGURE 2.62
Type 2 fairy ring with puffballs in a zoysiagrass lawn beginning to enter winter dormancy.

FIGURE 2.63
Type 3 fairy rings are distinguished by mushrooms or puffballs produced in rings or arcs without a stimulated zone. Inset shows a Type 3 birds nest (*Cyathus* spp.) fairy ring in a golf green.

A fairy ring is broken when its mycelium encounters an obstacle such as a rock, pathway or unfavorable soil condition. Rings may also disappear for no apparent reason. In general, two fairy rings will not cross one another; that is, at the point of intersection the growth of each ring stops (Figure 2.59). This obliteration at the point of contact is caused by the production of self inhibitory metabolites that also antagonize other members of the same or different fungal species. On slopes, the bottom of a ring usually is open, giving the appearance of an arc or crescent. This may be due to the downward movement of self-inhibitory metabolites that prevent development of the fungal mycelium in soil on the lower side of the ring.

When a plug of soil is removed from the edge of an active fairy ring a white thread-like network of mycelium sometimes may be seen in the thatch layer or clinging to soil and/or roots of grass plants (Figure 2.59). When environmental conditions are optimum for fungal growth, white mycelium may be seen on the surface of the thatch layer. Oftentimes, however, no mycelium is evident in the soil or thatch. Fairy ring infested soils, however, almost invariably have a mushroom odor, even if fungal mycelium is not evident. In golf greens, thatch layers in active rings are light-brown rather than the more normal dark-brown color.

Type 1 fairy ring fungi kill vegetation primarily by rendering infested soil impermeable to water. Very high concentrations of ammonium have been measured in the wilting zone of some Type 1 fairy rings and levels may be toxic enough to contribute to turf death. Some fairy ring fungi, including *M. oreades*, are known to parasitize roots and produce compounds toxic to roots such as hydrogen cyanide. It is likely, however, that most damage to turf can be attributed to the fungal mycelium rendering soil impermeable to water. Probing will reveal that soil in the dead zone is dry when compared to adjacent soil. The dead zone develops in response to an accumulation of fungal mycelium in such large amounts in soil that it prevents entry of rain or irrigation water, and thus plants die as a result of drought stress. It is quite characteristic for grass on the outer edge of the dead zone to display the blue-gray color of wilt before dying.

Management

Control of Type 1 fairy rings in native soil is made difficult by the water-impermeable nature of infested soil. Chemical control is frequently ineffective because some fairy ring fungi grow deeply into soil and lethal concentrations of fungicide do not come into contact with the entire fungal body. Good suppression of fairy rings, however, is obtainable in sand-based root zones of golf greens and tees. There are three approaches to combating Type 1 fairy rings: (1) suppression, (2) antagonism, and (3) eradication.

Suppression is the most practical approach to combating Type 1 fairy rings in most situations. This method of control involves a combination of core or solid tine aeration, deep watering, use of wetting agents, and proper nitrogen fertilization. Coring is beneficial since it aids in the penetration of air and water. The entire area occupied by the ring, to include a 2 ft (60 cm) periphery beyond the ring, should be cored 2.0–4.0 inch (5.0–10 cm) centers. The area should be treated with a soil wetting agent and then irrigated so that water penetrates to a depth of 4.0–6.0 inches (10–15 cm): Soil wetting agents aid water infiltration by breaking the surface tension of hydrophobic soil. The ring area should be re-treated in a similar fashion at the earliest indication of drought stress, that is, repeat the process whenever the dark-green grass turns blue-gray and begins to wilt. When a core aerator is not available, a deep root feeder with garden hose attachment may be useful to force water into dry soil. Even using a pitch fork to create openings is preferred to not creating any holes at all. Apply recommended amounts of nitrogen at the appropriate time of the year to help mask fairy rings.

In sand-based root zones, good fairy ring suppression has been reported by applying an appropriate fungicide in a high water volume (i.e., ≥100 gal water/acre; 935 L/ha) or by simply watering it off foliage. This may be because puff-ball fairy ring fungi common in golf greens grow closer to the surface. The depth of fungal growth is difficult to estimate. One way to check is to collect a cup cutter or profile sampler plug from within an active ring. Moisten the soil and place the sample in a plastic bag and incubate at a warm (>75°F; >13°C) temperature. Fungal mycelium may be evident within 24–48 h, thus giving an indication of the depth of fungal growth and, therefore, the depth to which a fungicide should be watered-in. Sometimes a fungicide is mixed with a wetting agent to facilitate entry and percolation, but caution is needed to ensure that chemical is not moved below the effective zone. If wetting agents are being used routinely, they may not be necessary unless soil is hydrophobic at the time of treatment. If effective suppression is still not achieved, then the previously described core aeration approach should be attempted prior to drenching in a fungicide. Effective suppression of fairy rings has been reported using azoxystrobin, flutolanil, polyoxin D, and several DMI/SI fungicides, including metconazole, propiconazole, tebuconazole, triadimefon, triticonazole, and possibly others. For best results fungicides should be applied preventively in spring when soils warm to about 55° to 60°F (13–16°C). A single fungicide application sometimes will control fairy rings on golf greens. In most cases, they reappear in a few weeks and require re-treatment.

The antagonism approach is based upon the observation that rings exhibit mutual antagonism, that is, elimination when they come into contact with one another. This method involves removal of sod. Soil must be roto-tilled repeatedly in several directions until the mycelium-infested soil has been thoroughly mixed. Soil then is prepared in the usual manner for seeding or sodding. This method has been shown to be promising, but has had only limited testing and obviously it is not practical. There are two methods of eradication: fumigation and excavation. Both methods are laborious, costly, and impractical.

Gray Leaf Spot

Pathogen: Pyricularia grisea (Cooke) Sacc. [synonym = *Magnaporthe grisea* (T.T. Hebert) Barr]

Primary hosts: St. Augustinegrass, kikuyugrass, perennial ryegrass, and tall fescue

Predisposing conditions: Prolonged periods of warm and humid weather in summer and autumn

Gray leaf spot is a common disease of St. Augustinegrass lawns in the southeastern United States. It can be particularly damaging to newly sprigged or sodded lawns, but may be a chronic problem in mature stands of St. Augustinegrass and kikuyugrass, and less commonly centipedegrass. The pathogen can also inflict significant damage to annual and perennial ryegrass, and tall fescue. Crabgrass (*Digitaria* spp.), foxtails (*Setaria* spp.), millet (*Panicum* spp.), and barnyardgrass (*Echinochola* spp.) are susceptible to *P. grisea* and these weeds may provide an important source of inoculum for triggering an epidemic.

Gray leaf spot in cool-season grasses historically was regarded as a seedling disease. Gray leaf spot did not become recognized as being an important disease in cool-season grasses until the 1990s, when it destroyed many acres of perennial ryegrass grown on golf courses in the mid-Atlantic region. During the epidemics of 1995 and 1998, the disease appeared in mid-summer during periods of extreme heat and drought stress. In 1998, the disease moved out of the mid-Atlantic region and was observed as far north as Rhode Island and Iowa, and as far west as Nebraska, Kansas, and Oklahoma. Gray leaf spot may also attack perennial ryegrass in overseeded bermudagrass in California and the southeastern United States. The disease may appear in late spring in overseeded perennial ryegrass during the bermudagrass transition period following heavy rainfall events in southern regions. Gray leaf spot is also a problem in perennial ryegrass sports fields, particularly stadiums with restricted air movement.

The most likely means of *P. grisea* spread is by wind dispersal of spores. Once the pathogen is established at a site, spores are rapidly dispersed by mowing. The pathogen survives winter as dormant mycelium in dead tissues. There are physiological races of *P. grisea* and the race that attacks perennial ryegrass is different from the race that attacks tall fescue. These races also differ from those associated with warm-season grasses.

Gray leaf spot can develop at any time from mid-summer to late autumn, but late summer and autumn are the most common times of the year for this disease to develop in cool-season grasses. Once blighting develops, the disease can remain active until there is a hard frost. Long leaf wetness durations (>9 h) and high humidity at night are required for infection at temperatures ranging from 68°F to 90°F (20–32°C). The fungus requires frequent cycles of leaf wetting and drying for spore production to become prolific.

Symptoms in Perennial Ryegrass

Gray leaf spot epidemics can hit rapidly and severely in summer, but usually progress more slowly in late summer and autumn. On golf courses, gray leaf spot is mostly destructive in perennial ryegrass areas not routinely receiving fungicides like roughs and tee box surrounds. From a standing position, the first observable symptom is the appearance of reddish-brown-, gray-, or tan-colored spots 1.0–2.0 inches (2.5–5.0 cm) in diameter, which could be easily confused with Pythium blight or dollar spot. There is, however, no cobweb-like foliar mycelium associated with gray leaf spot. During prolonged hot, humid, and/or dry weather, dead spots enlarge to 3–18 inches (7.5–45 cm) in diameter. At this point, disease symptoms mimic brown patch, but any kind of spot or patch symptom is short-lived when disease pressure is severe. Large areas of turf may collapse in 1–3 days. During severe epidemics, infected stands develop a bluish-gray hue, which is typical of drought stress symptoms (Figure 2.64). Hence, perennial ryegrass in roughs or fairways that appears wilted in the presence of adequate soil moisture is a good indicator of gray leaf spot. The disease is most severe in heat-sink areas, such as south-facing hillsides and knolls. Another feature observed in previous summer epidemics was that gray leaf spot frequently began and was more destructive in golf course roughs, particularly the intermediate rough where soil has been compacted by cart traffic. The higher canopy in roughs provides a more favorable microenvironment for the pathogen. This is supported by the observation that the disease generally is less severe in low-cut, perennial ryegrass approaches and collars.

Under less favorable environmental conditions, the disease usually requires several weeks to develop and become noticeable. This is normally the case when gray leaf spot develops in the autumn. In autumn, infected stands of perennial ryegrass appear chlorotic, thin, and unthrifty (Figure 2.65). Thereafter, turf may lose density gradually or rapidly depending on weather conditions.

FIGURE 2.64
Perennial ryegrass rough exhibiting the blue-gray, drought-like color symptoms of gray leaf spot.

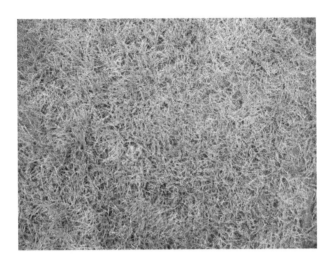

FIGURE 2.65
Slowly developing gray leaf spot in autumn in perennial ryegrass may only elicit a chlorosis and/or a thinning of stand density.

Gray leaf spot initially attacks leaves via spores, and within hours leaf tips may appear water-soaked and chlorotic. Thereafter, leaf spot lesions and leaf twisting occur. Sometimes, the youngest emerging leaf is twisted in the shape of a "fish hook." Plants with the fish hook symptom may not have any lesions. Spores can be disseminated rapidly by mowing, resulting in a streaking pattern (Figure 2.66). Within a very few hours, spores germinate, penetrate, and begin to ramify through cells before leaves dry in the morning.

FIGURE 2.66
Streak symptom of gray leaf spot caused by mowers redistributing spores.

Leaf lesions generally are circular to oblong, about 0.125–0.25 inch (3.1–6.2 mm) long and grayish-brown with a dark-brown border (Figure 2.67). Gray or brown lesions with or without a dark-brown border frequently develop along the edges or margins of leaf blades. A yellow halo can sometimes be observed bordering lesions. Lesions, however, may be the size of a pinhead and very dark-brown. In early morning hours, twisted leaf tips or lesions on the margins of leaves may appear felted, and infected tissues may be gray, dark-brown, purple, or yellow (Figure 2.68). The felted appearance is the result of the production of large numbers of *P. grisea* spores and their spore-bearing stalks known as conidiophores (Figure 2.68). Lesions produced by *P. grisea* can be dark brown in color and thus similar to those caused by the net-blotch pathogen, *D. dictyoides. Bipolaris sorokinana* and *Curvularia* spp. can also cause dark-brown leaf lesions and leaf twisting that mimic *P. grisea*; thus, proper disease diagnosis based on leaf lesions in the field is difficult (Figure 2.68). The most effective means of positively diagnosing gray leaf spot is to microscopically confirm the presence of the spores, which are distinctive and diagnostic by their morphology (Figure 2.68).

For unknown reasons, there have not been many reports of extremely rapid and devastating summer outbreaks of gray leaf spot in perennial ryegrass since the late 1990s. It is possible that some of this can be attributed to improved fungicide programming in perennial ryegrass sports turfs as well as conversions to resistance grasses. Autumn outbreaks are more common in perennial

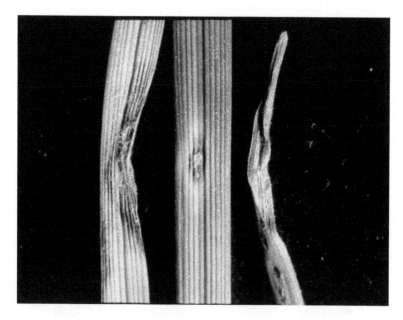

FIGURE 2.67

Gray leaf lesions on margins of leaves and a yellow halo caused by *Pyricularia grisea* in perennial ryegrass.

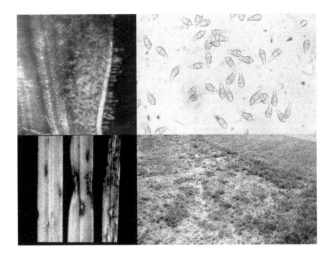

FIGURE 2.68
Felted appearance of an infected kikuyugrass leaf bearing numerous *Pyricularia grisea* spore-producing stalks called conidiophores. Brown *P. grisea* leaf lesions that are similar to those associated with Bipolaris, Curvularia, and Drechslera leaf spot diseases (lower left); distinctive tear-shaped morphology of *P. grisea* conidia (upper right); and rapid collapse of perennial ryegrass in a heat sink. (Top left inset photo courtesy of L. Stowell, PACE Turf, LLC.)

ryegrass in some regions (especially transition and southeastern regions of the United States). As previously noted, autumn outbreaks generally proceed more slowly and are not characterized by spots or circular patterns.

Symptoms in Tall Fescue

Gray leaf spot occurs in tall fescue mostly in the mid-Atlantic and southeastern United States. Leaf lesions are similar to those described for perennial ryegrass. Tall fescue is not as severely damaged compared to the potential in perennial ryegrass, and thinned-out stands generally recover when properly fertilized.

Symptoms in Warm-Season Grasses

Gray leaf spot primarily is a debilitating disease of St. Augustinegrass and less commonly centipedegrass and kikuyugrass. It can be especially damaging to newly sprigged or sodded stands. It develops during prolonged warm periods of high humidity from spring to autumn, but is generally most damaging in mid-to-late summer. Leaf lesions are oval to oblong and gray with purple or brown margins. Leaf lesions may have a yellow halo and range in size from 0.06 to 0.5 inches (1.5–12.5 mm) in length. During humid periods, lesions may have a velvet-gray appearance due to heavy sporulation by the pathogen (Figure 2.68). Leaves eventually turn yellow or grayish-tan, and badly infected stands have the appearance of having been subjected to severe drought stress. Gray leaf spot can cause severe loss of foliage, but seldom kills warm-season grass plants once they have become well established.

Cultural Management in Cool-Season Grasses

Under low disease pressure, reducing mowing height and removal of clippings helps to minimize damage, but under high pressure, these methods have little or no impact. Some nitrogen applied in autumn helps reduce gray leaf spot severity; however, during summer epidemics, nitrogen should be avoided since it can promote gray leaf spot during periods of heat stress. Since the pathogen requires long periods of leaf wetness to produce large numbers of spores, practices that minimize night irrigation should help reduce leaf wetness duration and thus spore production. Herbicide and plant growth regulator use can intensify the disease. Thus, some management practices that may help reduce gray leaf spot severity would include: avoid mowing when leaves are wet; collect clippings; avoid applying high rates of nitrogen fertilizer during summer periods when the disease is active; avoid herbicide and plant growth regulator use when the disease is active; and maintain adequate soil moisture levels by irrigating during daytime hours.

Severely damaged areas will need to be reestablished following a summer epidemic. Areas overseeded in late summer, however, often do not tiller or establish good density or quality playability because of high-temperature stress. During renovation, seed must make contact with soil. This is best achieved using a slicer seeder. Broadcasted seed does not establish well because either seed is directly attacked or very young seedlings are killed as they emerge by *P. grisea* and other damping-off pathogens such as *R. solani* and *Pythium* spp. Hence, tank-mixes of a broad-spectrum fungicide with a *Pythium*-targeted material need to be applied to seedlings if weather conditions are hot and humid. Low-rate, weekly sprays of a water-soluble nitrogen source like urea at 0.1–0.125 lb N/1000 ft^2 (5–6 kg N/ha) and syringing several times daily also help to improve seedling vigor. Commercial perennial ryegrass cultivars with improved gray leaf spot resistance are available; however, they may be damaged severely under the right conditions. Many of the current resistant perennial ryegrass cultivars (e.g., "Paragon" and "Palmer IV") have an open growth habit and may not maintain adequate density under low mowing. In extreme cases, renovation using a resistant species may become necessary.

Cultural Management in Warm-Season Grasses

Deep and infrequent irrigation during daytime hours, raising the mowing height, and removal of clippings can help alleviate the disease in St. Augustinegrass and other warm-season grasses. After disease subsides, an application of a complete fertilizer with half of the nitrogen in a slow-release form will assist in recovery. Alleviate soil compaction and reduce thatch by vertical cutting and coring when the disease is inactive and turf is growing vigorously. Avoid using herbicides when gray leaf spot is active. If possible, establish new lawns in the late spring and early summer to avoid the disease in mid-to-late summer in less mature turf.

Chemical Control

Where gray leaf spot is chronic in perennial ryegrass sports turfs and St. Augustinegrass lawns and sod fields, preventive applications of fungicide(s) are the most effective approach to controlling gray leaf spot. Gray leaf spot is especially destructive to perennial ryegrass seedlings in overseeded areas previously damaged by the disease. Vigilant scouting for gray leaf spot requires almost daily attention from mid-summer through late autumn (for cool-season grasses). Once disease develops, higher rates and more frequent fungicide applications are usually required. Among contact fungicides, chlorothalonil, maneb, and mancozeb are effective, but only for 5–7 days under high pressure. The QoI/strobilurin fungicides (e.g., azoxystrobin, fluoxastrobin, pyraclostrobin, and trifloxystrobin) are very effective and provide for long residual control. Thiophanate-methyl, polyoxin D, and DMI/SI (e.g., metconazole, propiconazole, tebuconazole, and triadimefon) are also effective, but the DMI/SI fungicides perform better when mixed with a contact fungicide. Once blighting appears, a high rate of a contact fungicide (i.e., chlorothalonil, mancozeb, or maneb) should be tank-mixed with one of the aforementioned penetrants. Affected areas should then be re-treated in 5–7 days with another application of a contact fungicide. Should a severe epidemic continue, tankmix combinations will likely be required on 10–21-day intervals, depending on fungicide and rate applied as well as environmental conditions. It is important to delay mowing for 24 h following a fungicide application.

Biotypes of the gray leaf spot fungus have developed resistance to azoxystrobin and possibly thiophanate-methyl. Hence, these and related fungicides should not be used continuously in gray leaf spot management programs. Because resistance is a threat, it is important to rotate frequently with other fungicide modes of action. Mixing a contact with a penetrant fungicide also reduces the potential for resistance. Fungicides like iprodione, vinclozolin, and flutolanil have little or no activity against gray leaf spot turf and can increase vulnerability of turf to the disease. More information on resistance management can be found in the Fungicides used to Control Turgrass Diseases section.

Leaf Spot, Melting-Out, Net-Blotch, and Red Leaf Spot

Pathogens: Bipolaris spp. and *Drechslera* spp.

Primary hosts: Colonial and creeping bentgrasses, bluegrasses, bermudagrass, fescues, and perennial ryegrass

Predisposing conditions: Warm to hot temperatures and frequent cycles of wet and dry weather

Leaf spot and melting-out often occur during cool and overcast periods and were previously discussed as spring and autumn diseases. Their Latin binomials and authorities (i.e., those who formally described the fungus) are shown in Table 2.4. In summer, Kentucky bluegrass, perennial ryegrass, fine-leaf fescues, and other grasses may decline due to invasion by *Bipolaris sorokiniana*. This fungus can cause leaf spot and melting-out phases. *B. sorokiniana* normally is most severe during overcast and/or wet periods in summer when temperatures exceed 80°F (>26.7°C) and humidity is high. Damage from this disease can be aggravated when infected stands are subsequently subjected to drought stress.

During warm and wet periods, *Bipolaris cynodontis* can cause a severe crown, stolon and root rot in bermudagrass. Red leaf spot (*D. erythrospila*) primarily is a late spring or summer disease of colonial bentgrass. Most creeping bentgrass cultivars have good red leaf spot resistance; however, the cultivars "Toronto" and "Penn A-1 and A-4" can contract the disease. Net-blotch (*D. dictyoides*) can cause leaf spot and melting-out in perennial ryegrass during or following rainy periods in the summer. Turfgrasses infected with *Bipolaris* spp. can be simultaneously attacked by *Curvularia* spp., resulting in a complex (i.e., two or more pathogens) that can be very destructive during periods of heat and/or drought stress.

Symptoms

The leaf spot and melting-out symptoms are the same as described previously for leaf spot and melting-out diseases of spring and autumn, which largely are caused by *Drechslera* species.

It is again noted that red leaf spot is mostly a colonial bentgrass disease. Lesions produced by the red leaf spot pathogen appear as small reddish spots. Infected colonial bentgrass leaves develop a brick-red color and dieback from the tip. In creeping bentgrass, red leaf lesions are present. During the melting-out phase, nonuniformly blighted areas develop. Many superintendents confuse red leaf spot with the red or purplish color of golf greens in late winter or early spring. This cold-weather-induced reddening or purpling in bentgrass is due to an increase in pigments, that is, anthocyanins in leaves. In bermudagrass, *B. cynodontis* causes dark-brown spots with a tan center on leaves and sheaths, which are oval shaped to elongate.

Management

For management of these diseases see "Leaf Spot and Melting-Out" in the Diseases Initiated in Autumn or Spring that may persist into summer section.

Localized Dry Spot

Pathogens: Unidentified basidiomycetes

Primary hosts: Turf grown in sand-based root zones

Predisposing conditions: Hot and dry periods of summer

Localized dry spots are common in sand-based greens and tees. Localized dry spots do not develop in clay, loam, or other native soils low in sand content. However, fairways that are routinely topdressed with sand develop localized dry spots. They normally develop in new golf course greens within a year or two of establishment. While they tend to decline in number and severity over time, some dry spots can be a persistent problem for indefinite periods. These dry spots develop with the advent of high temperatures and dry periods from late spring to autumn. They disappear during extended overcast and rainy periods.

A cause of localized dry spots has been attributed to decomposition of the mycelium of mushroom fungi and other sources of organic matter in soil by microbes. Microbial breakdown of organic matter releases organic substances (e.g., fulvic acid) that coat individual sand particles. Individually coated sand particles pack loosely together, rendering soil impervious to water infiltration. The water-repellent, hydrophobic condition is restricted to the upper few inches (3–6 cm) of soil. Removal of thatch alone will not significantly improve water infiltration.

Symptoms

Localized dry spot appears as solid patches of wilted or dried-out turf. Patches can be circular and range from a few inches (6–8 cm) to several feet (0.5–1.0 m) in diameter, or they may appear as large serpentine or irregularly shaped areas of wilted or dead turf (Figures 2.69 and 2.70). The soil within affected patches remains bone dry despite frequent irrigation. Water will penetrate the thatch, but not the thatch–soil interface, and will run off dry spot areas. Plants within affected areas develop a blue or purplish color that is indicative of wilt, and eventually the turf dies due to drought stress if no

FIGURE 2.69
Circular and serpentine-shaped localized dry spots on a golf green.

FIGURE 2.70
Serpentine pattern of localized dry spots with yellow borders on a golf green.

control measures are taken. In some cases, plants on the periphery of a dry spot appear yellow or chlorotic (Figure 2.70).

Management

Keeping the turf alive in localized dry spots requires numerous daily syringes and treatments with soil wetting agents during dry summer periods. Core aeration, water injection aeration, or hand pitch-forking in combination with frequent applications of a soil wetting agent helps to alleviate this condition. Where localized dry spot is chronic, the application of wetting agents should begin several weeks in advance of the time they normally appear. Once severe localized dry spots have appeared, the use of a soil wetting agent in conjunction with some form of aeration is the preferred control strategy. Isolated spots can be individually treated by frequent probing with a water-fork or a tree deep-root feeder that injects water. Fungicides have no known impact on the incidence, severity, or control of localized dry spots.

Pythium Blight

Pathogens: Pythium aphanidermatum (Edson) Fitzp., *Pythium ultimum* Trow, and less commonly *Pythium catenulatum* Matthews, others

Primary hosts: Annual bluegrass, bentgrasses, perennial ryegrass, and tall fescue. Bermudagrass in semitropical regions

Predisposing conditions: Hot and humid weather

Pythium blight develops rapidly during nighttime and is among the most feared and destructive turfgrass diseases. During periods of high relative humidity, night temperatures above 70°F (21°C), and abundant surface moisture, the disease progresses with remarkable speed. Large areas of turf can be destroyed within 24 h, particularly if there is rainfall at night. Pythium blight is often first observed in tree-lined, shaded, poorly drained, and low-lying areas adjacent to water with poor air circulation. A general misconception is that Pythium blight is a common disease in all turfgrasses. Although *Pythium* spp. can cause damping-off of any seedling species, it seldom attacks mature lawns or sports fields comprised of Kentucky bluegrass, fine-leaf fescue, bermudagrass, or zoysiagrass grown north of the transition zone. Pythium blight is most likely to attack creeping bentgrass, annual bluegrass, or perennial ryegrass grown under the intensive management (i.e., frequent night irrigation, low mowing, and high nitrogen fertility) conditions commonly found on golf courses and stadium athletic fields. Pythium blight also damages tall fescue, particularly in the transition zone and in the southeastern United States. Pythium blight attacks hybrid bermudagrass golf greens in Florida and the Gulf Coast region.

Symptoms

Pythium spp. are capable of producing an abundance of mycelium in just a few nighttime hours. When Pythium blight is active, a cottony web of grayish mycelium may be seen on or in the canopy during early morning hours when leaves are wet. Mycelia bridge leaf blades and are responsible for the cottony appearance seen on affected turf. The fungus primarily spreads through the turf canopy by rapid mycelial growth or by movement of mycelial fragments and motile spores (zoospores) in rain or irrigation water.

On closely mown creeping bentgrass and annual bluegrass golf greens, Pythium blight appears in spots, patches, rings, or in streaks that follow the water drainage pattern. Initial symptoms appear as reddish-brown- or orange-bronze-colored spots 1–2 inches (2.5–5.0 cm) in diameter (Figure 2.71). There may be a gray smoke-ring or grayish-white mycelium on the periphery of affected spots or patches. As the disease intensifies, patches increase in size and "frog-eye" or "satellite ring" symptoms may develop (Figure 2.72). Blighted areas can be found in low and high areas, but generally Pythium blight is most severe in low spots, in swales, in surface water drainage patterns, and in shaded or pocketed sites. Blighting can appear as streaks in swales and on slopes as a result of draining water carrying infectious mycelia and/or spores (i.e., propagules). Sometimes, streaks develop as a result of mowers dispersing propagules and the disease pattern may appear as small, orange or brown spots in a straight line (Figure 2.73). In low-lying areas where water collects, large nonuniform, reddish-brown areas collapse (Figure 2.74). Most plants are usually killed, and leaves are matted and have a greasy texture.

FIGURE 2.71
Pythium blight initially develops in small reddish-brown-colored infection centers in creeping bentgrass.

FIGURE 2.72
Satellite ring symptom of Pythium blight in creeping bentgrass. (Photo courtesy of M. Fidanza.)

In perennial ryegrass and tall fescue, infected foliage develops an oily or dark-gray color, and leaf blades have a water-soaked appearance. Other symptoms include small, brown-, reddish-brown-, orange-, or copper-colored dead spots 1–3 inches (2.5–7.5 cm) in diameter (Figure 2.75). Smoke-rings are more evident in tall fescue than perennial ryegrass. As the disease intensifies, spots or patches increase in size, coalesce, and large nonuniformly shaped areas die. Leaf blades collapse, mat together, and turn brown. Blighting is rapid and there are no distinctive leaf lesions.

FIGURE 2.73
Streak symptom of Pythium blight caused by the mower redistributing propagules (i.e., mycelium and spores) in a perennial ryegrass golf tee.

FIGURE 2.74
Pythium blight in a depression where water collected.

Management

Water management greatly influences Pythium blight severity. It is beneficial to irrigate early in the day to avoid moist foliage at nightfall. Removing leaf surface exudates by poling or dragging and installing fans is also beneficial. Improving air circulation and sunlight exposure by clearing brush and trees will help reduce disease potential, but these cultural measures are often expensive and difficult to achieve. Avoiding use of lime in alkaline soils and avoiding the application of nitrogen fertilizers at rates exceeding 0.25 lb N/1000 ft² (12.5 kg N/ha) during summer

FIGURE 2.75
Bright orange Pythium blight patches with smoke rings in tall fescue.

stress periods may help to reduce disease incidence and severity. Spoon-feeding nitrogen (i.e., 0.1–0.2 lb N/1000 ft^2; 5–10 kg N/ha) intermittently in the summer is agronomically important although it may slightly enhance Pythium blight. Cultural practices, however, will have only minimal beneficial effects on Pythium blight suppression during high disease pressure periods.

While fungicides are not often used in lawn care for Pythium blight control, they are considered a necessity on golf courses and stadium athletic fields in many regions of the United States. Because of the rapidity with which Pythium blight develops and progresses, fungicides are best employed preventively. Effective preventive control is provided by cyazofamid, fosetyl-aluminum, salts of phosphorous acid (i.e., phosphites), mefenoxam, and propamocarb. Chloroneb and ethazole are older fungicides that continue to be widely used for "knock-down" control of Pythium blight. Most *Pythium*-targeted fungicides can yellow or cause tip burn on golf greens when applied during warm to hot and humid weather, especially chloroneb and ethazole.

Widespread reliance and continuous usage of mefenoxam (and its isomer metalaxyl) on golf courses has led to reduced residual effectiveness and in some cases the selection of *Pythium* spp. biotypes resistant to this fungicide. To avoid the buildup of fungicide-resistant biotypes, and the reduction of residual effectiveness of compounds due to microbial buildup, *Pythium*-targeted fungicides should be rotated or applied in tank-mix combinations whenever economically feasible. Tank-mixing any two of the aforementioned compounds or tank-mixing with mancozeb also reduces the probability that resistant *Pythium* biotypes will dominate. More information on fungicide resistance management can be found in the section "Types of Fungicides."

Root Decline of Warm-Season Grasses

Pathogen: Gaeumannomyces graminis (Sacc.) Arx & D. Olivier var. *graminis*

Primary host: Bermudagrass, centipedegrass, kikuyugrass, seashore paspalum, St. Augustinegrass, and zoysiagrass

Predisposing conditions: Low mowing combined with warm, humid, cloudy, and wet periods in summer and early autumn

Root decline of warm-season grasses typically occurs during hot, humid periods marked by rainy weather from summer to early autumn. The pathogen (i.e., *G. graminis* var. *graminis*) produces darkly pigmented runner hyphae on roots, stolons, and basal leaf sheaths. Unlike other darkly pigmented pathogens (including those causing spring dead spot, summer patch, and take-all patch), it is not commonly associated with discrete circular patch symptoms. Bermudagrass root decline has been recognized in Florida for many years. With the advent of ultradwarf hybrid cultivars, root decline has become a widespread problem wherever these grasses are grown. The same pathogen is now recognized as causing severe damage to seashore paspalum and the disease sometimes is referred to as seashore paspalum decline. This pathogen also attacks centipedegrass, kikuyugrass, St. Augustinegrass, and zoysiagrass and the disease is sometimes referred to as take-all root rot. The disease is most common in golf greens and sometimes tees, which indicates that low mowing is a primary predisposing factor for the decline. Root decline, however, can occur in higher-cut lawns, fairways, and sod production fields.

Symptoms

Chlorosis (i.e., yellow) is a common early symptom of root decline (Figures 2.76 and 2.77). Lower leaves are the first to turn yellow and the yellowing progresses to younger leaves emanating from bud shoots. Leaves, roots, and stolons may have dark lesions and these tissues eventually turn brown or golden brown. Roots on stolons are typically shortened and blackened (Figure 2.78). Darkly pigmented runner hyphae and distinctive lobed infection structures (i.e., hyphopodia) of *G. graminis* var. *graminis* are prominent on stolons and roots (Figure 2.78). Wilted circular patches are sometimes associated with the disease. Leaves eventually become tan or bleached and turf generally dies in irregular patterns. Root decline of warm-season grasses grown on golf greens typically appears first in the perimeter cut where mowers turn causing mechanical stress. When turf is excessively soft and wet during hot and humid periods, infected stands are scalped in obvious mowing patterns (Figures 2.76 and 2.77).

FIGURE 2.76

Chlorosis (i.e., yellowing) is an early symptom of root decline in kikuyugrass. Note the turf in higher-cut rough is less affected. Eventually, turf dies in irregular shapes and often in mowing patterns as a result of mechanical injury due to scalping weakened turf. (Photos courtesy of L. Stowell, PACE Turf, LLC.)

In some regions, *G. graminis* var. *graminis* and *B. cynodontis* can simultaneously attack bermudagrass during wet periods following spring green-up. Affected plants exhibit a brilliant yellowing and the turf thins out in irregularly shaped areas. Stolons tend to dieback from younger tips, and yellowing progresses backwards to the crown. Roots on stolons are reduced to

FIGURE 2.77

Chlorosis in a seashore paspalum golf green during early stages of root decline (inset). Root decline in a bermudagrass green following the mowing pattern. (Photos courtesy of P. Harmon.)

FIGURE 2.78
Root decline in a low-maintenance bermudagrass turf. Note that roots on stolon have rotted (upper inset). Distinctive lobed hyphopodia (i.e., a flattened penetration structure) of *Gaeumannomyces graminis* var. *graminis* can be observed with a microscope on diseased roots, stolons, and stem bases (lower inset).

blackened nubs that can no longer anchor stolons in soil. Mild symptoms consisting of yellow spots 1.0–2.0 inches (2.5–5.0 cm) in diameter may appear in late summer.

Management

Root decline of bermudagrass and seashore paspalum golf greens is largely an environmental and mechanical stress-related disease. Thus, control focuses on cultural measures that reduce stress. Once symptoms appear, mowing height should be increased and mowing frequency reduced. Mowing should be avoided when turf is soft from excessive thatch and soil wetness. Golf greens should be mowed above 0.156 inches (3.9 mm) and tees and fairways above 0.5 inches (12.5 mm). Perimeter or "clean-up" areas that are damaged should be mowed no more than once every other day. Avoid overhead irrigation until soil is dry and promote cooling and drying of golf greens with fans. Core aerate, vertical cut, and topdress with sand when the turf is actively growing and symptoms have abated. Coring, vertical cutting, and other abrasive cultural practices, however, should be avoided when the disease is active. Improve surface water drainage and reduce shade. Spoon-feed golf greens with nitrogen and potassium in a 1:1 ratio and use ammonium sulfate as the primary nitrogen source. Ammonium sulfate is preferred because it acidifies soil and reduces the competitiveness of the pathogen. Soils low in manganese are believed to be a contributing factor. In near neutral or alkaline soils, avoid limestone and nitrate forms of nitrogen, which can cause increases in soil pH.

DMI/SI fungicides (e.g., propiconazole, triadimefon, others), QoI or stro-bilurin (azoxystrobin, pyraclostrobin, others) fungicides, or thiophanate-methyl should be applied preventively prior to the advent of high-temperature stress where the disease is commonly observed in golf greens. Higher rates and more frequent applications often are required and can be expected to provide only partial control. Avoid excessive usage of high rates of DMI/SI (i.e., propiconazole, triadimefon, others) as their growth-regulating effects can injure bermudagrass golf greens. Fungicides should be watered-in to a soil depth of about 1.0 inch (2.5 cm). Fungicide performance is most benefi-cial when mowing height has been increased.

Rust

Pathogen: Puccinia spp.

Primary hosts: Kentucky bluegrass, perennial ryegrass, and zoysiagrass

Predisposing conditions: Prolonged periods of overcast weather or shaded environments in late summer and autumn

There are many species and biotypes (known as races) of rust fungi that attack nearly all turfgrasses to include bluegrasses, bermudagrass, buffalo-grass, fescues, perennial ryegrass, St. Augustinegrass, and zoysiagrass. Stem rust (*Puccinia graminis* Pers.: Pers) of Kentucky bluegrass, crown rust (*Puccinia coronata* Corda) of perennial ryegrass, and zoysiagrass rust (*Puccinia zoysiae* Dietel) are the most common and important rust diseases of turf. Crown rust is sometimes found in Kentucky bluegrass. Leaf rust (*Puccinia recondite* Roberge ex Desmaz) is uncommon in Kentucky bluegrass, perennial rye-grass, and tall fescue, but is severe on strong creeping red fescue. In seed nurseries in the Pacific Northwest, both stem and crown rust can cause severe damage to perennial ryegrass and tall fescue.

Rust pathogens are obligate parasites, that is, they require living plant tis-sues in which to grow and complete their life cycles. Rust diseases are most commonly observed during cool, moist, and overcast periods of late summer and autumn. They are most damaging to poorly nourished turf and turfs grown under a low mowing height or in shade. In most regions of the United States, rusts do not often cause serious turf damage. However, in some envi-ronments marked by long periods of wet and overcast weather, such as coastal areas from northern California to Canada as well as New England and the Great Lakes region, rusts are chronic and debilitating diseases. They are par-ticularly severe in turfgrass seed-producing areas in the Pacific Northwest.

Symptoms

Rust-affected turfs exhibit a yellow, orange, or reddish-brown appearance from a distance (Figure 2.79). Close inspection of diseased leaves reveals the

FIGURE 2.79
Rusty red-brown-colored Kentucky bluegrass lawn with rust disease. Inset shows orange rust pustules on a Kentucky bluegrass leaf.

presence of conspicuous red, black, orange, or yellow pustules (Figure 2.79). These powdery pustules are comprised of large numbers of spores. Several types of spores are produced by rusts and these fungi have complicated life cycles. During most seasons, rust-affected plants appear healthy. Kentucky bluegrass turfs simultaneously infected with smut and rust can be severely thinned out in late summer.

Management

In most regions, rust-affected stands can be effectively maintained by employing sound cultural practices. A complete N + P + K fertility program is most often preferred to fungicides in situations where rust is damaging poorly nourished turfs. Irrigate early in the day to ensure leaf dryness prior to nightfall, irrigate deeply but infrequently, increase mowing height, and increase mowing frequency. By increasing mowing frequency, leaves bearing immature spores are removed and this reduces the potential for more leaf infections. Improve sunlight penetration and air circulation in shaded environments by removing brush and trees.

DMI/SI (e.g., propiconazole, triadimefon, and others) and QoI or strobilurin (e.g., azoxystrobin, pyraclostrobin, and others) fungicides effectively control rust diseases in a single spring or autumn application. Contact fungicides (e.g., chlorothalonil and mancozeb) are not very effective and multiple applications are required to reduce rust injury.

Slime Mold

> *Pathogens: Mucilago crustacea* A. Wigg., *Physarum cinereum* (Batch) Pers. and *Fuligo* spp.

Primary hosts: All turfgrasses

Predisposing conditions: Warm, rainy periods in summer

Symptoms

Slime molds are primitive fungal-like organisms known as myxomycetes. Taxonomically, they are considered Protista's (i.e., primitive unicellular organisms) rather than fungi. Slime molds move about as protoplasmic amoebae and feed on dead organic matter in thatch. During prolonged periods of warm, wet overcast weather in summer, they migrate onto turfgrass leaves to reproduce. Slime mold appears overnight as black-, yellow-, orange-, or purple-colored patches, arcs, or rings in the turf canopy (Figure 2.80). Once on leaves, they differentiate into fruiting bodies called sporangia. These sporangia appear as small, rounded, ball-like clusters about the size of a pin head, and are initially white, yellow, orange, purple, or black in color (Figure 2.80). *P. cinereum* is the most common slime mold found in turf. When *P. cinereum* sporangia mature and dry on leaves, they are typically gray, and appear similar to cigarette ashes (Figure 2.80). Spores are discharged and fall into thatch where they develop into tiny, shapeless masses of protoplasm. Slime molds are not parasitic and basically are harmless curiosities.

Management

Slime mold protoplasm and sporangia are removed by mowing or by irrigating turf. Sporangia can also be easily brushed off leaves. Slime molds are harmless and affected turfs do not need to be treated with fungicides.

FIGURE 2.80

Black slime mold growth in an arc on a lawn following a summer rain. Insets show immature fruiting bodies (i.e., sporangia; upper left) and dried cigarette-like fruiting bodies discharging spores (lower right inset).

Southern Blight or Sclerotium Blight

Pathogen: Sclerotium rolfsii Sacc.

Primary hosts: Most cool-season grasses and bermudagrass

Predisposing conditions: Warm to hot and humid weather

Southern blight is primarily a disease of annual bluegrass, creeping bent-grass, and bermudagrass grown on golf courses. The disease is uncommon in sites where fungicides (especially QoI/strobilurins, flutolanil, and poly-oxin D) are routinely used. Kentucky bluegrass, perennial ryegrass, and tall fescue lawns, fairways, and roughs, however, may also be attacked. The disease is more prevalent in warm and humid regions, particularly Southern California and southeastern states. The disease has been reported as far north as Washington D.C. in the east, and northern Illinois in the mid-west. Warm or hot temperatures (usually >90°F; >32°C), high soil moisture and thatch favor development of southern blight. In Southern California, the disease may begin in May and continue throughout the summer, whereas elsewhere it tends to be more prevalent during hot and humid periods of summer.

Symptoms

Southern blight initially appears as small, circular, yellow, orange or brown patches or crescents. Patches increase in size rapidly and may produce frog-eyes or crescents 6 inches to 3 ft (15–90 cm) in diameter (Figures 2.81 and 2.82).

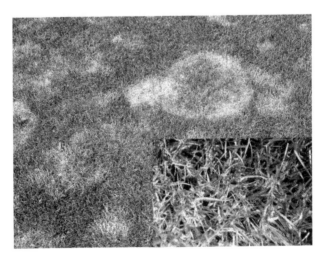

FIGURE 2.81
Southern blight in Kentucky bluegrass. Inset shows foliar mycelium at the periphery of a blighted patch.

FIGURE 2.82
Southern blight in a mixed perennial ryegrass and annual bluegrass rough. (Photo courtesy of
L. Stowell, PACE Turf, LLC.)

Turf may turn a reddish color as it dies. At this stage, the disease mimics
summer patch and necrotic ring spot in Kentucky bluegrass. Patches 6 to 9 ft
(2–3 m) in diameter have been observed. On golf greens, patches, rings and
crescents are yellow or reddish-brown in color. Patches may be uniformly
blighted or contain healthy plants (i.e., frog-eyes).

Inspection of wet foliage or thatch during early morning hours normally
reveals the presence of grayish-white mycelium and/or sclerotia (i.e., resistant structures of the fungus) (Figure 2.83). An abundance of foliar mycelium
may be observed on hot and humid mornings attacking both the turf and
broadleaf weeds, especially clovers. Sclerotia are produced in abundance at
or near the thatch layer and are initially white. In time, sclerotia develop a
mustard-brown color and they appear similar to mustard seeds or sulfur
coated urea granules in color and size (Figure 2.83).

Management

Southern blight is culturally managed by acidifying soil with ammonium
sulfate as a nitrogen source, and by reducing thatch through vertical cutting
and core aeration. Southern blight has become less common on golf greens
since the advent of flutolanil, polyoxin D and QoI/strobilurin products,
which are highly effective when used preventively. For large areas that are
chronically affected, renovation with a less susceptible turfgrass species may
be more cost effective than the use of fungicides.

Summer Patch

Pathogen: Magnaporthe poae Landschoot & Jackson

FIGURE 2.83
Sclerotia of *Sclerotium rolfsii* are initially small and white, but darken to a mustard-brown color as they mature and can be as large as sulfur-coated urea prills. (Inset courtesy of L. Stowell, PACE Turf, LLC.)

Primary hosts: Annual bluegrass, Kentucky bluegrass, creeping red fescue

Predisposing conditions: High temperature stress, moist soils and low mowing

Summer patch is a destructive disease of Kentucky bluegrass, creeping red fescue, and annual bluegrass turf. Under low mowing, Chewings fescue, sheep fescue, and hard fescue can also develop the disease. While uncommon, summer patch can occur in creeping bentgrass grown on golf greens in the southeastern United States and Southern California. Summer patch most commonly occurs in older, mature turfs.

Summer patch generally appears when daytime air temperatures above 88°F (>31°C) prevail. The most important environmental factors required for summer patch development are for soils to be warm (≥75°F; ≥12.8°C) to hot and moist. It is most severe on sunny, exposed slopes or other heat-stressed areas such as those adjacent to paved walks and driveways, and bunker faces on golf courses. The disease most frequently occurs during periods of drought stress that were preceded by wet weather in late spring or early summer. Summer patch may flare up following rainy periods in late summer. Low and frequent mowing and light and frequent irrigation are the primary cultural practices leading to severe outbreaks of summer patch. Other predisposing factors include spring applications of high levels of nitrogen fertilizer; use of nitrate forms of nitrogen (e.g., calcium nitrate, sodium nitrate, and potassium nitrate); accumulation of thatch; frequent summer thunderstorms; and soil compaction. Kentucky bluegrass grown in sites prone to flooding, as well as sodded bunker faces, and tee box and green surrounds are especially prone to summer patch.

Symptoms

Symptoms of summer patch in Kentucky bluegrass initially appear as wilted, gray-green or pale yellow-green areas of turf. These areas rapidly turn into straw-brown, dead patches that initially resemble those of dollar spot (Figure 2.84). These patches soon increase in size and may become crescent-shaped or remain circular. Fully developed patches generally range from 6 to 18 inches (15–45 cm) in diameter (Figures 2.84 and 2.85). Plants at the periphery of affected patches display a yellow, bronze, or copper color when the disease is active. The yellow- or copper-colored plants at the edge of patches only remain evident for a few days, and they are most conspicuous under low mowing (Figure 2.85). Healthy turf may persist in the center of patches producing rings or frog-eye symptoms. Frog-eye symptoms are observed only occasionally, while the circular patch with only a few or no living plants in the center is more common. Depressions in turf called "crater pits" are common. Patches may coalesce, and large nonuniformly shaped areas of turf can be destroyed within a 10–20-day period (Figure 2.86). There are no distinctive leaf or sheath lesions associated with this disease, but leaves dieback from the tip.

In creeping red and other fine-leaf fescues, dead patches generally are 3.0–6.0 inches (7.5–15 cm) in diameter, but some can exceed 12 inches (30 cm) in diameter. Turf initially develops a yellow, bronze, or reddish-brown color,

FIGURE 2.84
Initial infection centers caused by *Magnaporthe poae* in Kentucky bluegrass appear as small, straw-colored spots of blighted turf similar in appearance to dollar spot in Kentucky bluegrass.

FIGURE 2.85
Active summer patch frog-eyes with bronzed peripheries in annual bluegrass. (Photo courtesy of S. McDonald.)

but dead leaves invariably turn a straw-brown color. "Crater pits" are common and all plants within a patch usually are killed. Dead, circular voids in fine-leaf fescue stands often remain evident for years, unless damaged areas are renovated.

In annual bluegrass grown on golf greens, patches range from a few inches (5–8 cm) to about 1 ft (30 cm) in diameter. In mixed greens, infected annual

FIGURE 2.86
Severe summer patch pitting in Kentucky bluegrass.

FIGURE 2.87
Active summer patch in annual bluegrass golf greens may appear as yellow- or reddish-brown-colored patches. Creeping bentgrass eventually fills areas left void by the disease, giving greens a patch-work quilt appearance.

bluegrass plants initially turn yellow or bronze in ribbons adjacent to healthy creeping bentgrass. Death of plants may be nonuniform rather than in discrete patches, particularly when annual bluegrass is mixed with creeping bentgrass. Affected plants develop a reddish-brown, bronze, or yellow color before dying in either a patch or irregular pattern (Figure 2.87). Yellow color symptoms developing in irregular patterns can be confused with anthracnose (Figure 2.88). Stem bases typically are water-soaked and brown in color (Figure 2.88). While summer patch is occasionally found in creeping bentgrass golf greens in the southeastern United States and Southern California, it is unaffected by the disease elsewhere. Creeping bentgrass plants adjacent to dying or dead annual bluegrass plants fill into dead areas vacated by annual bluegrass in autumn, giving greens a patch-work quilt appearance (Figure 2.87). Darkly pigmented runner hyphae are generally not produced in abundance on roots, but dark-brown growth cessation structures may be present and are diagnostic. Summer patch in annual bluegrass greens and fairways can also appear as yellow spots or patches or as a generalized yellowing. The feeding activity of black turfgrass Ataenius grubs and annual bluegrass weevils produce symptoms similar to summer patch in annual bluegrass.

Management

Low mowing, frequent irrigation, and soil compaction are major factors that intensify summer patch. For higher-cut Kentucky bluegrass and creeping red fescue grown on lawns and roughs, increase mowing height to 3 inches (7.5 cm) in late spring, and apply water deeply and only at the onset

FIGURE 2.88
Summer patch symptoms include yellowing in irregular patterns in annual bluegrass, which mimic anthracnose or feeding of annual bluegrass weevils. Infected stem bases are water-soaked and brown (right inset).

of wilt. Similarly, the height of cut on annual bluegrass golf greens should be increased to the maximum acceptable level, preferably >0.156 inches (>3.9 mm). Core aeration alleviates summer patch severity in compacted soils and thatched turfs, but aeration should be performed in the spring or autumn when the disease is inactive. Divert the concentration of cart and other traffic. Avoid use of Kentucky bluegrass in areas that chronically flood or heat sinks such as southwest-facing slopes and bunker faces.

On sunny days, when soils are wet, it is not uncommon for temperatures in the upper 2 inches (5 cm) of the soil to exceed ambient air temperature. Irrigating during sunny periods will elevate soil temperature because water in thatch efficiently absorbs and conducts heat. Hence, syringing golf greens to avoid wetting of soil is preferred to irrigation on hot and sunny days. In lawns, use slow-release acidifying nitrogen fertilizers, such as sulfur-coated urea, and irrigate deeply but infrequently to avoid wilt. Acidification with ammonium sulfate also reduces summer patch severity over time. Conversely, nitrate forms (i.e., calcium, potassium, or sodium nitrate) of nitrogen and limestone applications should be avoided as they can intensify summer patch by increasing soil pH.

Preventive applications of a QoI/strobilurin fungicide (i.e., azoxystrobin, fluoxastrobin, pyraclostrobin) are the most effective chemical approach to summer patch control. DMI/SI fungicides (e.g., metconazole, myclobutanil, propiconazole, tebuconazole, and triadimefon) are also effective, but they can discolor golf greens and at high rates can cause a leafy texture to develop. Thiophanate-methyl drenches may provide a satisfactory level of

curative control. To control the disease in chronically affected annual bluegrass golf greens, fungicides should be applied beginning mid-spring, that
is, several weeks prior to the anticipated onset of summer patch when soil
temperatures reach 65°F (18°C). On chronically affected golf greens, fungicides should be applied on roughly a 21–28-day interval until late summer.
Even on higher-cut Kentucky bluegrass, two or more fungicide applications
may be required in hot spots such as south-facing slopes. Most fungicides
do not require watering-in to be effective, but should be applied in about 100
gallons water per acre (935 L/ha). Fungicides are ineffective if turf is allowed
to enter drought-induced dormancy.

Superficial Fairy Ring

> *Pathogens: Coprinus kubickae* Pilat & Svrcek, *Trechispora cohaerens*
> (Schwein.) Julich & Staplers, others
>
> *Primary hosts:* Golf greens
>
> *Predisposing conditions:* Warm, rainy periods in summer

Superficial fairy rings (SFRs) are caused by mushroom fungi in the same
group associated with fairy ring. The disease appears during warm and
rainy periods in summer, and is most commonly observed under conditions
of low nitrogen fertility. In some regions, SFR can be evident in the autumn
and winter. These fungi develop principally in thatch and do not penetrate
more than 0.5–1.0 inches (1.2–2.5 cm) into the underlying soil.

Symptoms

SFR is sometimes referred to as "white patch" since they appear primarily as
white, circular patches that range from 3.0 inches (7.5 cm) to 3.0 ft (0.9 m) in
diameter in golf greens (Figure 2.89). At the edge of these well-defined, circular patches is a 1.0–2.0 inch (2.5–5.0 cm) fringe of dense, white mycelium.
Many SFRs are hidden under the canopy and only become evident when turf
is treated with a nonselective herbicide for renovation purposes.

Older leaves of plants in the whitish fringe die prematurely and have a
bleached-white appearance. Although SFRs appear unsightly, these fungi
do not infect leaves. Instead, they cause premature senescence and death of
leaves and sheaths by blocking incoming sunlight, which results in the rapid
breakdown of chlorophyll in these tissues. Some SFR fungi may be weakly
pathogenic, but most appear to be saprophytes or possibly senectophytes
(i.e., only pathogenic on senescent tissues). Once leaves and sheaths die, the
fungus uses the necrotic tissues as a source of nutrition. Turf may lose density and thatch degradation by the fungus may cause shrinkage, which dis-
figures and/or interferes with trueness of putting surfaces.

FIGURE 2.89
Superficial fairy ring in a golf green.

Management

SFRs are more common in poorly nourished turf; therefore, an application of 0.25 to 0.5 lb nitrogen per 1000 ft^2 (12.5–25 kg N/ha) would mask their appearance. Mechanical disruption of thatch and fungal mycelium by coring, vertical cutting, or spiking helps to minimize adverse effects of these fungi. Fungicides listed for fairy ring control would likely have activity on SFR.

White Blight or Melanotus White Patch

Pathogen: Melanotus phillipsii (Berk. & Broome) Singer

Primary hosts: Tall fescue, creeping red fescue, and Chewings fescue

Predisposing conditions: Warm, rainy, or humid weather in summer

Tall fescue and fine-leaf fescues are the primary hosts for this pathogen, but the disease may occur in Kentucky bluegrass and perennial ryegrass. This disease develops during very hot and/or humid periods, and primarily in sunny places. Prolonged overcast rainy weather in the summer either preceding or following a period of heat and/or drought stress also appears to favor white blight, particularly in Kentucky bluegrass and perennial ryegrass. Morning shade may favor the disease in creeping red fescue. White blight tends to be most severe in tall fescue, particularly in stands less than 1 year old. White blight seldom occurs north of the transition zone.

FIGURE 2.90
White blight in tall fescue. Inset shows stalkless mushrooms of *Melanotus phillipsii* growing on creeping red fescue leaves.

Symptoms

Circular patches of blighted leaves range from 3 to 12 inches (7.5–30 cm) or greater in diameter (Figure 2.90). Leaves die back from the tip and become light tan to white in color. Leaf blades become matted and foliar mycelium may be evident. Small, tan-white, stalkless mushroom caps 0.125–0.250 inch wide (3–6 mm) develop on blighted leaves (Figure 2.90). The presence of stalkless mushrooms on blighted foliage is the key diagnostic sign for white blight. They first appear as small round balls on blades, but open at the bottom to expose their gills. There is, however, a saprophytic fungus (*Marasmius* sp.) that produces tiny mushrooms with stalks on senescent turfgrass leaves.

Management

Avoid drought stress by watering deeply but infrequently. Blighting is restricted to leaves, therefore, late summer and autumn applications of a complete N-P-K fertilizer will stimulate growth and promote recovery. There are no known chemical control measures; however, fungicides that effectively control brown patch and southern blight may reduce the severity of white blight.

Yellow Ring

Pathogen: Trechispora alnicola (Bourd. & Galzin) Liberta

Primary host: Kentucky bluegrass

Predisposing conditions: Warm and wet weather from spring to autumn

FIGURE 2.91
Yellow ring in Kentucky bluegrass. Large amounts of *Trechispora alnicola* mycelium can be seen in the thatch layer (upper inset).

Symptoms

This pathogen also belongs to the class of mushroom fungi. As its name implies, this disease causes characteristically yellow rings to appear in Kentucky bluegrass (Figure 2.91). There are no other known hosts, and the rings may come and go. Normally, there are large amounts of white fungal mycelium present in the thatch layer, which produces a strong mushroom odor (Figure 2.91). The fungus lives saprophytically in thatch and during cool weather the pathogen infects epidermal root cells, but does not kill roots. Yellow ring may be evident from spring to autumn.

Management

Given the ephemeral nature of this disease, and the fact that it causes no permanent damage to turf, control measures are seldom recommended. Vertical cutting or core aeration disrupt fungal growth, reduce thatch, and may speed up the disappearance of the yellow rings. Flutolanil, polyoxin D, and QoI/strobilurin fungicides would be expected to reduce yellow ring severity.

Seedling Diseases or Damping-Off

Pathogens: Pythium spp., *Rhizoctonia solani, Bipolaris* spp., *Drechslera* spp., *Curvularia* spp., *Fusarium* spp., *Pyricularia grisea*, others

Primary host: All seeded species

Predisposing conditions: Seeding during warm, humid, or wet periods

Many common soil fungi can parasitize seeds and seedlings. Seed decay or damping-off of seedlings occurs most often when cool-season grasses are seeded in spring or summer when high temperatures and high humidity prevail. Conditions of high temperature and humidity, especially at night, slow seed germination and reduce seedling vigor. Furthermore, these are the same conditions that are conducive to growth of *Pythium* spp. and *R. solani*. *Bipolaris* spp., *Curvularia* spp., and *Fusarium* spp. are destructive under conditions of high temperature and/or drought stress. During the winter and spring following an autumn seeding, young plants can be damaged by Microdochium patch or leaf spot and melting-out diseases (*Drechslera* spp.). Hence, temperature and moisture extremes that enhance fungal growth, but reduce growth and vigor of seedlings, are the major factors that predispose seeds and seedlings to damping-off diseases.

Symptoms

Damaged stands often appear patchy with severely thinned-out areas non-uniformly distributed throughout areas of good density. Stands may also appear spotty with small patches of yellow or collapsed plants distributed nonuniformly. Close inspection of yellowed plants may reveal the presence of brown leaf spot lesions, which may be caused by *Bipolaris, Curvularia, Drechslera, Fusarium* spp., or *P. grisea* (especially perennial ryegrass seedlings). *Pythium* spp. and *R. solani* progress rapidly (24–48 h) during warm and humid weather, initially causing seedlings to appear darkened and water-soaked. Gray- or brown-colored foliar mycelium may be present (Figure 2.92). Infected seedlings soon shrivel collapse, turn brown, and die. Dead seedlings may be matted and have a greasy appearance.

Other factors that can enhance damping-off are as follows: old and slow-to-germinate seed; seed planted too deeply; poor seed to soil contact; excessive nitrogen fertilization; high soluble salts; excessive irrigation; poor surface and/or internal water drainage; compacted soil that puddles; and poor air circulation.

Management

Cultural approaches to minimize damping-off are listed in Table 2.5. Fungicides may be used preventively or curatively, but preventive measures are most effective. A preventive approach would include the use of fungicide-treated seed. Mefenoxam (metalaxyl or Apron®) or ethazol-treated seed will protect seeds from *Pythium* spp., but not other pathogens. Fungicidal dusts can be prepared in the absence of commercially treated seed by mixing wettable powder formulations at a rate of 1–2% by weight with seed (e.g., 1–2 lb

FIGURE 2.92
Pythium damping-off in cool-season turfgrass seedlings.

of fungicide per 100 lb of seed; 454–908 g fungicide/45 kg seed). Seed dusting should be performed in a sealed container, in a well-vented room or outside, and workers should wear a respirator, gloves, and a spray suit.

Foliar sprays are also effective in controlling seedling diseases. These sprays are best applied right after it is apparent that most seedlings have emerged. A *Pythium*-targeted fungicide (e.g., cyazofamid, fosetyl-aluminum, mefenoxam, and propamocarb) should be tank-mixed with a broad-spectrum fungicide (e.g., chlorothalonil, iprodione, and mancozeb) and applied to seedlings and/or seedbed and allowed to dry prior to the first irrigation. A foliar spray of a tank-mix combination may need to be reapplied on a 14–21-day

TABLE 2.5

Cultural Approaches to Reduce Seed and Seedling Diseases

Delay renovation or seeding of cool-season grasses until cooler periods in autumn.
Prepare the seedbed properly, which may require roto-tilling and use of soil amendments.
Provide for surface water drainage by proper grading before seeding.
Ensure proper seed-to-soil contact by rolling and avoid planting seed too deeply. In older turfs, remove thatch prior to renovation or overseeding.
Use certified seed of regionally adapted cultivars, which recently has been tested for percent germination.
During establishment, use a complete N-P-K fertilizer that has at least half of the nitrogen component in a slow-release form.
Avoid fertilizing seedlings with mostly water-soluble nitrogen sources in the spring and summer.
Syringe the seedbed frequently, but do not allow water to puddle, inundate, or wash seeds or seedlings.
Avoid nighttime irrigation and time afternoon watering so that foliage is dry prior to nightfall.
Maintain a proper cutting height and remove clippings where practical.

interval if hot, humid, and rainy conditions persist. *Pythium*-targeted chloroneb and ethazole fungicides are potentially too phytotoxic to be applied to seedlings in warm and humid weather. Chloroneb and ethazole, however, can be applied to the seedbed just before or after seed is planted.

Bacterial Diseases

Bacterial diseases are uncommon in turfgrasses. Plant pathogenic bacteria are single-celled, usually rod-shaped, and have rigid cell walls. They reproduce by binary fission and they may or may not be mobile. Bacteria have no means of penetrating cells so they must enter plants through natural openings such as stomates and hydathodes, or through wounds. Once inside plants, they cause injury by producing toxins or they can plug vascular tissues. By occluding vessels, for example, they prevent the movement of water and nutrients, which causes plants to die primarily due to lack of sufficient water. There are two reported bacterial disease of turfgrasses in the United States.

Bacterial Wilt

> *Pathogen:* Xanthomonas translucens pathovar *poannua*
>
> *Primary hosts:* Annual bluegrass
>
> *Predisposing conditions:* Rain and overcast weather from spring to autumn

Bacterial wilt in the United States was first recorded on "Toronto" (also known as C-15) creeping bentgrass. The cause of the disease was initially unknown and the malady was referred to as "C-15 Decline." Other vegetatively propagated cultivars of creeping bentgrass such as Cohansey and Nimislia also proved to be susceptible. Since the original report of C-15 Decline in the early 1980s in the mid-western United States, there have been no other authenticated cases of bacterial wilt in creeping bentgrass turf. Thus, annual bluegrass is the primary host for bacterial wilt.

Annual bluegrass turf is predisposed to bacterial wilt by various stresses such as intensive mowing, grooming, and other mechanical stresses; poor growing conditions (i.e., shade, poor air and water drainage, soil compaction, etc.); and environmental stresses. Heavy rains and thunderstorm activity often trigger and intensify the disease. The disease may abate during dry periods and flare up following rainfall. In most cases, bacterial wilt is problematic on small annual bluegrass greens on older golf courses. The etiolated leaf symptom described below commonly appears in annual bluegrass growing in fairways and approaches, but the bacterium causes little or no damage under high mowing heights. The incidence of the disease may be in large part related to the trend for very low mowing heights (<0.125 inches; <3.1 mm), more frequent mowing, lower nitrogen fertility, and aggressive grooming practices (i.e., sand topdressing, vertical cutting, brushing, grooved rollers, etc.) to improve green speed.

Symptoms

Bacterial wilt tends to develop first on pocketed or shaded greens. Damage appears in high and low areas but tends to be most severe on the outer periphery or clean-up areas, on walk-on and walk-off areas, and in impeded water drainage patterns that are subject to ice formation during winter months. In annual bluegrass grown on golf greens, bacterial wilt generally first appears in late spring, but may remain active into early autumn. Infected annual bluegrass plants on golf greens initially turn yellow, lime-green, or blue-gray in spots. Soon, individual infected plants turn reddish-brown and die in white- or tan-colored spots about the size of a dime (1.8 cm), giving greens a speckled appearance (Figures 2.93 and 2.94). In severe cases, whitish-tan dead spots ranging from 0.50 to 1.0 inch (1.2–2.5 cm) in diameter coalesce and the pattern of injury appears as a general blighting, rather than a speckled appearance. The spot and speckle symptoms of bacterial wilt can mimic active or residual anthracnose basal rot. To add to the confusion, both diseases may occur simultaneously. Infection by one of these pathogens probably predisposes plants to infection by others.

There appears to be no consistent leaf or sheath symptoms that can be used reliably to diagnose bacterial wilt. However, etiolation and yellowing of leaves is a common field symptom (Figure 2.95). When infected plants are

FIGURE 2.93
Bronze-colored spot symptoms of bacterial wilt in an annual bluegrass golf green.

FIGURE 2.94
Inactive, tan-colored speckles and spots of bacterial wilt in an annual bluegrass golf green.

incubated at high temperature in a lab for about 2 days, the youngest leaf of some infected plants may become yellowed and elongated (i.e., etiolated) (Figure 2.95). Infected leaves are sometimes bent in the shape of a shepherd's crook (Figure 2.95). Leaf etiolation, however, is not uncommon in apparently healthy turf (especially creeping bentgrass and perennial ryegrass) and is known as etiolated tiller syndrome, which has no known cause. Older leaves

FIGURE 2.95
Etiolated leaves and leaves bent in the shape of a shepherds crook are symptoms of bacterial wilt.

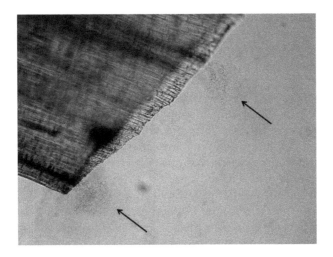

FIGURE 2.96
Bacteria streaming from two vascular bundles in an annual bluegrass leaf.

of infected plants can appear darker green or mottled and stunted, but infected leaves are usually yellow at the tip or at the base of the lamina. Stem bases of infected plants may be variously discolored and water-soaked. The variable leaf symptoms of bacterial wilt are not stable characteristics and are of little value in field diagnosis.

Bacterial wilt is often difficult to diagnose if there is no leaf etiolation. In the laboratory, a diagnostician will place leaves or roots of plants suspected of being infected into a drop of water on a microscope slide, sever them with a razor blade and look for streaming of bacterial cells from vascular bundles. Slow oozes from senescent tissues are common, but rapid streaming of bacterial cells from vascular bundles of infected leaves, crowns, or roots is the best indicator of bacterial wilt (Figure 2.96). Unfortunately, it is difficult and time consuming to isolate and positively identify *Xanthomonas* spp. and other plant pathogenic bacteria.

Management

Bacteria are spread by moving water and mowers. Obviously, mowing creates wounds and contaminated equipment will distribute bacteria onto wounded tissue. Mowing the turf when leaves are dry later in the morning may lower the progression of the disease. Increasing mowing height reduces bacterial wilt severity, but this practice also slows green speed. It is very important to avoid mowing when greens are excessively wet and soft, and to only mow every other day, particularly the perimeter clean-up cut. When the disease is restricted to one or a few greens, a mower should be dedicated to those greens. It is best to use a lightweight, walk-behind greens mower. Grooved rollers should be replaced with solid rollers and grooming, core aeration, and sand topdressing should be avoided when the disease is active. This is because grooming, coring, and topdressing practices abrade and wound tissue, and create openings for easy entry of the bacteria.

Currently, there are no known highly effective chemical control measures for bacterial wilt. Most information is anecdotal. Agricultural antibiotics (e.g., oxytetracycline and streptomycin) suppress bacteria wilt, but they are expensive, difficult to handle, are sometimes phytotoxic, and must be applied in high volumes of water at dusk on a frequent basis. Currently, however, no antibiotics are labeled for use on turf. Some superintendents report positive, short-term benefits from applications of hydrogen dioxide, which may act as a disinfectant killing bacteria on leaf surfaces. Products containing copper hydroxide may slow disease progression, but they do not control bacterial wilt. Superintendents have observed poor to good results with copper hydroxide. There are, however, no research-based recommendations for using copper-based products for bacterial wilt control in turf. Acibenzolar is a chemical that activates natural plant defense mechanisms. Preventive applications of acibenzolar are helpful in suppressing bacterial diseases in some vegetable crops. Currently, acibenzolar is available in a prepackaged mixture with chlorothalonil (i.e., Daconil Action®) and is labeled for suppression of bacterial wilt in turfgrasses.

The most important cultural practices for managing bacterial wilt include increase mowing height; reduce mowing frequency; replace grooved rollers with solid rollers; avoid mowing when greens are soft and wet; avoid use of plant growth regulators; and cease all grooming, coring, and topdressing practices when the disease is active. Weekly spray applications of 0.125 lb N/1000 ft² (5–6 kg N/ha) from urea are recommended, particularly where there is a mix of annual bluegrass and creeping bentgrass. On greens where creeping bentgrass is the dominant species, the disease is best managed by employing cultural practices that promote bentgrass growth. Water management is also important. Soils should not be allowed to become excessively wet. This necessitates frequent syringing and hand-watering of areas that are prone to rapid drying. Longer-range planning should include modifications to the growing environment. This may involve tree and brush removal, use of fans to promote air circulation, and improving surface and internal water drainage. In situations where the disease is a chronic problem, greens composed primarily of annual bluegrass may have to be regrassed with creeping bentgrass.

Bacterial Decline

Pathogen: *Acidovorax avenae* pathovar *avenae*
Primary host: Creeping bentgrass

Predisposing conditions: Prolonged periods of extreme heat stress and sustained high humidity

In 2010, a new bacterial disease of creeping bentgrass caused by *A. avenae* pathovar *avenae* was reported (Giordano et al., 2012). This disease should not be confused with bacterial wilt of annual bluegrass caused by *X. translucens*. Bacterial decline (aka Acidovorax bacterial disease and bacterial etiolation) is a malady of creeping bentgrass golf greens, which normally is triggered by prolonged periods of high-temperature stress and sustained high humidity. The disease may begin in late spring, but intensifies in response to prolonged periods of heat stress in summer. Daily air temperatures above 88°F (31°C) and night temperatures above 80°F (26.7°C) appear to be optimal for this disease.

Symptoms

Symptoms of bacterial decline are variable and can include etiolated growth and yellow leaves. Over time golf greens develop a generalized chlorosis, however, there may be healthy darker green biotypes or clones of creeping bentgrass dispersed throughout areas of the putting surface (Figure 2.97). Annual bluegrass growing in affected greens does not develop bacterial decline. Etiolated or elongated creeping bentgrass leaves may curl and twist and these leaves are not easily removed by mowing. Chlorosis can be mild or pronounced with gradual turf loss or there can be rapid and widespread decline. In mild cases, affected creeping bentgrass can recover rapidly

FIGURE 2.97
Bacterial decline develops during periods of extreme heat and initially causes chlorosis in creeping bentgrass golf greens and sometimes a rapid collapse of turf. (Photos courtesy of C. Sain.)

following a shift to a less stressful weather pattern. Presumptive identifications for this disease in creeping bentgrass are based on leaf chlorosis and heavy streaming of bacteria from cut leaves.

Management

The cultural approaches for managing this disease are the same as described for bacterial wilt. Control is aimed at alleviating strees by increasing mowing height and reducing mowing frequency. In particular, greens should not be mowed when soft from heavy rainfall. As previously noted, there are no known highly effective chemical control materials for turfgrass bacterial diseases. Acibenzolar is a chemical that activates natural plant defense mechanisms. Preventive applications of acibenzolar are helpful in suppressing some bacterial diseases. Currently, acibenzolar is available in a prepackaged mixture with chlorothalonil (i.e., Daconil Action) and is labeled for suppression of bacterial decline.

Plant Parasitic Nematodes

Plant parasitic nematodes are known to cause extensive injury to turfgrasses in warm, temperate, and subtropical areas of the United States. Parasitic nematodes also can be troublesome in the transition zone and northern regions, especially in creeping bentgrass and annual bluegrass grown on golf greens. The overall importance of parasitic nematodes in higher-cut turf in northern regions is unknown. Warm-season grasses grown in southern regions, however, are very susceptible to injury from plant parasitic nematodes. In general, parasitic nematodes populations increase as soil temperatures rise in spring, and high populations are usually present throughout summer months. The feeding activity of parasitic nematodes on turfgrass roots combined with environmental stress debilitates turf. Except for sting nematode, which largely is restricted to the southeastern United States, it is likely that parasitic nematodes by themselves do not directly kill cool-season turfgrasses in the absence of other stress factors. In the presence of summer stresses, parasitic nematodes further weaken and debilitate turf, causing chlorosis, proneness to wilt, and turf thinning. Although not commonplace, however, extraordinarily high population densities (e.g., ≥15,000 parasitic nematodes per 250 cc soil) can develop in clusters that are damaging to cool-season grasses.

The Anguina seed and gall nematode (*Anguina pacificae*) is unlike other nematodes of turfgrasses since it attacks above-ground plant parts. This nematode produces galls at the base of annual bluegrass stems and is considered the most important parasitic nematode on northern California golf courses. There are other foliar damaging nematodes (e.g., *Anguina agrosi-tis*), which attack floral parts of bentgrasses in seed production fields in the Pacific Northwest. Only the most common, root-feeding nematodes will be considered below (Table 2.6).

All plant parasitic nematodes are microscopic (<0.02 inch in length; <0.5 mm) round worms (Figure 2.98). Nematodes reproduce by eggs. The emerging juveniles molt four times before reaching adult size. Each female is capable of producing large numbers of eggs (especially root knot nematode) and the entire life cycle for most species is completed in 5–6 weeks under suitable conditions. Because nematodes that attack plants are obligate parasites, they must feed on living tissues to grow and reproduce. Most parasitic nematodes are capable of attacking a wide range of plant species, and can survive on weeds in the absence of turfgrasses. Most nematodes store large quantities of food, which enables them to survive long periods in soil in the absence of suitable plants. Many survive in frozen soils and may overwinter in living roots or in dead plant tissues. Nematode activity is favored by warm and moist soil conditions. Their activity is enhanced in light-textured soils and reduced in compacted or heavy soils where aeration becomes restricted. Hence, the sandy soils of golf greens provide an ideal environment for nematodes. Nematodes are unable to move more than a few millimeters in soil, but they may be transported over longer distances via coring, moving water, soil, sod, sprigs, and equipment.

Literally millions of nematodes can inhabit a few square feet (or a square meter) of soil, but most nematodes are nonpathogenic. Most soil-dwelling nematodes actually perform a beneficial service in soil by helping to degrade organic matter. Although there are numerous nematode species, only about 15 genera are known to be common turfgrass parasites. All plant parasitic nematodes bear a hollow, spear-like structure called a stylet. The stylet is similar to a hypodermic syringe, and is used to inject enzymes into plant cells. Simultaneously, partially digested food is withdrawn. Plant parasitic nematodes are grouped according to their feeding habit (Table 2.6). Endoparasitic nematodes (i.e., root knot, lesion, and lance) partially or totally burrow into plant tissues and feed primarily from within, whereas ectoparasitic nematodes (i.e., awl, dagger, needle, pin, ring, sheath, spiral, sting, stubby root, and stunt) feed from root surfaces, although a small portion of the body may be embedded in the root. Most genera encountered in turf are ectoparasitic and, therefore, are more commonly found than endoparasitic nematodes. However, the endoparasitic root knot and lance (lance behaves as both endoparasitic and ectoparasitic) nematodes are among the most common and injurious parasitic nematodes, particularly in annual bluegrass, creeping bentgrass, and bermudagrass grown on golf greens. As noted previously, parasitic nematodes are more commonly destructive to warm-season grasses grown in warm and moist climates such as found in the southeastern United States. The sting nematode, an ectoparasite, is among the most destructive nematode species in warm-season grasses, especially

TABLE 2.6

Common Plant Parasitic Nematodes Known to Be Injurious to Turfgrasses in the United States

Common Name/ Genus	Feeding Group	Turfgrasses Injured/Root Symptoms
Lesion (*Pratylenchus* spp.)	Endoparasitic	Cool- and warm-season grasses are injured. Root lesions initially minute and brown, but enlarge and may prune the root system.
Lance (*Hoplolaimus* spp.)	Ectoparasitic and endoparasitic	Warm-season grasses, creeping bentgrass, and annual bluegrass are injured. Common in golf greens. Causes swelling of roots followed by necrosis and sloughing of cortical tissues.
Root-knot (*Meloidogyne* spp.)	Endoparasitic	Warm-season (especially bermudagrass and zoysiagrass) and cool-season grasses (especially golf greens) injured. Galls (i.e., swellings or knots) develop on roots. Galls may be small and difficult to see.
Ring (*Criconemella* spp.)	Ectoparasitic	Especially important in centipedegrass, also injures Kentucky bluegrass, bentgrass, bermudagrass, and zoysiagrass. Roots are stunted; brown lesions on roots.
Spiral (*Helicotylenchus* spp.)	Ectoparasitic	Cool- and warm-season grasses injured, especially seashore paspalum and St. Augustinegrass. Roots are poorly developed with premature sloughing of cortical tissues.
Stunt or stylet (*Tylenchorhynchus* spp.)	Ectoparasitic	Warm- and cool-season grasses are injured. Roots shortened, shriveled, and brown lesions evident on roots.
Stubby root (*Paratrichodorus* spp. and *Trichodorus* spp.)	Ectoparasitic	Warm- and cool-season grasses. Large brown lesions on roots; swelling on root tips. *Trichodorus obtusus* induces damage similar to sting nematode in bermudagrass and St. Augustinegrass.
Sting (*Belonolaimus* spp.)	Ectoparasitic	Bermudagrass, centipedegrass, zoysiagrass, St. Augustinegrass, bentgrass, and ryegrass in southeastern United States are injured. Most damaging species in turfgrasses. Root shortened and lesions evident, especially on root tips.
Dagger (*Xiphinema* spp.)	Ectoparasitic	Bermudagrass, zoysiagrass, and perennial ryegrass are injured. Root lesions are reddish-brown to black and sunken.
Pin (*Paratylenchus* spp.)	Ectoparasitic	Kentucky bluegrass, fine-leaf fescue, and tall fescue are injured. Fewer lateral roots with distinct lesions.
Awl (*Dolichodorus* spp.)	Ectoparasitic	Bentgrass and warm-season grasses are injured, primarily in wet sandy soils in the southeastern United States, but it is rare.
Sheath (*Hemicycliophora* spp.)	Ectoparasitic	Tall fescue and warm-season grasses are injured, primarily in the southeastern United States.

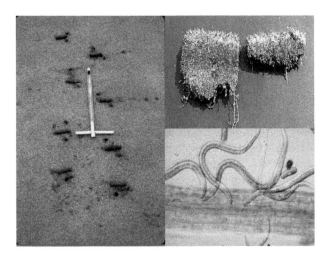

FIGURE 2.98
Parasitic nematodes are microscopic round worms that feed on roots of turfgrasses diminishing the size and function of root systems. Parasitic nematodes should be sampled in random patterns to root zone depth. (Photos courtesy of P. O'Brien [upper right] and L. Krusberg [lower right].)

bermudagrass. Sting nematode also is injurious to creeping bentgrass grown on golf greens in the southeastern United States. Lance, stubby root, spiral, ring, and other parasitic nematodes can be very destructive to centipedegrass and St. Augustinegrass when populations are high. Very high populations of the aforementioned parasitic nematode species can weaken cool-season grasses, especially in sandy soils.

Symptoms

Symptoms of parasitic nematode injury include yellowing, stunting, wilting or early signs of drought stress, thinning of the stand, and diminished root systems (Figure 2.98). Commonly, weed and blue-green algae infestations are more problematic in soils infested with damaging populations of nematodes. These symptoms are related to the injury nematodes inflict upon the root system, and may not become noticeable until water becomes limiting. Owing to similarities between environmental stress symptoms and nematode injury, the source of the problem is often difficult to diagnose. Like many so-called weak or secondary fungal pathogens, parasitic nematodes may not cause much of a problem until environmental extremes, poor growing conditions, mechanical injury, insect pests, or a fungal disease reduces turf vigor. Parasitic nematodes cluster and affected areas may appear as streaks or oval-shaped areas called "hot spots." Often, however, there is no pattern to nematode injury in cool-season grasses. Severe infestations can result in a dramatic reduction in turf density, vigor, and color (Figure 2.99).

Inspection of roots may or may not reveal some indication of nematode feeding. Roots may exhibit one or more of the following symptoms: knot-like swellings, red or brown lesions, excessive root branching, short stubby root tips, necrotic root tips, and root rot (Table 2.6).

Turfgrass areas damaged by nematodes are especially prone to wilt and do not respond readily to an application of fertilizers or fungicides. The search for undiagnosed problems of golf greens between spring and autumn by superintendents and diagnosticians almost always ends in assaying soil for parasitic nematodes. One problem with assay interpretation is that parasitic nematode populations fluctuate with and influence root growth dynamics. In most regions, parasitic nematode populations are usually found in high populations during the growing season after soils warm in late spring. Proper soil sampling is required to get an accurate picture of nematode population densities. Sending one or two cup cutter plugs of soil to a lab will provide incomplete and/or unreliable information. Sampling from severely thinned areas will also yield unreliable results because these obligate parasites will not survive in large populations in the absence of living plants. Instead, soil should be collected at root zone depth from numerous areas at the edge or interface between healthy and injured turf. For more information on how to obtain a proper soil sample for a plant parasitic nematode assay, see the section "Collecting and Sending Diseased Samples to a Lab."

Various methods for nematode extraction are used in the laboratory. Unfortunately, there are no reliable data correlating parasitic nematode number per sample and expected degree of turf injury in the field. Thresholds (i.e., the number of parasitic nematodes that are viewed as damaging) have

FIGURE 2.99
A severe infestation of lance and other parasitic nematodes caused a loss of density in this mixed annual bluegrass–creeping bentgrass fairway. (Photo courtesy of N. Mitkowski.)

been established by some nematologists. Thresholds are of limited value for many nematode species because growing conditions greatly influence the potential importance of parasitic nematode population densities. For example, low mowing, wear or mechanical damage, other fungal diseases (especially anthracnose and summer patch), and environmental stresses weaken turf thus predisposing a stand to more potential damage by fewer parasitic nematodes. Furthermore, the threshold for individual species may be less important than the total number of parasitic nematodes found. While there is no composite threshold consensus among pathologists, a combined count of over 1000 lance, root knot, spiral, stubby root, or stunt per 250 cc of soil (note labs vary in soil sample size) is a cause for concern in most cool-season grasses. Annual bluegrass is considered more susceptible to injury from plant parasitic nematodes, when compared to creeping bentgrass. For annual bluegrass golf greens, however, the combined threshold would be considerably lower in summer, especially if another disease is involved. For bermudagrass and creeping bentgrass grown on golf greens in the southeastern United States, sting nematode counts above 50 per 250 cc of soil may be cause for action. Where parasitic nematodes are an annual concern, turf should be sampled monthly between spring and autumn to provide valuable information on natural variations in the rise and fall of population densities. Pathologists ultimately will make management recommendations based on the species and the total number of parasitic nematodes present in a sample.

Cultural and Biological Management

Because parasitic nematodes may only be injurious during summer stress periods, cultural practices that alleviate stress are the wisest approach to minimize injury. Such practices would include increasing mowing height, reducing mowing frequency, correct nutrient deficiencies, improve air circulation, and alleviate compaction. A healthy root system will support higher nematode populations, but ultimately a well-developed root system is the best defense against these parasites. Deep and infrequent irrigation promotes a deeper root system and thus improves the stress tolerance of most turfgrasses. On golf greens, spoon-feeding (0.1–1.5 lb N/1000 ft^2; 5–7.5 kg N/ha) and avoiding mechanical injury during summer months can improve or help maintain vigor. Applications of biostimulants and micronutrients and other materials that promote rooting are recommended.

Applying beneficial species of nematodes (e.g., *Steinernema carpocapsae* and *Steinernema riobrave*) as biological agents have provided mixed results, but have generally failed to significantly impact parasitic nematode populations. In addition to predatory nematodes, nematode-trapping fungi, bacterial pathogens of nematodes (e.g., *Pasteuria* spp.), and bacterial antagonists (i.e., *Bacillus firmus*) are being evaluated as biological control agents. For example, the bacterium *B. firmus* applied to turfgrasses colonizes roots and its presence in the rhizosphere (i.e., microenvironment surrounding roots) discourages the feeding activity of selected parasitic nematode species in warm-season grasses. Granular chitin–protein materials do not have any observable activity against plant parasitic nematodes found in turfgrasses. Mustard seed extract has shown promise for reducing populations of sting and possibly other parasitic nematodes (Cox et al., 2007).

Chemical Management

Currently there are few nematicides labeled for use on turfgrasses. The soil fumigant 1,3-dichloropropene (i.e., Curfew®) has been labeled for use on golf greens (bermudagrass and creeping bentgrass) in some southern states. This fumigant (a liquid under pressure) is applied directly into the soil using a knife injection system. It quickly volatilizes to a gas, which is toxic to nematodes and some insects. The fumigant kills foliage on contact and the knife injection causes mechanical damage to golf greens. This material is toxic to turf it contacts, can be injected only once annually, and can only be handled by custom applicators. Regardless, this fumigant can be highly effective when applied properly (Figure 2.100). Parasitic nematode populations, however, recover over time.

FIGURE 2.100
The soil fumigant 1,3-dichloropropene (Curfew) was injected into this bermudagrass turf chronically affected by sting nematodes about 30 days prior to the photograph being taken. (Photo courtesy of B. Martin.)

Special local need labels have been approved for the insecticide abamectin (i.e., Avid®). Currently, labels are only available in some southern states for ring and sting nematode control. Abemectin should be sprayed in a high water volume with a soil penetrant surfactant and further incorporated by overhead irrigation. This method of application is nondestructive to turf and the material can be applied several times in a season to achieve suppression of sting nematode populations.

In some situations, such as extremely high and persistent parasitic nematode population densities in chronically injured golf greens, fumigation and regrassing may ultimately become the best solution. Unfortunately, parasitic nematodes eventually will reinvade these sites.

Virus Diseases

Viruses are submicroscopic and those that infect grasses are pathogenic entities, which consist of nucleic acid (usually RNA) covered by a protective protein coat. Viruses can only be seen with expensive and powerful electron microscopes, which generally are not available to diagnostic labs. Viruses are unable to grow, are immobile, and exhibit no metabolic activity outside of living cells. They have no means of penetrating cells when acting alone. Despite these limitations, they are common inhabitants of grasses.

Viruses gain entry into grasses primarily by insect vectors or they may be seedborne. They may also be mechanically transmitted in sap, which is rare in nature yet it is a primary means of transmission of Panicum mosaic virus in St. Augustinegrass and centipedegrass. When a virus gains entry into a cell, it biochemically triggers the plant to produce nucleic acid and protein identical to that of the invading virus. As large amounts of virus particles are produced, they are spread throughout the plant and cause it to lose vigor, turn chlorotic, and in some cases die. While viruses are common in many grasses, little is known about virus diseases of turfgrasses. St. Augustine decline and centipede mosaic are the only described virus diseases of turf.

St. Augustine Decline and Centipede Mosaic

Pathogen: Panicum mosaic virus

Primary host: St. Augustinegrass and centipedegrass

Predisposing conditions: During periods of active turf growth in spring and summer

This disease is primarily a problem in Gulf Coast states from Florida to Texas. Panicum mosaic virus causes a decline and sometimes death of

St. Augustinegrass. In centipedegrass, symptoms are mild and there is generally no decline or death of plants. The disease is transmitted in sap during mowing. Panicum mosaic virus can predispose turf to more damage from insect pests, parasitic nematodes, and cold temperatures.

Symptoms

Early symptoms include a mild yellowing and leaf stippling that mimics mite damage (Figure 2.101). Leaves may develop a mottled yellow-green mosaic, which becomes more pronounced or severe the longer plants are infected. In advanced stages, plants are stunted, lack vigor, and exhibit very slow stolon growth. In extreme cases, turf appears severely chlorotic, and leaves become necrotic and large areas thin out or die out completely. It is uncommon, however, for the Panicum mosaic virus to kill large areas of turf.

Management

Virus diseases cannot be controlled with chemicals. Virus diseases are managed primarily by imposing sound cultural practices and alleviating environmental stresses such as drought and shade. Cultural practices, such as increasing mowing height, avoiding mowing when leaves are wet, preventing drought stress and applying the proper rates of complete N-P-K fertilizers assist in maintaining turf vigor and density. Renovating with a Panicum mosaic virus-resistant cultivar or conversion to another species that is unaffected by the disease may become necessary.

FIGURE 2.101
Chlorosis, stunting, leaf stippling, and leaf mosaics are symptoms of Panicum mosaic virus in St. Augustinegrass. (Photos courtesy of N. Jackson.)

Blue-Green Algae, Moss, and Black-Layer

Blue-green algae and moss contain chlorophyll and are very simple plants. These plants do not parasitize grasses, but they are included here because they are highly invasive and often outcompete grasses for space in wet or shaded environments. Black-layer is a soil physical malady that acts continuously to damage plants and develops under conditions similar to those that encourage algae and moss.

Blue-Green Algae (aka Cyanobacteria)

Pathogen: Nostoc spp., *Oscillatoria* spp., *Phormidium* spp., others

Primary hosts: All turfgrasses grown in wet and/or shaded sites, particularly golf greens

Predisposing conditions: Prolonged periods of warm, overcast, and rainy weather or excessive irrigation

Algae contain chlorophyll and are often described as primitive plants because they lack roots, stems, and leaves. The types of algae found in turf are primarily blue-green, filamentous, and have a mucus-like coating. The blue-green algae are capable of fixing atmospheric nitrogen, which may enable them to contribute some nitrogen to turf. While this may seem beneficial, blue-green algae can seal surfaces, creating an anaerobic condition that is very harmful to the grass. They are referred to as cyanobacteria because their nuclei are not bound by a membrane (i.e., prokaryotic), and as such some books classify them as bacteria. The true algae (i.e., eukaryotic) have membrane-bound nuclei, and can be green or brown in color.

Common genera of blue-green algae found in turf include *Nostoc* spp., *Phormidium* spp., and *Oscillatoria* spp. Blue-green algae filaments are long and are comprised of a single cell. These algae reproduce by binary fission (i.e., a splitting in half), and individual cells may develop thick walls and behave like spores. This enables blue-green algae to survive unfavorable environmental conditions. These algae secrete a gelatinous substance (i.e., mucilage) and eventually can form a black scum in thatch or on the surface of wet soils. The mucilage protects cells by helping to maintain a moist microenvironment around individual filaments.

Blue-green algae can be a chronic problem in golf greens, especially those with poor internal water drainage and excessively irrigated. Growth of blue-green algae is promoted during extended periods of warm, overcast, and rainy weather in summer and affected golf greens have a blackened appearance (Figure 2.102). Black algal scums are particularly common on golf greens that are compacted, low-lying, adjacent to ponds, and shrouded by trees or brush. Algae are less troublesome on higher mowed collars, tees,

FIGURE 2.102
Blue-green algae (aka cyanobacteria) proliferate in wet environs causing blackening within the canopy of golf greens.

or fairways, but they can be invasive and form scums in low and wet areas of higher-cut turf (Figure 2.103). They are particularly a problem on greens because the very low mowing heights expose a much greater thatch surface. Algal cells and filaments can bind sand particles and plug pore spaces thereby impeding water percolation. Algal scums also impede oxygen and other gas diffusion into and out of soils, and may play a role in the formation of some black-layers. Algal filaments can accumulate on older, lower leaves causing senescence of tissues, and thus reduce turf density.

FIGURE 2.103
Blue-green algal scum sealing the surface of a wet golf tee.

Management

Blue-green algae are not effectively controlled for prolonged periods unless the conditions that predispose turf to their growth are corrected. Control, therefore, is aimed at alleviating wet soil conditions by improving surface and internal water drainage and aeration as well as by employing proper irrigation practices. This ultimately may require removal of trees and shrubs; cleaning, repairing, or installing new drains; and possibly the rebuilding of greens. Increasing mowing height is essential; persistent low mowing will only continue to aggravate the condition.

Applications of ground agricultural limestone, iron sulfate, and/or fatty acid soaps provide some short-term suppression by desiccating algae. Prior to application of any desiccant, the scum should be manually broken up by hand-held garden tools (e.g., "Weed Weasel") or by vertical cutting, spiking, or narrow tine aeration, and the debris removed. Simultaneously, mowing height should be increased. The higher turf canopy will shade blue-green algae, reducing their photosynthetic capacity and, therefore, their ability to grow and compete with turf. A spray application of a low rate of water-soluble nitrogen (i.e., 0.10–0.25 lb/1000 ft^2; 5–12 kg N/ha) in combination with iron sulfate (1.0–2.0 oz/1000 ft^2; 3–6 kg/ha) is beneficial to golf greens that exhibit slow growth or poor vigor. During autumn, nitrogen fertility should be increased to stimulate tillering, which improves stand density. Soil acidification using ammonium sulfate can help suppress blue-green algae.

For golf greens, the most effective chemical approach to managing blue-green algae is to apply chlorothalonil, fosetyl-aluminum or mancozeb prior to the advent of overcast and rainy weather in summer. Higher rates of chlorothalonil and mancozeb are needed once the blackening from blue-green algae has appeared. Copper hydroxide and copper hydroxide plus mancozeb are also effective curative materials for controlling blue-green algae. Copper-based products should be used with caution as they can burn or yellow turf and their excessive use can result in copper toxicity. Frequent applications of DMI/SI fungicides (e.g., propiconazole, triadimefon, others) and phosphorus fertilizers applied in the summer can encourage blue-green algal growth on golf greens.

Moss

Pathogen: Byrum argenteum Hedw. *Funaria* spp., *Mnium* spp., *Polytrichum* spp., others

Primary host: Turfgrasses growing in wet and/or shady sites and golf greens

Predisposing conditions: Prolonged periods of warm or cool, overcast, and rainy weather

Mosses are more advanced plants than algae, but like algae they have no vascular system. Common moss species found in turf include *Amblystegium* spp., *Brachythecium* spp., *Byrum* spp., *Funaria* spp., *Mnium* spp., and *Polytrichum* spp. *Byrum argentum* (silvery thread moss) is among the most common species found on golf greens. Moss plants are composed of sheets of cells with leafy stems and filamentous, root-like rhizoids. Rhizoids form at the base of shoots, anchor moss in soil, and function in water and nutrient uptake. Mosses have a complicated life cycle and water is required for reproduction to occur. They produce spores in a sporangium, which is borne on the top of a stalk. Although mosses usually proliferate in shady and moist places, they are capable of growing in full sun while surviving periods of drought. They can grow in open areas during extended periods of overcast and wet weather. Moss may also be found on slopes and other droughty areas that are subjected to frequent scalping.

Moss colonies growing in golf greens disrupt uniformity and playability by creating bumpy putting surfaces (Figure 2.104). The presence of moss in golf greens is often an indicator that greens are being cut too low, are underfertilized, and/or are overwatered. Like blue-green algae, moss problems on greens often are linked to excessive irrigation, particularly in poorly drained soils. Mosses seldom are a problem on collars, tees, or a fairway, demonstrating that low mowing is a major cause for moss invasion. Mosses are very aggressive and opportunistic, and they can eventually dominate. Mosses are especially invasive in the Pacific Northwest, New England, and the United Kingdom because weather conditions remain cool and overcast for long periods. They are also a problem in lawns in other regions, particularly where lawns are heavily shaded and where turf density is poor (Figure 2.104).

FIGURE 2.104
Silvery thread moss colonies in a golf green. Inset shows moss growing in a shady lawn.

Management

As is the case with blue-green algae, control is aimed at correcting those soil or cultural conditions that provide moss their competitive edge. Increasing mowing height while increasing fertility are key cultural approaches to reducing moss invasiveness. Other cultural measures include improving water and air drainage, avoiding excessive irrigation, selectively pruning trees and shrubs, maintaining proper soil fertility and soil pH levels, and alleviating soil compaction. Frequent sand topdressing and core aeration help to reduce moss competitiveness on golf greens by disrupting their growth, injuring cells, and by improving water and air movement into soil. Ammonium sulfate applications to supply 3–4 lb N/1000 ft^2 (150–200 kg N/ha) per year also reduces moss. Higher rates of nitrogen, however, increase mowing frequency and reduce green speed. Spoon-feeding with ≤0.125 lb N/1000 ft^2 (6 kg N/ha) would be expected to improve the competitiveness of golf green turf. Fertilizing frequently in summer with relatively high nitrogen rates (≥0.3 lb N/1000 ft^2; 15 kg N/ha) can encourage moss (Thompson et al., 2011).

The herbicide carfentrazone is likely the most effective material available in the United States for controlling moss on golf greens. Carfentrazone should be mixed with a surfactant and used in conjunction with a spoon-feeding program. Sodium bicarbonate (i.e., baking soda) also is effective when used as a spot treatment. Copper hydroxide and copper hydroxide plus mancozeb are sometimes effective, but are risky since they can scorch and discolor turf. Soaps and other desiccants are erratic and seldom consistently effective. All of the aforementioned products are most effective when used in spring and autumn. Summer applications are generally erratic. High label rates of chlorothalonil, however, shrink moss colonies when applied in hot weather. Repeated treatments of the aforementioned materials are often needed. One product may work well in one year or at one location, but fail in other years or locations. Chemicals are unlikely to provide long-term control unless the conditions that cause moss to develop are addressed.

Black-Layer

Pathogen: Abiotic

Primary hosts: Turf grown on golf greens

Predisposing conditions: Anaerobic conditions developing in response to a water-saturated soil

Black-layer is a physical condition that is primarily associated with high-sand-content golf greens. Black-layer can develop in native soils and is commonplace in wetlands. Anaerobic conditions in soil are the cause of black-layer. Anaerobic conditions develop as pore spaces become filled with water (i.e., saturated), thus displacing oxygen. Under anaerobic conditions (i.e., void of oxygen), sulfur is converted by sulfate-reducing bacteria to form hydrogen sulfide gas, which is directly toxic to roots. The sulfurous odor of hydrogen sulfide gas is normally evident.

Black-layer develops during periods of excessive rain or because of excessive irrigation, especially in golf greens with poor internal soil drainage or where soil water perches. The condition is frequently encountered where sand topdressing is layered over a greens mix of a different sand particle size, a heavier soil type, or a subsurface organic layer. Black-layers also develop on sodded greens where soil from the sod forms a layer-effect. Both situations can cause water to become trapped in the upper layer, resulting in soil becoming anaerobic. This trapping of water between soil layers is referred to as a "perched water table." Root development is usually poor below a black-layer or between layers of different soil types. Liners in new golf greens can act as dams on low sides where water collects and drives oxygen from the soil. Black-layer can also develop when a superintendent irrigates every night to the same depth in soil. This keeps the upper soil profile almost continuously wet, which can cause an anaerobic condition. Blue-green algae may play a role in the development of some black-layers by plugging pore spaces and/or by sealing putting surfaces.

Symptoms

Symptoms usually first appear in low spots of golf greens as turf develops a yellow, blue-purple, or reddish-brown appearance, and eventually turf thins in irregular patterns (Figure 2.104). Loss of density generally is most severe in low-lying, shaded areas where air circulation is poor. Black-layer detection is easy. Examination of the soil profile reveals the presence of a surface or subsurface black-layer, and a foul, sulfurous odor. Where layers are restricted to the surface, thatch may have a black color. Subsurface layers are coal-black in color and may develop several inches (3–12 cm) below the surface (Figure 2.105). Bands of blackened sand usually range from 0.25 to 1.0 inch (0.6–2.5 cm) in width, but bands greater than 1 inch (>2.5 cm) in width are not uncommon. Sometimes, back metal precipitates can be observed on stem bases, indicating that the thatch layer had become anaerobic (Figure 2.105). There is a foul, sulfurous, rotten-egg odor associated with affected soils due to the production of hydrogen sulfide gas by bacteria under anaerobic conditions. The actual blackening is caused by the precipitation of metal sulfides, usually iron, which are black in color.

Management

Increasing soil aeration and reducing soil wetness are the only approaches to alleviating anaerobic conditions. In the short term, spiking and solid or hollow tine coring can be performed quickly to increase soil aeration and drying with minimal disruption (Figure 2.106). Fans are also helpful in promoting

FIGURE 2.105
Black-layer (i.e., anaerobic conditions) in golf greens turns turf a reddish-brown or purple color. Distinct black bands can be seen where water perches (upper right inset) and sometimes black metal precipitates form on stem bases (upper left inset).

drying. In the long term, it may be necessary to install sand channel drainage lines to speed up water drainage. Drains through liners in new greens may also be needed. Correcting compaction or creating bypasses through layers with deep core aeration with the verti-drain or using drill-and-fill machines in the spring and autumn also greatly improves soil aeration. These operations, however, may have to be performed several times before their overall benefits can be objectively assessed.

FIGURE 2.106
Black-layer bypassed by coring, which allows for improved soil aeration. (Photo courtesy of S. McDonald.)

Trees and brush should be removed where necessary to improve air movement and sunlight penetration. It is especially beneficial to ensure morning sunlight exposure of greens. Furthermore, it is helpful to control blue-green algae and thatch. Avoid applying materials containing elemental sulfur. Nitrate forms of nitrogen may help alleviate black-layer.

Collecting and Sending Diseased Samples to a Lab

Fungi cause most disease problems in turf (Tables 2.7 through 2.9). Many fungal diseases are easily distinguishable because of unique signs (i.e., visible parts of a pathogen such as mycelium, spores, etc.) or symptoms that accompany their activity. For example, the red thread fungus produces a pink-gelatinous mycelium and characteristic red, antler-like sclerotia, making this disease easy to diagnose. Unfortunately, many diseases do not always produce textbook symptoms. Certain signs, such as foliar mycelium can be deceiving. Summer pathogens that cause dollar spot, brown patch, southern blight, and Pythium blight often produce large amounts of foliar mycelium. Most turfgrass managers equate foliar mycelium with Pythium blight. In many situations, this is a diagnostic error. These four diseases, however, can be easily separated by the microscopic observation of the mycelium. Other pathogens, particularly those that attack roots, can be very difficult to identify. Take-all patch, summer patch, necrotic ring spot and spring dead spot are root and stem diseases. With few exceptions, these pathogens cannot be identified by mycelial characteristics. To be absolutely certain, a pathologist must isolate the pathogen on an agar medium. Isolation techniques, however, are not consistently successful. Once isolated, some of these fungi can be presumptively identified by colony characteristics, but many must be manipulated to produce their spore-producing structures. These techniques often require a significant waiting period, and frequently they are unsuccessful. Molecular techniques are fast and reliable, but currently are not routinely available in most turf disease diagnostic labs. The problem with molecular diagnostics is that it is expensive to hire, equip, and train qualified specialists to do this type of work.

Many fungal diseases, or the activity of parasitic nematodes, may produce symptoms that mimic environmental stress or insect pest damage. Similar symptomatologies of numerous diseases, as well as environmental stresses, make diagnostics a challenge for even the experts. When perplexed or exasperated, turf managers eventually send samples to a diagnostic laboratory. Most land-grant universities provide plant disease diagnostic services. Many of these laboratories, however, are not staffed by pathologists trained to diagnose turfgrass diseases. Few labs are able to identify bacteria or virus diseases of turfgrasses and as previously noted, molecular techniques are not routinely employed. Most university and all private labs charge a fee. It is prudent to choose a laboratory that you have confidence in, but it is equally important that samples are properly collected and shipped. Since time is always an essential factor in treating turf diseases, samples must be transported as rapidly as possible. Either direct transport or overnight express shipment is preferable. Most turf pathologists at universities have little or no daily responsibilities in plant disease labs or clinics. Hence, if you wish for a particular specialist to handle your problem, it is wise to contact that individual by telephone or e-mail prior to shipping samples. Otherwise, the specialist may be out of town or have other commitments that preclude a rapid handling of your sample. If shipping by express mail, do not ship on Thursday, Friday, or Saturday, as samples arriving over a weekend, holiday, or late on Friday are not likely to be processed until the following Monday or workday. Samples that cannot be properly handled in a timely manner deteriorate rapidly.

Samples must be collected while the disease is actively injuring or causing turf decline. Samples collected even a few days after disease activity has subsided may yield negative or misleading information. Avoid sending dead sod or samples from areas recently treated with fungicides. Sod pieces the size of a book or cup cutter size plugs are ideal for disease diagnostics.

Two or more plugs should be carefully selected from areas where both early and advanced symptoms are evident. Each plug should contain some dead tissue (if present), and some healthy-appearing plants. The majority (>75%) of the sample, however, should contain various stages of unhealthy, declining, or blighted tissue. Plugs should be taken to a soil depth of about 2 inches (5 cm). Send only soil that is clinging to roots; do not send deep plugs where there is a conspicuous cleavage zone beyond which roots do not penetrate. The only exception would be if black-layers are evident. Plugs should be collected when soil is moist, but well drained and not in a muddy or soggy condition. Samples then should be wrapped in an inner, damp paper towel, followed by an outer wrapping of aluminum foil (Figure 2.107). Some labs prefer foil wrapping of roots and soil only, thereby leaving foliage exposed and unwrapped. Aluminum foil is preferred because it helps ensure that soil and roots remain intact during transport. Do not box plugs that are loose or unbound in paper bags or newspaper. Plugs wrapped in paper or crowded into large plastic bags normally arrive as a loose mass of soil mixed on top and within the turfgrass canopy. Broken plugs with soil mixed into diseased foliar tissue greatly impede the pathologist, and may be the cause for an improper or delayed diagnosis. Before packing foil-wrapped plugs, ensure that field identification information is securely fixed to the outside of the foil. Obviously, if diseased turf samples are being hand delivered, they would not have to be as carefully wrapped. However, samples shipped by express or overnight mail should be foil wrapped and packed with newspaper or appropriate packing material to prevent tumbling and break-up of plugs. Package the samples for transport as if they were glass or some other breakable object.

FIGURE 2.107
Cup cutter plugs with only 1 or 2 inches of soil should be wrapped in a damp paper towel and then in aluminum foil to keep the soil moist and intact during shipment to a diagnostic lab.

In the laboratory, most samples are incubated overnight in a humidity chamber, particularly if a foliar pathogen is suspected. Within about 24 h under high humidity, foliar mycelium and sometime spores may be present to assist the diagnostician. Otherwise leaves, stems, and roots are inspected for signs or symptoms of pathogenic activity. For most diseases, a diagnosis is generally made within 48 h of receiving a fresh, properly collected and shipped sample. If the pathologist must attempt to isolate a possible pathogen on a sterile media, the process may require a waiting period of one or more weeks. All techniques that involve an attempt to isolate a pathogen are tedious and time consuming to perform. During the interim, however, a turf pathologist is likely to suggest cultural and/or chemical approaches to alleviate stress and reduce further turf damage.

Bacterial diseases are uncommon in turf, and are primarily diagnosed by sectioning tissue and searching for heavy streaming of bacterial cells. Bacteria are identified based on colony growth and color characteristics on specialized bacteriological media; staining (i.e., Gram stain) character; ability to grow in the absence of oxygen (anaerobic growth) or on special substrates; by molecular (i.e., DNA testing); and other methods.

Samples should be clearly marked on the outside of packaged turf plugs and should be accompanied with a sample submission form from the lab or a letter that provides the information outlined below. The form or letter should be placed in a separate plastic bag. Information helpful to the diagnostician and some commonly asked questions are as follows:

1. Turf species and site affected
   a. Green, tee, fairway, lawn, athletic field, and so on

    b. Full sun, partial shade, heavy shade

    c. Soil often wet or dry

    d. Name of cultivar and age of turf, if known

2. Symptoms and environmental conditions

    a. Describe symptoms (e.g., circular patches, diffuse thinning, and color changes) and distribution (e.g., widespread, localized, and scattered)

    b. Describe weather conditions just prior to and during the period when injury became evident

    c. Date when symptoms first appeared and date of sample collection

3. Cultural and chemical

    a. List all pesticides applied to the site within 14 days prior to the appearance of symptoms

    b. Note the total amount nitrogen applied in the last 3 months

    c. Irrigation frequency, amounts and time of day

4. Digital photographs

    a. Close-up photos and photos taken from a standing position are particularly useful

    b. E-mail photos to the diagnostic lab immediately

Parasitic Nematode Assay

All too often, turfgrass managers send a shovel-full of soil or a cup cutter plug from a single-site to a lab and request a parasitic nematode assay. This type of sample is unacceptable since the results will likely provide meaningless information. A good sample from a single golf green, lawn, athletic field, and so on will require several minutes of work. It is best to collect soil a day after rain or irrigation when it is moist, but not wet or muddy. About one pint of soil (480 cc) is necessary from each site, rather than a composite sample from two or more greens, tees, lawns, and so on. Where parasitic nematodes are a chronic problem, the soil should be collected in late spring or early summer and again in late summer. Nematode counts from monitored sites will provide baseline information on population densities of parasitic nematodes for each site. This baseline information is particularly useful for interpreting problems with turfs that are poorly rooted and tend to wilt rapidly in summer.

    Samples should be collected in the upper 3 inches (7.5 cm) of soil since this area represents the zone where most turfgrass roots are found. If roots are confined only to a 1–2 inch (2.5–5.0 cm) soil depth, this would be the appropriate zone from which to sample. The best tool for the job is the common "soil probe," which extracts cores that range from 0.75 to 1.0 inch (2.0–2.5 cm) in diameter (Figure 2.98). Soil cores should be taken randomly

TABLE 2.7

Summary of Common Diseases of Major Cool-Season Turfgrass Species in the United States

Bentgrasses
Creeping, Colonial, Velvet

Anthracnose

Bacterial decline

Brown patch

Copper spot (especially velvet)

Dead spot

Dollar spot

Fairy ring

Leaf and sheath spot (transition and southern United States)

Leptosphaerulina blight

Localized dry spot

Microdochium patch

Parasitic nematodes

Pink patch

Pythium blight

Pythium-induced root dysfunction

Pythium patch/snow blight

Rapid blight

Red leaf spot (especially colonial)

Red thread

Typhula blight

Southern blight

Superficial fairy ring

Summer patch (southern United States)

Take-all patch

Yellow patch

Yellow tuft

Bluegrasses
Annual

Anthracnose

Ascochyta leaf blight

Bacterial wilt

Brown patch

Brown ring patch

Dollar spot

Fairy ring

Leaf spot/melting-out
 (*Bipolaris* and *Drechslera* spp.)

Leptosphaerulina blight

Localized dry spot

Microdochium patch

Necrotic ring spot

Parasitic nematodes

Pink patch

Pythium blight

Pythium-induced root dysfunction

Rapid blight

Red thread

Southern blight

Summer patch

Superficial fairy ring

Typhula blight

Yellow patch

Yellow tuft

Kentucky

Ascochyta leaf blight

Brown patch

Dollar spot

Fairy ring

Leaf spot/melting-out
 (*Bipolaris* and *Drechslera* spp.)

Leptosphaerulina blight

Microdochium patch

Powdery mildew

Red thread

Rust

Summer patch

Smut, flag or stripe

Necrotic ring spot

Parasitic nematodes

Pink patch

TABLE 2.7 (continued)

Summary of Common Diseases of Major Cool-Season Turfgrass Species in the United States

Southern blight	Yellow tuft
Typhula blight	
Yellow ring	

Fine-Leaf Fescue

Chewings and Creeping Red

Anthracnose	Pink patch
Ascochyta leaf blight	Red thread
Dollar spot	Summer patch
Fairy ring	Typhula blight
Leaf spot/melting-out	White blight
(*Bipolaris* and *Drechslera* spp.)	
Microdochium patch	
Necrotic ring spot	

Hard and Sheep Fescue

Dollar spot	Pink patch
Fairy ring	Red thread
Leaf spot/melting-out	Summer patch
(*Bipolaris* and *Drechslera* spp.)	Typhula blight
Microdochium patch	
Necrotic ring spot	

Tall Fescue

Ascochyta leaf blight	Pythium blight
Brown patch	Red thread
Leaf and sheath spot (transition and southern United States)	Rust
	Stripe smut
Dollar spot	Typhula blight
Fairy ring	White blight
Gray leaf spot	
Microdochium patch	
Net-blotch	

Perennial Ryegrass

Bipolaris and *Drechslera* spp. diseases

Brown blight	Microdochium patch
Leaf spot/melting-out	Pink patch
Net-blotch	Pythium blight
Brown patch	Rapid blight
Dollar spot	Red thread
Fairy ring	Rust
Gray leaf spot	Stripe smut
Leaf and sheath spot	Southern blight
Leptosphaerulina blight	Typhula blight

TABLE 2.8

Summary of Common Diseases of Major Warm-Season Turfgrass
Species in the United States

Bermudagrass

Bipolaris leaf spot/melting-out	Parasitic nematodes
Dead spot	Pythium blight (subtropical regions)
Dollar spot	Root decline
Fairy ring	Rust
Large patch	Southern blight
	Spring dead spot

Centipedegrass

Centipede mosaic	Large patch
Dollar spot	Parasitic nematodes
Fairy ring	Root decline

Kikuyugrass

Bipolaris leaf spot/melting-out
Fairy ring
Gray leaf spot
Large patch
Root decline

Seashore Paspalum

Dollar spot	Microdochium patch
Fairy ring	Parasitic nematodes
Large patch	Root decline

St. Augustinegrass

Dollar spot	Large patch
Downy mildew	Parasitic nematodes
Fairy ring	Root decline
Gray leaf spot	Rust
	St. Augustine decline

Zoysiagrass

Dollar spot	Parasitic nematodes
Bipolaris leaf spot/melting-out	Root decline
Fairy ring	Rust
Large patch	Spring dead spot

from the interface between damaged or declining turf and healthy turf. The
thatch layer, and plant stems and leaves should be discarded. Only soil and
roots are needed. Samples from dead areas, or where few plants persist, will
provide false population density information. This is because plant parasitic
nematodes are obligate parasites that feed only on living plant tissue. Hence,
parasitic nematodes naturally migrate outwards from severely thinned

TABLE 2.9

Diseases Arranged by Tissues Attacked and Pathogen

Disease	Pathogen(s)
I. Patch Diseases Caused by Root-invading Pathogens	
Necrotic ring spot	*Ophiosphaerella korrae*
Pythium-induced root dysfunction	*Pythium volutum*, possibly others
Spring dead spot	*Ophiosphaerella korrae, O. herpotricha,* or *O. narmari*
Summer patch	*Magnaporthe poae*
Take-all patch	*Gaeumannomyces graminis* var. *avenae*
II. Foliar Diseases	
Ascochyta leaf blight	*Ascochyta* spp.
Brown ring patch	*Waitea circinata* var. *circinata*
Leaf spot	*Drechslera* or *Bipolaris* spp.
Leptosphaerulina blight	*Leptosphaerulina australis*
Pink patch	*Limonomyces roseipellis*
Red thread	*Laetisaria fuciformis*
Leaf and sheath spot	*Rhizoctonia zeae*
Typhula blight	*Typhula incarnata* and *Typhula ishikariensis*
White blight	*Melanotus phillipsii*
Yellow patch/cool	*Rhizoctonia cerealis*
Temperature brown patch	
III. Foliar, Stem, or Root Diseases	
Anthracnose	*Colletotrichum cereale*
Dead spot	*Ophiosphaerella agrostis*
Dollar spot	*Sclerotinia homoeocarpa;* (new Latin binomial forthcoming)
Gray leaf spot	*Pyricularia grisea*
Large patch	*Rhizoctonia solani* (AG 2-2 LP)
Leaf spot and melting-out	*Drechslera* or *Bipolaris* spp.
Microdochium patch	*Microdochium nivale*
Parasitic nematodes	Many species
Pythium-induced root dysfunction	*Pythium aristoporium, P. arrhenomanes, P. graminicola, P. torulosum,* others
Root decline of warm-season grasses	*Gaeumannomyces graminis* var. *graminis*
IV. Diseases Caused by Obligate Parasites	
Flag smut	*Urocystis agropyri*
Powdery mildew	*Blumeria graminis*
Rust	*Puccinia* spp.
Stripe smut	*Ustilago striiformis*
Yellow tuft or downy mildew	*Sclerophthora macrospora*

continued

TABLE 2.9   (continued)

Diseases Arranged by Tissues Attacked and Pathogen

Disease	Pathogen(s)
V. Diseases That Cause Hydrophobic Thatch or Soil, or Thatch Shrinkage	
Fairy ring	Many basidiomycetes
Brown ring patch	*Waitea circinata* var. *circinata*
Localized dry spot	Microbial decomposition of organic matter
Superficial fairy ring	*Coprinus kubikae*, others
Yellow ring	*Trechispora alnicola*
VI. Seedling Disease	
Damping-off	*Pythium* spp., *Fusarium* spp., *Rhizoctonia solani*, *Bipolaris* spp., *Drechslera* spp., *Curvularia* spp., *Pyricularia grisea* others
VII. Others	
Bacterial decline	*Acidiovorax avenae* pathovar *avenae*
Bacterial wilt	*Xanthomonas translucens* pathovar *poannua*
St. Augustine decline	Panicum mosaic virus
Centipede mosaic	Panicum mosaic virus

areas, and populations generally are highest in the region where unhealthy and healthy plants meet. About 15–20 randomly selected soil cores, or one pint (480 cc) of soil, will provide a satisfactory sample size. Soil cores should be uniformly mixed and placed in a plastic bag or specially lined bags provided by the testing laboratory. Avoid using paper lunch bags, as moist soil is likely to cause them to fall apart during transport. It is very important that site information be marked on the outside of the bag with a waterproof marker. Do not place small scraps of paper that identify the collection site inside the bag. These paper scraps deteriorate rapidly, and will probably be unreadable within a few hours of contact with moist soil.

Once collected, soil should be kept out of direct sunlight. Soil placed on the dashboard or in the trunk of a car will heat up rapidly, killing nematodes. If samples must be held for a few days, they should be refrigerated or placed in a cool and dark room. Extremes in temperature, high or low, will kill nematodes. This must be avoided because only living nematodes can be effectively extracted from soil. In the laboratory, various methods for extraction are used. Many labs use a glass funnel technique. Soil is placed in a funnel lined with porous paper. The funnel base is clamped, and soil is flooded with water. Within 24 h most of the living nematodes will swim from soil through the paper and be trapped at the base of the funnel. Water in the base of the funnel stem is drawn-off into a shallow dish. The dish is placed under a dissecting microscope, and the nematodes are identified to genus level and counted. As previously discussed, there are no reliable data correlating nematode number per sample and expected degree of turf injury in the field. The nematologist, however, generally will be able to provide a good management recommendation based not only on the total number of plant parasitic nematodes but also on the species of nematodes that are present.

Fungicides Used to Control Turfgrass Diseases

Promoting vigorous growth through sound cultural practices is the first step in minimizing disease injury. Frequently, however, environmental stresses, poor growing environments, traffic, and inappropriate management practices weaken plants, predisposing them to invasion by fungal pathogens. When disease symptoms appear, it is imperative that a rapid and accurate diagnosis be made. The prudent manager also attempts to determine those environmental and cultural factors that have led to the development or contributed to the intensity of the disease. Management practices that promote diseases include frequent and close mowing, nightly programmed or otherwise excessive irrigation, and applications of inadequate or excessive amounts of nitrogen fertilizer. Poor growing environments include excessive thatch and/or mat layers, shade, poor air or water drainage, and soil compaction. Despite hard work and adherence to sound management practices, diseases may become a serious problem. This normally occurs when environmental conditions favor disease development, but not plant growth and vigor.

Fungicides may be applied preventively (i.e., preplant infection and before disease symptoms appear) or curatively (i.e., when disease symptoms first become evident). Preventive fungicide treatment is recommended for chronically damaging diseases. Curative applications are more economically wise for less severe and/or less rapidly developing diseases such as red thread, leaf spot, rust, and stripe smut. The key to a successful curative fungicide program is vigilant scouting.

Contact fungicides protect only surfaces contacted, but can be used effectively in managing several diseases. Contact fungicides, however, only may provide 7–10 days of control under conditions of high disease pressure. Penetrant fungicides move into tissues and may or may not move from the point of uptake. Since some molecules are protected from environmental forces, they generally provide 14–18 days protection during moderately high-pressure disease periods. In general, once a disease appears, the application of a contact plus a penetrant fungicide is preferred. Tank-mixing a contact plus a penetrant fungicide provides a quicker knockdown, a longer residual effect, and a wider spectrum of diseases controlled. Frequently, a fungicide may only be needed to help the turf better survive a high pressure disease period. Favorable changes in weather conditions such as a shift from hot and humid conditions to an extended cool and dry period, however, often reduce and sometimes eliminate a disease problem.

Professional Fungicide Use Considerations

Lawns and School Sports Fields

Proper utilization and selection of fungicides is too difficult and complicated for amateurs. Because of this, only well-trained employees of lawn care, sports turf, and landscape management companies can provide the most reliable disease services for lawns and community sports fields. Fungicides, however, should not become a part of a normal application schedule. As a general rule, use of fungicides is not recommended in most lawn or sports field situations because (a) proper diagnosis and proper fungicide selection are difficult, (b) generally it is too late to achieve the economic and aesthetic benefits of a fungicide once extensive injury has occurred, (c) management companies capable of only dry or granular applications do not have the proper spray equipment or they cannot obtain the desired fungicide(s) in granular form, and (d) it may be less expensive and better in the long run to overseed or renovate damaged turf areas with disease-resistant species or cultivars.

There are several disease situations of lawn and sports turfs that are best controlled through a preventive fungicide application. Some notable situations include (1) Kentucky bluegrass turfs injured in previous years by summer patch, necrotic ring spot, and perhaps dollar spot; (2) perennial ryegrass turfs injured in previous years by Pythium blight, brown patch, dollar spot, or gray leaf spot; (3) tall fescue turfs chronically damaged by brown patch; and (4) warm-season grass turfs chronically damaged by root decline, large patch, and possibly dollar spot. Many diseases, however, are effectively controlled with curative fungicide applications when disease symptoms first appear. For example, leaf spot, smut, and rust in Kentucky bluegrasses, rust and red thread in perennial ryegrass, or gray leaf spot in St. Augustinegrass can be controlled effectively with one or two curative fungicide applications. Dollar spot disease is extremely common and if allowed to go unchecked, it may cause extensive injury to many turfgrass species. When diagnosed in its early stages, however, dollar spot is also controlled effectively in lawns and sports fields by curative fungicide applications. Given these situations, it becomes obvious that effective fungicide programs hinge upon (a) knowledge of past disease problems in a particular lawn, neighborhood, sports field, and so on; (b) the ability to distinguish between turfgrass species; and (c) the ability to diagnose turfgrass diseases at an early stage in their development. Hence, in addition to the expense and logistical problems associated with applying fungicides, lawn care, sports turf, and landscape management companies must also educate its employees to scout and effectively diagnose diseases. This educational process is best achieved by in-house training programs and attending turf workshops and conferences.

Remember, once a disease has severely reduced stand density, overseeding or renovating with resistant species or cultivars normally is suggested. Fact sheets describing different diseases and a list of cultural practices that help minimize disease injury should be provided by lawn care, sports turf, and landscape management companies. Fact sheets on managing turf diseases are available online at many land-grant university extension websites.

Golf Courses and Stadium Athletic Fields

Where extremely high-quality turf is required, fungicides will be needed in most years in many areas of the United States. The indiscriminate use of fungicides or employment of numerous, preventive applications of fungicides for all potential disease problems is wasteful and discouraged. Fungicides are expensive and complicated compounds. Unlike most herbicides and insecticides, multiple annual applications often are required for a diverse group of potential pathogens. To be economically and environmentally responsible, professional turfgrass managers should familiarize themselves with the strengths and weaknesses, behavior in and on plants, mode of action, and target diseases of fungicides.

Types of Fungicides

Fungicides are somewhat mysterious chemicals because their behavior on or inside of plants, and how they physiologically affect microorganisms, is complicated and imperfectly understood. Fungicides are divided into three mobility types: contact, penetrant, and systemic. Fungicides also are classified by mode of action or chemical composition (Table 2.10). These classification systems or groupings, and some other terms associated with fungicides, are confusing and sometimes misleading. For example, most fungicides are not fungicidal, and penetrants can provide activity on the surface of tissues as well as from within tissues. Most fungicides are fungistatic. That is, most only prevent growth or development of fungi (i.e., fungistatic) and they do not actually kill them. Several contact fungicides can kill fungal spores as they germinate, but even most contacts are fungistatic. Another contradiction centers on the use of the word "systemic." A systemic chemical by definition is capable of moving throughout a plant from leaves to stems to roots and vice versa. In fact, the only truly systemic turf fungicides are fosetyl-aluminum and phosphites. Most other so-called systemics are better referred to as penetrants because they either remain localized inside tissue or primarily move upward in xylem via the transpiration stream. The translocation behavior and mode of action of penetrants are described below and are summarized in Table 2.10.

The largest chemical groups of penetrant fungicides in the turf market are the DMI/SI and QoI/strobilurin fungicides. Most DMI/SI fungicides are chemically classified as triazoles, which include difenoconazole metconazole, myclobutanil, propiconazole, tebuconazole, triadimefon, and triticonazole. When applied to a turfgrass plant, a DMI/SI fungicide will penetrate tissue and move upward (i.e., acropetal penetrant) via the xylem from the

TABLE 2.10

Common Chemical Name, Some Trade Names, Chemical Class, Mobility, and FRAC Codes for Common Turfgrass Fungicides Used in the United States

Common Name	Some Trade Name(s)	Chemical Class	Mobility[a]	FRAC Code[b]
Azoxystrobin	Heritage®	Strobilurin or QoI	AP	11
Boscalid	Emerald®	Carboximide[c]	AP	7
Chloroneb	Terraneb SP®	Aromatic hydrocarbon	C	14
Chlorothalonil	Daconil Ultrex®, Concorde®, Echo®, others	Chloronitrile	C	M3
Cyazofamid	Segway®	Cyanoimidazole	LP	21
Ethazole/etridiazole	Koban®, Terrazole®	Triadiazole	C	14
Fenarimol	Rubigan®	DMI/SI	AP	3
Fluazinam	Secure™	Pyridinamine	C	29
Fludioxonil	Medallion®	Phenylpyrolle	C	12
Fluoxastrobin	Disarm®	Strobilurin or QoI	AP	11
Fosetyl-aluminum	Chipco Signature®, Prodigy®	Phosphonate	XS	33
Flutolanil	ProStar®	Carboximide[c]	AP	7
Iprodione	Chipco 26 GT®, Iprodione Pro®	Dicarboximide	AP	2
Maneb	Pentathalon®	Ethylenebis-dithiocarbamate	C	M3
Mancozeb	Dithane®, Fore Rainshield®, Protect®	Ethylenebis-dithiocarbamate	C	M3
Mefenoxam	Subdue MAXX®	Phenylamide	AP	4
Metalaxyl	Subdue®, Apron® Seed Treatment	Phenylamide	AP	4
Metconazole	Tourney®	DMI/SI	AP	3
Myclobutanil	Eagle®	DMI/SI	AP	3
Phosphite	Alude®, Magellan®, Vital®, others	Phosphonate	S	33
Penthiopyrad	Velista®	Carboximide	AP	7
Polyoxin D	Affirm®, Endorse®	Peptidic nucleoside antibiotic	LP	19
Propamocarb	Banol®	Carbamate	LP	28
Propiconazole	Banner MAXX®	DMI/SI	AP	3
Pyraclostrobin	Insignia®	Strobilurin or QoI	LP	11
Quintozene	PCNB®, Terraclor®	Aromatic hydrocarbon	C	14
Tebuconazole	Torque®	DMI/SI	AP	3
Thiophanate-methyl	3336 Plus®, T-Storm	Benzimidazole	AP	1
Thiram	Spotrete®	Dialkyl dithiocarbamate	C	M3
Triadimefon	Bayleton®	DMI/SI	AP	3

TABLE 2.10 (continued)

Common Chemical Name, Some Trade Names, Chemical Class, Mobility, and FRAC
Codes for Common Turfgrass Fungicides Used in the United States

Common Name	Some Trade Name(s)	Chemical Class	Mobility[a]	FRAC Code[b]
Triticonazole	Trinity®, Triton®	DMI/SI	AP	3
Trifloxystrobin	Compass®	Strobilurin or QoI	LP	11
Vinclozolin	Curalan®, Touche®	Dicarboximide	LP	2

[a] Contact (C) = Contact protectant fungicides are only active on leaf and sheath surfaces.
Acropetal penetrant (AP) = Fungicide is absorbed, moves upward (acropetal) in the xylem and can provide activity both on the outside and on the inside of plant tissues.
Local penetrant (LP) = Fungicide moves into leaves and may move across leaves (translaminar) providing activity both on the outside and on the inside of plant tissues.
Systemic (S) = True systemic fungicides are absorbed and translocated throughout the plant via phloem and xylem.

[b] Fungicide Resistance Action Committee (FRAC), 2012.

[c] Chemical class also known as succinate dehydrogenase inhibitor.

point of tissue contact. The DMI/SIs are also capable of lateral diffusion from the upper to lower leaf surface (i.e., translaminar) and vice versa. Chemical that contacts basal leaf sheaths, or runs down between leaf sheaths, may be transported into axillary buds and possibly stems.

Strobilurin or QoI fungicides were initially derived from a mushroom fungus (*Strobilurus tenacellus*) and stabilized to prevent rapid deterioration in the environment. Members of this group include azoxystrobin, fluoxastrobin, pyraclostrobin, and trifloxystrobin. Azoxystrobin and fluoxastrobin are acropetal penetrants that move across leaves from upper to lower leaf surfaces and vice versa and also move upwards in the xylem from the point of contact. Pyraclostrobin and trifloxystrobin are localized penetrants. That is, when contact is made with tissue, the molecules move into leaves and may move from upper to lower leaf surface, but otherwise are not translocated in the xylem.

Other acropetal penetrants include boscalid, flutolanil, penthiopyrad, thiophanate-methyl, propamocarb, and mefenoxam. Other localized penetrants include cyazofamid (weak uptake, mostly contact activity), iprodione, polyoxin D, and vinclozolin. Iprodione and vinclozolin are dicarboximides and boscalid and penthioprad are carboximides (aka succinate dehydrogenese inhibitors). Polyoxin D is an antibiotic produced by the bacterium *Streptomyces cacaoi* var. *asoensis*.

Contact fungicides provide activity on the outside surfaces of plants and protect only those tissues they contact. Because contact fungicides are subjected to removal by mowing and the forces of the environment (i.e., wash-off, UV light degradation, microbial breakdown, etc.), they tend to provide a relatively short period of protection. The contact fungicides used on turfgrasses include: chloroneb, chlorothalonil, ethridiazole or ethazol fluazinam, fludioxonil, maneb, mancozeb, PCNB (or quintozene), and thiram. Contact fungicides are extremely important in disease management programs because they generally are free of resistance problems. Contacts also tend to have more rapid knockdown activity on foliar pathogens than penetrants and, therefore, are preferred in a tank-mix combination with a penetrant once disease symptoms have appeared. Contact fungicides normally interfere with several, mostly unknown biochemical or physiological processes in susceptible fungi.

The DMI/SI and QoI/strobilurin fungicides as well as boscalid, flutolanil, mefenoxam, penthiopyrad, propamocarb, and thiophanate-methyl are single-site specific. That is, they only disrupt one specific biochemical or physiological process in a susceptible fungus. The single process disrupted is often controlled by a single gene and therefore the probability of a resistant biotype developing is increased. For example, DMI/SI fungicides interrupt the production of ergosterol in sensitive fungi by blocking a single demethylation reaction. In the absence of this sterol, membrane form and function are impaired and the fungus cannot grow. The QoI/strobilurin fungicides provide disease control by interfering with respiration, which depletes a key energy compound (i.e., ATP) needed by living cells to grow and function. Thiophanate-methyl prevents spindle fiber development during mitosis, so cell division does not occur. Mefenoxam blocks RNA synthesis; boscalid, flutolanil, and penthiopyrad inhibit a respiratory enzyme; and propamocarb interferes with cell membrane function. Because these fungicides interrupt only one of thousands of biochemical reactions occurring in fungi, the probability of the existence of a small subpopulation of tolerant biotypes in the ecosystem is likely. Continuous use of fungicides with the same mode of action can selectively remove susceptible biotypes, which allows resistant biotypes to proliferate and eventually dominate in a turf.

Nontarget Effects of Fungicides

Other than cost, some reasons why repeated fungicide applications may be undesirable include (a) development of fungicide-resistant pathogens, which is most likely to occur with those fungi causing dollar spot, gray leaf spot, anthracnose, and Pythium blight; (b) continuous fungicide usage may lead to a buildup of microorganisms that degrade the active ingredient, resulting in reduced residual control; (c) disease resurgence, a phenomenon in which a disease recurs more rapidly and causes more injury in turfs previously treated with fungicides, when compared to nonfungicide-treated sites; (d) a fungicide may control one disease, but encourage another disease; (e) phytotoxicity or objectionable plant growth regulator effects; and (f) encouragement of blue-green algae on golf greens.

Fungicide Resistance, Reduced Sensitivity, and Residual Effectiveness

The development of fungal biotypes resistant to fungicides is well documented. Resistant biotypes of the dollar spot fungus (i.e., *S. homoeocarpa*) first developed as a result of repeated usage of cadmium and benzimid-azole (e.g., benomyl and thiophanate) fungicides on golf courses. Biotypes of the dollar spot fungus resistant to iprodione, and DMI/SI fungicides (e.g., propiconazole, triadimefon, others) have also been reported. *Pythium aphanidermatum* biotypes resistant to metalaxyl/mefenoxam are well documented. The anthracnose (i.e., *C. cereale*) pathogen has developed resistance to thiophanate-methyl and azoxystrobin. Gray leaf spot (*P. grisea*) resistance to azoxystrobin is also common. It is important to note that when resistance develops to a fungicide, all fungicides in the same chemical class theoretically will exhibit cross resistance (see Table 2.10). For example, *S. homoeocarpa* biotypes resistant to one DMI/SI fungicide will also be resistant to all other DMI/SI fungicides. After many years of use (usually ≥8 years), the dollar spot pathogen loses its sensitivity (i.e., less residual control) to DMI/SI fungicides, which may be a prelude to resistance.

The buildup of resistant biotypes of fungi occurs in response to a selection process that eventually enables a small but naturally occurring subpopulation of resistant biotypes to dominate in the turfgrass microenvironment. Resistance problems can be delayed or averted by rotating fungicides with different modes of action, by tank-mixing fungicides of difference modes of action, or by tank-mixing known synergists. Synergistic combinations are those where two or more fungicides with different modes of action are tank-mixed together at low rates. A synergistic tank-mix provides a level of control equivalent to or better than the normal use rate of either component applied alone. There are, however, few well-documented studies demonstrating synergism among tank mixtures. The only evidence that supports synergism in turfgrass pathology was demonstrated by mixing one-half the low label rate of two *Pythium*-targeted fungicide (fosetyl-aluminum, propamocarb, or mefenoxam). Such a mixture would be expected to provide a level of Pythium blight control equivalent to or better than either component applied alone at the full rate.

A simple rule is to avoid the use of fungicides with confirmed resistance, as well as their biochemical mode of action relatives, at times when high-risk diseases are active. The Fungicide Resistance Action Committee (FRAC) is an international industry-based organization. A code system was developed by FRAC to assist end users in choosing compounds having different chemical modes of action. Each fungicide is assigned a number and the user simply chooses fungicides that have different FRAC codes when tank-mixing or rotating fungicides (Table 2.10).

Perhaps a more common negative phenomenon associated with frequently used fungicides, which may be confused with resistance, is reduced residual effectiveness. For example, when metalaxyl (now mefenoxam) was first introduced in the early 1980s, it was common for it to provide 18 or more days of residual Pythium blight control. Today, on numerous golf courses where metalaxyl/mefenoxam has been used for many years, the fungicide may provide only 5 or 10 days of Pythium blight control. Microorganisms are largely responsible for breaking down pesticides in the environment.

Some microbes can rapidly build up in response to the continuous use of certain fungicides from the same chemical class. The microbes use the active ingredient as an energy source. As a result of the fungicide being more rapidly degraded, residual effectiveness becomes less and less over time. Loss of residual effectiveness can be an indicator that resistant biotypes are building in the turf. In many cases, however, loss of residual effectiveness is likely due to a buildup of high populations of microbes that use the fungicide as an energy source. The improper application of fungicides, use of nozzles that do not atomize droplets (e.g., floodjet), rainfall before chemical has time to dry on leaves, and mowing within 24 h of spraying also contribute to reduced residual effectiveness.

Other Nontarget Effects of Fungicides

Some diseases may recur more rapidly and severely in turfs previously treated with fungicides, when compared to adjacent untreated areas (e.g., treated fairways vs. untreated roughs). Dollar spot, brown patch, and gray leaf spot are probably the most common diseases to exhibit this phenomenon. The mechanism of disease resurgence is unknown but has been attributed to a fungicide reducing populations of beneficial microorganisms, which naturally antagonize and keep disease-causing fungi in abeyance. It is also conceivable that nonfungicide-treated turf that was blighted, yet able to recover due to a shift in environmental conditions, is better prepared to resist future infections due to natural defense systems in plants having been activated by the initial attack.

Fungicides applied to control one disease may encourage other diseases. Thiophanate-methyl may increase rust (*Puccinia* spp.) in perennial ryegrass and Bipolaris/Drechslera leaf spot; iprodione can increase yellow tuft; some formulations of azoxystrobin and flutolanil may enhance dollar spot; and chlorothalonil can increase summer patch and stripe smut in Kentucky bluegrass. Azoxystrobin has been shown to intensify Pythium blight in tall fescue. The encouragement of disease in these situations is again attributed to offsetting the balance between antagonistic and pathogenic microorganisms in the ecosystem. It is important to note that using a selected fungicide will not invariably result in an increase in a nontarget disease. These problems are sporadic and enhancement of nontarget diseases cannot occur unless environmental conditions are conducive for the disease to occur naturally. Hence, when dollar spot is active, fungicides like azoxystrobin and flutolanil should be avoided or if they are needed, they should be tank-mixed with a fungicide that targets dollar spot.

The phytotoxicity that accompanies the usage of some fungicides generally is not severe. Most phytotoxicity problems occur when fungicides are applied to annual bluegrass and creeping bentgrass golf greens during periods of high-temperature stress. Fungicides formulated as emulsifiable concentrates are most likely to cause a foliar burn when applied during periods of high temperature and humidity. Ethazole/ethridazole and chloroneb can burn golf greens when applied on warm and humid mornings. A missapplication of excessive rates of ethazole/ethridazole or chloroneb can be very injurious to golf greens. Some formulations of triticonazole injure annual bluegrass. Chlorothalonil can severely injure some cultivars of creeping red fescue and Chewings fescue when mowed low (<2.0 inches; 5.0 cm) in summer. Repeated applications of DMI/SI fungicides (e.g., propiconazole, triadimefon, others) elicit a blue-green color in creeping bentgrass and a yellow color in annual bluegrass and suppress the foliar growth of most turfgrass species. High rates and frequent applications of DMI/SI fungicides can be particularly injurious to annual bluegrass, creeping bentgrass, and bermudagrass golf greens. Applying DMI/SI fungicides in conjunction with some plant growth regulators (e.g., paclobutrazol and flurprimidol) may cause objectionable levels of discoloration or injury, particularly in turf grown on golf greens.

The DMI/SI fungicides may promote the growth of blue-green algae (aka cyanobacteria) on golf greens. The mechanism for this phenomenon is unknown. Azoxystrobin use on greens has also been linked to enhanced algal growth. Conversely, chlorothalonil, fosetyl-aluminum, and mancozeb have been shown to suppress blue-green algae growth on golf greens. Fungicides also frequently improve turf color and quality in the absence of disease and mancozeb has been shown to mitigate some types of mechanical injury on golf greens.

It should be noted that the harmful side effects just described are often isolated events or occur only after repeated use of one chemical class of a fungicide over the course of several years. As a general rule, nontarget effects are sporadic and they do not invariably occur in most situations. Experienced turfgrass managers have long recognized that tank-mixing or rotating fungicides with different modes of action greatly minimize these potential problems.

Fungicide Application

Most fungicides are diluted in water and sprayed onto turfgrasses. Nearly all efficacy research with fungicides involves sprayable formulations. Little effort has been devoted to comparing sprayable formulations with granular forms. In general, granular forms of fungicides are more expensive and provide a shorter period of residual control than their sprayable counterpart. Granular fungicides that penetrate plant tissue provide somewhat effective control of foliar blighting pathogens, but generally have much reduced activity against root pathogens. Granules can move in surface water if a heavy rain occurs soon after application. This may leave turf in surface water drainage patterns unprotected. Granular fungicides, however, have an important place in disease management programs. They can be used rapidly without the logistical problems associated with mixing and spraying. They are particularly useful in small units where diseases are localized and spraying is impractical. For example, if only a portion of one or two tees is showing a disease symptom on a Sunday morning, it is more prudent to quickly spot-treat with a granular fungicide rather than to prepare a tank for broadcast spraying.

Aside from improper sprayer calibration, perhaps the single greatest error in using fungicides is applying them in insufficient amounts of water to provide good plant coverage. Sprayable fungicides perform well when applied in 1.0 or 2.0 gallons of water per 1000 ft² (44–88 gal/A; 407–814 L/ha). Fungicides targeting foliar diseases provide effective control when applied in 1.0 gallon of water per 1000 ft² (44 gal/A; 407 L/ha). A higher water dilution of 2.0 or 3.0 gallons per 1000 ft² (88–130 gal/A; 814–1221 L/ha), however, may be recommended for targeting root diseases.

For most diseases, fungicides must be allowed to dry on leaves prior to rainfall or irrigation to provide maximum effectiveness. Contact fungicides can lose most of their effectiveness if a rain storm occurs prior to the fungicide drying on leaves. Even fungicides that penetrate tissues exhibit reduced effectiveness if rain or irrigation occurs before the chemical completely dries on leaves. There are a few exceptions to this no post-application irrigation principle, and they largely apply to fungicides used to target root diseases. For example, thiophanate-methyl provides better summer patch control if watered-in before it has time to dry on leaf surfaces. With the exception of fosetyl-aluminum and phosphites, fungicides that target Pythium-incited root dysfunction should be lightly watered-in, but only to a soil depth of 0.5–1.0 inches (1.5–2.5 cm). Some fungicides targeting fairy ring perform better when lightly watered-in, whereas others may need to be drenched into coring holes.

Fungicides should be sprayed through nozzles (110° preferred) that atomize droplets. Flat-fan nozzles are generally more efficient than nozzles that deliver a large droplet, such as flood jet and hollow cone nozzles. Air-induction nozzles are preferred if wind and drift are a concern. Low pressure produces larger droplets and can be another cause of reduced effectiveness. Pressure in the spray boom at delivery should be in the range of 30–60 psi (207–414 kPa). In short, it is important to use enough water and pressure to move fungicide(s) into the turf canopy so that the chemical(s) can wash down between leaf sheaths and contact stem bases.

Sprayers need to be accurately calibrated prior to mixing fungicides. Calibration should be rechecked often. Screens and nozzles should be visually checked prior to each spray to ensure uniform fungicide delivery. Turn on the agitation system before adding fungicides, and allow it to run continuously. Poor agitation may result in settling-out of insoluble formulations, which will result in less effective control. Spray tanks should be filled half way with water before adding any fungicide(s). When tank-mixing products, place water-insoluble materials (i.e., those formulated wettable powders, emulsifiable concentrates, dry dispersible granules, or flowables) into the tank first. Soluble materials such as liquids or soluble powders are added to the tank after insolubles. Do not tank-mix more than one emulsifiable concentrate as turf burning may occur, particularly when treating golf greens. In general, fungicides should not be tank-mixed with insecticides or herbicides unless otherwise stated on labels. For example, insecticides targeted for white grubs and some preemergence herbicides targeted for annual grass weeds should be watered-in immediately and depending on the disease, this practice could reduce fungicide performance. Whenever in doubt, apply materials separately rather than in tank-mix combination. Thoroughly clean the spray tank, hose, boom, and nozzles after each use. All too often, disasters have occurred when a fungicide was applied through an improperly cleaned sprayer that was previously used for a nonselective herbicide application.

Little information exists regarding chemical interactions of tank-mixes. Most well-known chemical incompatibilities are noted on pesticide labels. There are two general types of incompatibilities: chemical and physical. Chemical incompatibilities generally occur when the pH of the final solution or the presence of one of the compounds reduces the efficacy or increases the phytotoxicity of a pesticide. Some examples of chemical incompatibilities are as follows: mixing lime, alkaline reacting fertilizer or water having a high pH with thiophanate-methyl or an ethylenebis-dithiocarbamate (e.g., mancozeb, maneb, and thiram) fungicide can reduce their effectiveness; tank-mixing nonchelated iron sulfate with an emulsifiable concentrate may cause phytotoxicity; and tank-mixing a DMI/SI fungicide with some plant growth regulators (especially paclobutrazol and flurprimidol) may discolor or damage annual bluegrass, creeping bentgrass, and bermudagrass grown on golf greens. Fosetyl-aluminum or acid reacting fertilizers (especially phosphorous acid) can dramatically drop pH of the mixture. Hence, fosetyl-aluminum and phosphite fungicides may not be compatible with some fertilizers or copper-based pesticides. The pH of the final tank-mixture should be between 6.5 and 7.0. Additives are available for adjusting the pH of spray solutions. A pH meter should be used by managers who spray pesticides more than a few times per year. These meters require frequent calibration and stock buffer solutions should be purchased for the purpose of recalibration.

Physical incompatibility is normally associated with excessive foaming or settling-out of particles. Mixing with some wettable powder fungicides may cause the formation of a precipitate (i.e., solid particles that separate out of the suspension or solution to form a solid material at the bottom of the tank). Mixing flowable formulations of chlorothalonil or mancozeb with fosetyl-aluminum can cause a precipitate to form. Physical incompatibility can indicate that there is an equipment problem. For example, wettable powders mixed without sufficient agitation or without a sufficient amount of water will clog screens. Prewetting and creating slurry is helpful in getting wettable powders into suspension, especially when spraying with a small quantity of water. It is important to always keep the agitation system running, even during breaks or when in transit.

Only enough material that can be sprayed in one day should be prepared. Chemicals will interact in the tank and if enough time elapses, the effectiveness of pesticides may diminish. Temperature also influences pesticide effectiveness. As the temperature in the tank is increased, the reaction rate of chemicals will increase and the likelihood of reduced efficacy is enhanced.

As previously noted, many incompatible combinations are listed on pesticide labels. Frequently, however, compatibility questions arise especially when dealing with new formulations of pesticides or when unusual combinations are being considered. It, therefore, becomes necessary to test the compatibility of a mix yourself. This is best achieved through a simple two-step test. Step 1 involves placing a mixture of the precise dosage of pesticides plus the appropriate amount of water (i.e., proportional to normal water dilution) in a quart jar for 30 min. If the separation of chemicals occurs or if materials settle out or form scums or flakes, it is unwise to use the mixture. Also, if the jar begins to feel warm, chemical reactions are occurring and the mix should be considered incompatible. Step 2 should be performed regardless of problem-free results acquired in Step 1. In Step 2, the mixture is applied in a test strip to turf. Preferably, the mixture should be applied during adverse environmental conditions such as hot and humid weather and intentionally overlapped to ensure that phytotoxicity does not occur with double concentration of chemicals. A minimum of 72 h should elapse before the response can be properly evaluated.

Acknowledgments

I am grateful to Bruce Martin, Clemson University and Larry Stowell, PACE Turf, LLC for reviewing and editing sections dealing with diseases of warm-season grasses. I thank Michael Fidanza, Penn State University; John Kaminski, Penn State University; Steven McDonald, Turfgrass Disease Solutions, LLC; Nathaniel Mitkowski, University of Rhode Island; and Derek Settle, Chicago District Golf Association for their valuable insights and contributions.

Bibliography

Anderson, M., A. Guenzi, D. Martin, C. Taliaferro, and N. Tisserat. 2002. Spring dead spot: A major bermudagrass disease. *USGA Green Section Record* 40(1):21–23.
Brent, K.J. and D.W. Hollomon. 2007. *Fungicide Resistance: The Assessment of Risk.* FRAC Monograph No. 2, revised. CropLife International, Brussels.
Burpee, L. and R. Latin. 2008. Reassessment of fungicide synergism for control of dollar spot. *Plant Dis.* 92:601–606.

Cook, P.J., P.J. Landschoot, and M.J. Schlossberg. 2009. Inhibition of *Pythium* spp. and suppression of Pythium blight of turfgrasses with phosphonate fungicides. *Plant Dis.* 93:809–814.

Couch, H.B. 1995. *Diseases of Turfgrasses.* Krieger Publishing Co. Malabar, FL.

Cox, C.L., B. McCarty, and S.B. Martin. 2007. Suppressing sting nematodes using botanical extracts. 2007. *Golf Course Manage.* 75(9):94–97.

Davis, D.B. and P.H. Dernoeden. 1991. Severity of summer patch and turf quality as influenced by mowing height, irrigation and nitrogen source. *Agron. J.* 83:670–677.

Davis, J.G. and P.H. Dernoeden. 2001. Fermentation and delivery of *Pseudomonas aureofaciens* strain Tx-1 to bentgrass affected by dollar spot and brown patch. *Int. Turfgrass Soc. Res. J.* 9:655–664.

Davis, J.G. and P.H. Dernoeden. 2002. Dollar spot severity, tissue nitrogen and soil microbial activity in bentgrass as influenced by nitrogen source. *Crop Sci.* 42:480–488.

Dernoeden, P.H. 2013. *Creeping Bentgrass Management.* 2nd Edition. CRC Press, Boca Raton, FL.

Dernoeden, P.H., J.N. Crahay, and D.B. Davis. 1991. Spring dead spot and Tufcote bermudagrass quality as influenced by nitrogen source and potassium. *Crop Sci.* 31:1674–1680.

Dernoeden, P.H. and J. Fu. 2008. Fungicides can mitigate injury and improve creeping bentgrass quality. *Golf Course Manage.* 76(4):102–106.

Ervin, E.H., D.S. McCall, and B.J. Horvath. 2009. Efficacy of phosphate fungicides and fertilizers for control of Pythium blight on a perennial ryegrass fairway in Virginia. Online. *Appl. Turfgrass Sci.,* doi: 10:1094/ATS-2009-1019-01-BR.

Fidanza, M.A. 1999. Conquering fairy ring disease with new tools. *Golf Course Manage.* 67(3):68–71.

Fidanza, M.A., J.L. Cisar, S.J. Kosta, J.S. Gregos, M.J. Schlossberg, and M. Franklin. 2007. Preliminary investigations of soil chemical and physical properties associated with type-I fairy ring symptoms in turfgrass. *J. Hydrol. Process.* 21:2285–2290.

Fidanza, M.A. and P.H. Dernoeden. 1997. A review of brown patch forecasting, pathogen detection, and management strategies for turfgrasses. *Int. Turfgrass Soc. Res. J.* 8:863–874.

Fidanza, M.A., J.E. Kaminski, M.L. Agnew, and D. Shepard. 2009. Evaluation of water droplet size and water-carrier volume on fungicide performance for anthracnose control on annual bluegrass. *Int. Turfgrass Soc. Res. J.* 11:195–205.

Fungicide Resistance Action Committee. 2012. *FRAC Code List: Fungicides Sorted by Mode of Action (Including FRAC Code Numbering).* Available at http:frac.info/frac/index.htm. CropLife International, Brussels.

Giordano, P.R., A.M. Chaves, N.A. Mitkowski, and J.M. Vargas, Jr. 2012. Identification, characterization, and distribution of *Acidovorax avenae subsp. avenae* with creeping bentgrass etiolation and decline. *Plant Dis.* 96: 1736–1742.

Heckman, J.R., B.B. Clarke, and J.A. Murphy. 2003. Optimizing manganese fertilization for the suppression of take-all patch disease on creeping bentgrass. *Crop Sci.* 43:1395–1398.

Hill, W.J., J.R. Heckman, B.B. Clarke, and J.A. Murphy. 2001. Influence of liming and nitrogen on the severity of summer patch of Kentucky bluegrass. *Int. Turfgrass Soc. Res. J.* 9:388–393.

Hodges, C.F. and L.W. Coleman. 1985. *Pythium*-induced root dysfunction of secondary roots of *Agrostis palustris*. *Plant Dis.* 69:336–340.

Hsiang, T. and S. Cook. 2001. Effect of *Typhula phacorrhiza* on winter injury in field trials across Canada. *Int. Turfgrass Soc. Res. J.* 9:669–673.

Hsiang, T., A. Liao, and D. Benedetto. 2007. Sensitivity of *Sclerotinia homoeocarpa* to demethylation-inhibiting fungicides in Ontario, Canada, after a decade of use. *Plant Pathol.* 56:500–507.

Inguagiato, J.C., J.A. Murphy, and B.B. Clarke. 2008. Anthracnose severity on annual bluegrass as influenced by nitrogen fertilization, growth regulators, and vertical cutting. *Crop Sci.* 48:1595–1607.

Inguagiato, J.C., J.A. Murphy, and B.B. Clarke. 2009. Anthracnose disease and annual bluegrass putting green performance affected by mowing practices and light-weight rolling. *Crop Sci.* 49:1454–1462.

Jordon, K.S. and N.A. Mitkowski. 2006. Population dynamics of plant-parasitic nematodes in golf course greens turf in Southern New England. *Plant Dis.* 90:501–505.

Jo, Y.-K, A.L. Niver, J.W. Rimelspach, and M.J. Boehm. 2006. Fungicide sensitivity of *Sclerotinia homoeocarpa* from golf courses in Ohio. *Plant Dis.* 90:807–813.

Jo, Y.-K., S.W. Chang, M. Boehm, and G. Jung. 2008. Rapid development of fungicide resistance by *Sclerotinia homoeocarpa* on turfgrass. *Phytopathology* 98:1297–1304.

Jung, G. and Y. Jo. 2008. New challenge to an old foe, dollar spot fungicide resistance. *Golf Course Manage.* 76(2):117–121.

Kaminski, J.E. and P.H. Dernoeden. 2002. Geographic distribution, cultivar susceptibility and field observations on bentgrass dead spot. *Plant Dis.* 86:1253–1259.

Kaminski, J.E. and P.H. Dernoeden. 2005. Dead spot of creeping bentgrass and hybrid bermudagrass. Online. *Appl. Turfgrass Sci.*, doi: 10.1094/ATS-2005-0419-01-DG.

Kaminski, J.E. and P.H. Dernoeden. 2005. Nitrogen source impact on dead spot (*Ophiosphaerella agrostis*) recovery in creeping bentgrass. *Int. Turfgrass Soc. Res. J.* 10:214–223.

Kerns, J.P., M.D. Soika, and L.P. Tredway. 2009. Preventive control of Pythium root dysfunction in creeping bentgrass putting greens and sensitivity of *Pythium volutum* to fungicides. *Plant Dis.* 93:1275–1280.

Kerns, J.P. and L.P. Tredway. 2008. Pathogenicity of *Pythium* species associated with Pythium root dysfunction of creeping bentgrass and their impact on root growth and survival. *Plant Dis.* 92:862–869.

Kerns, J.P. and L.P. Tredway. 2010. Pythium root dysfunction of creeping bentgrass. Online. *Plant Health Prog.*, doi: 10.1094/PHP-2010-0125-01-DG.

Kim, Y., E.W. Dixon, P. Vincelli, and M.L. Farman. 2003. Field resistance to strobilurin (QoI) fungicides in *Pyricularia grisea* caused by mutations in the mitochondrial cytochrome b gene. *Plant Dis.* 93:891–900.

Koch, P.L., C.R. Grau, Y. Jo, and G. Jung. 2009. Thiophanate-methyl and propiconazole sensitivity of *Sclerotinia homoeocarpa* isolates collected from golf course putting greens, fairways, and roughs in Wisconsin and Massachusetts. *Plant Dis.* 93:100–105.

Latin, R. 2011. *A Practical Guide to Turfgrass Fungicides*. The American Phytopathological Society of America. St. Paul, MN.

Martin, D.L., G.E. Bell, J.H. Baird, C.M. Taliaferro, N.A. Tisserat, R.M. Kuzmic, D.D. Dobson, and J.A. Anderson. 2001. Spring dead spot resistance and quality of seeded bermudagrasses under different mowing heights. *Crop Sci.* 41:451–456.

McDonald, S.J., P.H. Dernoeden, and C.A. Bigelow. 2006. Dollar spot and gray leaf spot severity as influenced by irrigation, paclobutrazol, chlorothalonil and a wetting agent. *Crop Sci.* 46:2675–2684.

McDonald, S.J., P.H. Dernoeden, J. Kaminski, and M. Agnew. 2009. Curative control of yellow tuft in creeping bentgrass. *Golf Course Manage.* 77(5):106–111.

Melvin, B.P. and J.M. Vargas, Jr. 1994. Irrigation frequency and fertilizer type influence necrotic ring spot in Kentucky bluegrass. *HortScience* 29:1028–1030.

Miller, G.L., L.F. Grand, and L.P. Tredway. 2011. Identification and distribution of fungi associated with fairy rings on golf putting greens. *Plant Dis.* 95:1131–1138.

Mocioni, M., M. Gennari, and M.L. Gullino. 2001. Reduced sensitivity of *Sclerotinia homoeocarpa* to fungicides on some Italian golf courses. *Int. Turfgrass Soc. Res. J.* 9:701–704.

Monteith J. Jr. and A.S. Dahl. 1932. Turf diseases and their control. *Bull. U.S. Golf Assoc. Green Section* 12(4):85–188.

Nelson, E.B. 1997. Microbial inoculants for the control of turfgrass diseases. *Int. Turfgrass Res. Soc. J.* 8:791–811.

Nikolai, T.A. 2002. More light on lightweight rolling. *U.S. Golf Assoc. Green Section Record* 40:9–12.

Pigati, R.L., P.H. Dernoeden, and A.P. Grybauskas. 2010. Early curative dollar spot control in creeping bentgrass as influenced by fungicide spray volume and application timing. Online. *Appl. Turfgrass Sci.*, doi: 10.1094/ATS-2010-0312-03-RS.

Pigati, R.L., P.H. Dernoeden, A.P. Grybauskas, and B. Momen. 2010. Simulated rain and mowing impact fungicide performance when targeting dollar spot in creeping bentgrass. *Plant Dis.* 94:596–603.

Putman, A.I. and J.E. Kaminski. 2011. Mowing frequency and plant growth regulator effects on dollar spot severity and duration of dollar spot control by fungicides. *Plant Dis.* 95:1433–1442.

Roberts, J.A., J.C. Inguagiato, B.B. Clarke, and J.A. Murphy. 2011. Irrigation quantity effects on anthracnose disease in annual bluegrass. *Crop Sci.* 51:1244–1252.

Ryan, C.P., P.H. Dernoeden, and A.P. Grybauskas. 2012. Seasonal development of dollar spot epidemics in six creeping bentgrass cultivars in Maryland. *HortScience* 47:422–426.

Sanders, P.L., W.J. Houser, P.J. Parish, and H. Cole, Jr. 1985. Reduced-rate fungicide mixtures to delay fungicide resistance and to control selected turfgrass diseases. *Plant Dis.* 69:939–943.

Settle, D.M., J.D. Fry, G.A. Milliken, N.A. Tisserat, and T.C. Todd. 2007. Quantifying the effects of lance nematode parasitism in creeping bentgrass. *Plant Dis.* 91:1170–1179.

Settle, D.M., J.D. Fry, and N.A. Tisserat. 2001. Development of brown patch and Pythium blight in tall fescue as affected by irrigation frequencies, clipping removal and fungicide application. *Plant Dis.* 85:543–546.

Settle, D.M., J.D. Fry, T.C. Todd, and N.A. Tisserat. 2006. Population dynamics of the lance nematode (*Hoplolaimus galeatus*) in creeping bentgrass. *Plant Dis.* 90:44–50.

Shantz, H.L. and R.L. Piemeisel. 1917. Fungus fairy rings in eastern Colorado and their effects on vegetation. *J. Agric. Res.* 11:191–245.

Shepard, D., M. Agnew, M. Fidanza, J. Kaminski, and L. Dant. 2006. Selecting nozzles for fungicide spray applications. *Golf Course Manage.* 74(6): 83–88.

Smiley, R.W., P.H. Dernoeden, and B.B. Clarke. 2005. *Compendium of Turfgrass Diseases.* 3rd Edition. APS Press, St. Paul, MN.

Smith, J.D., N. Jackson, and A.R. Woolhouse. 1989. *Fungal Diseases of Amenity Turf Grasses.* E. & F.N. Spon, London.

Thompson, C., M. Kennelly, and J. Fry. 2011. Effect of nitrogen source on silvery-thread moss on a creeping bentgrass putting green. *Appl. Turfgrass Sci.*, doi: 10.1094/ATS-2011-1018-02-RS.

Toda, T., T. Mushika, T. Hayakawa, A. Tanaka, T. Tani, and M. Hyakumachi. 2005. Brown ring patch: A new disease on bentgrass caused by *Waitea circinata* var. *circinata*. *Plant Dis.* 89:536–542.

Tredway, L.P. 2005. First report of summer patch of creeping bentgrass caused by *Magnaporthe poae* in North Carolina. *Plant Dis.* 89:204.

Tredway, L.P. and E.L. Butler. 2007. First report of spring dead spot of zoysiagrass caused by *Ophiospharella korrae* in the United States. *Plant Dis.* 91:1684.

Tredway, L.P., E.L. Butler, M.D. Soika, and M.L. Bunting. 2008. Etiology and management of spring dead spot of hybrid bermudagrass in North Carolina USA. *Acta Hortic.* 783:535–546.

Tredway, L.P., M.D. Soika, and B.B. Clarke. 2001. Red thread development in perennial ryegrass in response to nitrogen, phosphorus, and potassium fertilizer applications. *Int. Turfgrass Soc. Res. J.* 9:715–722.

Todd, T.C. and N. Tisserat. 1993. Understanding nematodes and their impact. *Golf Course Manage.* 61(5):38, 42, 44, 48, 52.

Vargas, J.M., Jr. 2005. *Management of Turfgrass Diseases*. John Wiley & Sons, Inc., Hoboken, NJ.

Vincelli, P. and E. Dixon. 2002. Resistance to QoI (strobilurin-like) fungicides in isolates of *Pyricularia grisea* from perennial ryegrass. *Plant Dis.* 86:235–240.

Vincelli, P. and D.W. Williams. 2011. *Chemical Control of Turfgrass Diseases 2011.* Cooperative Extension Service, University of Kentucky, Lexington, KY.

Walker, N.R. 2009. Influence of fungicide application timing on the management of bermudagrass spring dead spot caused by *Ophiosphaerella herpotricha*. *Plant Dis.* 93:1341–1345.

Westerdahl, B.B., M.A. Harivandi, and L.R. Costello. 2005. Biology and management of the nematodes of turfgrass in northern California. *USGA Green Section Record* 43(5):7–10.

Williams, D.W., P.B. Burrus, and P. Vincelli. 2001. Severity of gray leaf spot in perennial ryegrass as influenced by mowing height and nitrogen level. *Crop Sci.* 41:1207–1211.

Williams, D.W., A.J. Powell, Jr., P. Vincelli, and C.T. Dougherty. 1996. Dollar spot on bentgrass influenced by displacement of leaf surface moisture, nitrogen, and clipping removal. *Crop Sci.* 36:1304–1309.

Wong, F.P. and S.L. Midland. 2007. Sensitivity distributions of California populations of *Colletotrichum cereale* to the DMI fungicides propiconazole, myclobutanil, tebuconazole, and triadimefon. *Plant Dis.* 91:1547–1555.

Wong, F.P., S.L. Midland, and K.A. de la Cerda. 2007. Occurrence and distribution of QoI-resistant isolates of *Colletotrichum cereale* from annual bluegrass in California. *Plant Dis.* 91:1536–1546.

Yuen, G.Y. and O. Kilic. 2001. Evidence of induced resistance in the control of *Bipolaris sorokiniana* in tall fescue by *Stenotrophomonas maltophilia* C3. *Int. Turfgrass Soc. Res. J.* 9:736–741.

3

Turfgrass Insect and Mite Management

Goal of Insect and Mite Management

Generally, the goal of insect and mite management in turf is to suppress the damage to a level where the desired aesthetics and health of the turf are maintained. Obviously, this implies that pests will be present and some damage will occur. The amount of pest damage that is allowable will be different for home lawns, sports turf, golf courses, and sod farms. In fact, the amount of pest damage allowable will differ significantly within each of these turf usage areas, depending on the needs and desires of the individual owner and/or manager!

Currently, to achieve insect and mite management in turfgrass, the integrated pest management (IPM) approach is highly recommended. IPM is accomplished through a *process* of identifying and understanding turfgrass pests that are present or could infest the turf, understanding the turf species being used and its needs, systematically monitoring of the pests, establishing population thresholds that require action, and selecting appropriate control methods.

Pest Management Process

The term "process" is being used to suggest that there is no single program or management procedure that will effectively control turfgrass pests, within a season or between seasons! Weather conditions change. The turf stand, itself, changes over time. Insect and mite populations change over time. Management tools (e.g., equipment, pesticides, and biological controls) change over time. With the knowledge that all of these variables change over time, the smart turfgrass manager can still adopt some general processes that will allow for flexibility in maintaining effective management of pests.

1. **Pest identification.** The turf habitat provides suitable living space for a multitude of animals. These animals can be beneficial, damaging, or

of no real importance. Turf managers must be able to identify each of these animals and determine if any control action is necessary. March fly larvae feeding on dead turf killed by a disease may be mistaken for the cause of the dead turf. In this case, disease management is needed, not insect control. As broad-spectrum insecticides are being replaced with more targeted products, beneficials (e.g., ground beetles, spiders, and predatory bugs) and nuisance animals (e.g., earthworms, slugs, millipedes, and ground-nesting bees and wasps) are on the increase. In most cases, these can be tolerated, in fact, encouraged.

2. **Pest life cycles.** Each insect and mite found in turf has a unique life cycle. Turf managers must become familiar with these life cycles because certain stages within these life cycles are often resistant to controls. Other stages are quite vulnerable to control and these should be the target of control actions. More importantly, the life cycles can differ significantly, depending on the geographic location of the turf. Cutworms generally have two to three generations in cool-season turf, but may have three to six generations in warm-season areas.

3. **Determine the zone of activity: The "target principle"**—Turfgrass is a special environment with unique attributes. Thinking of turf as a regular field crop will result in pest management failures. Turf is a perennial plant cover with distinctive subsurface (soil-inhabiting) and surface-inhabiting pests. Whether using pesticides, bio-based products, or biological controls, having knowledge of where your "target" is at any given time is often a major key to successful management!

   Turf-infesting caterpillars feed on grass blades, so having an insecticide in the thatch or soil is not of much use! On the other hand, leaving pesticide residues on leaf blades will not control white grubs that are feeding in the soil–thatch interface. Thatch and the high organic matter formed in the upper inch of the soil can bind and restrict the movement of chemical controls. Pesticides that work in agricultural soils with 1–3% organic matter simply do not move well in the 3–10% organic matter soils under turf.

   In this chapter, the insects will be rather artificially broken into groups, according to the turfgrass zone that they inhabit. Pests in the upper, *foliage/stem zone* of turf are often conspicuous and fairly easy to control because of their exposure. Pests in the *stem/thatch zone* can more easily evade detection for some time. Pests located in the *thatch/soil zone* also evade detection until their damage to turf roots results in significant turf loss! These pests are often referred to as "soil-inhabiting" pests. While they often do inhabit the soil, they come to the soil–thatch interface to feed on organic matter, including turfgrass roots, stolons, and crowns. Pests located in the stem/thatch zone and thatch/soil zone are often the most difficult to manage with pesticides unless irrigation is available to move the pesticide residues into the areas occupied by the pests.

Finally, we must acknowledge that there are *nuisance pests* associated with turf. Nuisance pests rarely damage the turf directly, but their presence can concern the manager and users of the turf. Insects such as bees, ants, earwigs, fleas, flies, midges, and others are "people problems," which rarely damage the turf directly. Other nuisances are not insects, but relatives of insects: millipedes, centipedes, snails, slugs, and earthworms. Where nuisance pests are not tolerated, turf users often demand that the turf manager control the nuisances!

4. **Monitoring**. Pest monitoring is one of the most important principles in pest management. By using regular monitoring, the turf manager can detect new pests and determine when, or if, existing pests will reach damaging numbers. Monitoring is also used to determine if a pest is in a vulnerable stage of its life cycle. Monitoring also includes record keeping! Pests often reoccur in an area that they damaged previously. There may be a particular microhabitat that the insect prefers and they will generally show up in these places first. Excellent area maps are generally available on websites, and turf managers should use these to mark where and when pests occurred.

5. **Select appropriate controls**. Pesticides (chemical controls) are powerful pest management tools but using pesticides as your only tool will result in eventual failure. IPM approaches pest control as a system of decisions and control tactics. In pest management, we must constantly use cultural (e.g., selecting resistant turf species/cultivars, water and thatch management, and mowing) and biological controls (e.g., parasites, predators, and pathogens) to their fullest potential. This can help reduce the unnecessary usage of pesticides, preserve their usefulness, and develop a more diversified and resilient turf habitat.

Pest Identification

Formal training in entomology, the study of insects, can certainly help the turf manager make correct identifications of the various insects, mites, and other invertebrates encountered in the turf environment. The pictures and illustrations provided in this book can help. However, even seasoned managers can make errors in identification. Therefore, managers should learn to make contact with experts located at state land-grant universities (usually through their cooperative extension services), consultants associated with professional turf organizations (e.g., golf course, lawn and landscape care, and sports field associations), product suppliers (be cautious that these folks have a vested interest in selling you products), and other turfgrass managers.

Be a questioning person! If you encounter a pest or turf damage that is not exactly like the ones described in books, factsheets, and webpages, do not be afraid to ask for assistance in identification. Fall armyworm larvae cause the same types of "pock marks" on golf greens that black cutworms make, but they are different species. Most annual white grub species look the same from the side so you have to carefully inspect their bristle patterns to determine what species is present. This is important because some species are less susceptible to certain controls. Chinch bugs, billbugs, and white grubs can all cause "brown turf," so careful sampling and inspection is needed to determine which of these is causing the problem. In some cases, the brown turf may actually be the result of disease, drought stress, or other factor.

No single book can answer all pest identification problems. Turf managers are encouraged to build a library of information. See the general references located at the end of this chapter for additional information on insects, mites, and other animals associated with turf.

Insects and Mites Associated with Turf: An Introduction

Many invertebrate and vertebrate animals can be associated with turf. The turfgrass environment provides a sizeable source of food and living space for a variety of plant feeders, predators, scavengers, and decomposers. The vast majority of invertebrate species associated with the turf habitat can be considered beneficial or inconsequential in habits. However, most turfgrass is fed upon by a limited number of herbivores that eat leaves, stems, or roots. When in sufficient numbers, these herbivores become significant pests. Others, such as fleas, spiders, ants, and wasps, can cause concern because of their nuisance ability to sting, bite, or simply get in the way or annoy.

The invertebrates most commonly encountered by the turf manager are as follows:

1. **Phylum Nematoda.** Nematodes or roundworms are micro- to macroscopic in size and include numerous plant pests, decomposers, and animal parasites. Nematodes that attack turf roots are generally too small to see without the aid of a microscope. Larger nematodes appear to be worms without segments, are pointed at each end, and are constantly coiling and twisting (Figure 3.1). These larger species are usually beneficial decomposers or parasites of other animals. A group of insect parasitic nematodes (entomopathogenic) have been commercially developed and several species are useful tools to manage insect pests, including some turfgrass pests. Plant parasitic nematodes that attack turfgrass are difficult to manage and experts on these invertebrates are often aligned with departments

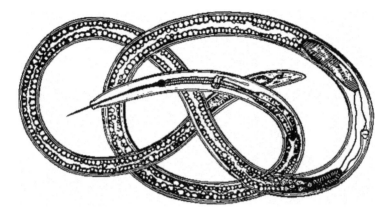

FIGURE 3.1
Diagram of a free-living nematode, *Xiphinema* spp. (Redrawn from Cobb.)

of plant pathology, entomology, or plant protection at universities. Nematicides are pesticides designed to manage these invertebrates.

2. **Phylum Mollusca**. Snails and slugs are small- to large-sized animals without segmentation of the body (Figures 3.2 and 3.3). The slugs are merely snails that have secondarily lost or reduced the shell. Most are plant or fungal feeders or decomposers. They are usually considered a nuisance in turf. Molluscicides are pesticides developed to control snails and slugs.

FIGURE 3.2
Gray garden slugs commonly feed on soft plant tissues, fungi, and decaying plant materials, including grass clippings.

FIGURE 3.3
Garden snails, *Helix* spp., are common inhabitants of gardens where they feed on plant leaves.

3. **Phylum Annelida**. Segmented worms include earthworms
   (Figure 3.4) as well as numerous smaller, closely related worms.
   Though considered beneficial because of their soil aeration habits
   and ability to aid in the decomposition of organic matter, certain
   earthworm species can become nuisance pests when they create "lumpy" turf in lawns or leave muddy castings on golf course
   greens and tees. There are no pesticides specifically designed to
   manage segmented worms in turf.

4. **Phylum Arthropoda**. Arthropoda is the largest group of animals
   in terms of the total number of species known, and often in terms
   of the number of individuals per unit area. They are characterized
   by having segmented bodies with segmented appendages and an
   exoskeleton that has to be shed during the growth of the animal.
   The group includes the crustaceans (sowbugs and pillbugs as well
   as shrimps and crabs), arachnids (spiders and mites), chilopods
   (centipedes), diplopods (millipedes), symphylans (garden centipedes), and the insects. Each of these groups has a special set of
   characters that are used in identification. However, all arthropods
   have an exoskeleton divided into segments and body regions (e.g.,
   head and trunk; cephalothorax and abdomen; and head, thorax,
   and abdomen) that must be shed at periodic intervals to accom-
   modate growth; paired, jointed appendages (legs, mouthparts, and
   occasional abdominal appendages); a ventral nerve cord; and dorsal pumping heart.

FIGURE 3.4
Common earthworm or nightcrawler prefers soils with high organic matter. They often come to the surface at night to feed on decaying plant material and they leave soil and organic matter feces, called castings, on the surface.

Classes of Arthropods

The six major groups (classes) of arthropods are the Arachnida, Crustacea, Chilopoda, Diplopoda, Symphyla, and Insecta.

The *Arachnida* includes spiders, mites and ticks, scorpions, daddy longlegs (= harvestmen), and other less common groups (Figures 3.5 through 3.9).

FIGURE 3.5
Wolf spiders often make burrows in turf and emerge at night to search for insects.

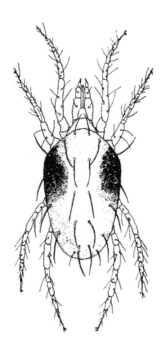

FIGURE 3.6
Diagram of a twospotted spider mite, *Tetranychus urticae*, adult female. (USDA.)

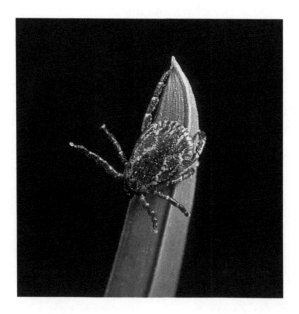

FIGURE 3.7
Ticks, like the American dog tick, *Dermacentor variabilis*, feed on the blood of animals, but they are often found in taller grasses.

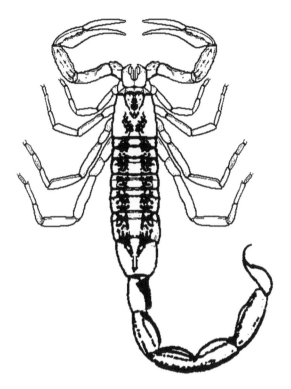

FIGURE 3.8
Diagram of a common scorpion, *Centruroides vittatus* (CDC).

FIGURE 3.9
A "daddy longlegs" or "harvestmen" is another group of arachnids, the Opiliones, that are harmless though they look like spiders.

All normally have four pairs of legs, a pair of near-oral appendages called pedipalps (antenna-like, leg-like, or pincher-like), a pair of pincher-like mouthparts called chelicerae, no antennae, and two body regions (the cephalothorax and abdomen). Spiders have the two body regions separated by a constriction while most of the other groups have the two regions broadly joined. Since most arachnids are terrestrial, they use tracheal tubes, book lungs, or simple diffusion (small mites) to respire. Though arachnids are often feared, in turfgrass systems, most arachnids are considered to be beneficial predators or nuisance pests. However, some mite groups attack turf plants and ticks (large, blood-sucking mites) are commonly found in turf where their host animals reside.

The *Crustacea* are primarily aquatic (crabs, shrimp, and crayfish), though a few groups have adapted to terrestrial life (sowbugs and pillbugs, the order Isopoda). Crustaceans have two pairs of antennae and the body is divided into two regions—a head and trunk or cephalothorax and abdomen. The isopods (Figure 3.10) have the second pair of antennae reduced to a small stub and have the head/trunk body design. Isopods are considered beneficial decomposers, though they may feed on grass seedlings. Crayfish (= crawdads) (Figure 3.11) often live in soils with high water tables. In such habitats, they construct numerous mud "chimneys" up to the surface (Figure 3.12). These mud tubes extend above the turf and can be unsightly. Controls are not recommended as these chimneys extend directly to the water table and the burrows may also connect to nearby streams or wetland habitats.

The *Chilopoda* or centipedes (Figure 3.13) are terrestrial and have head and trunk regions with a single pair of antennae. Each trunk segment has only one pair of legs attached to the side of the body. The first pair of legs has

FIGURE 3.10
A common sowbug, *Armadillium* spp., that is found in places where decaying plant material is found.

FIGURE 3.11
Crayfish can be aquatic or semiterrestrial in habits.

FIGURE 3.12
Semiterrestrial crayfish species often deposit soil around burrow openings. These are called mud chimneys.

FIGURE 3.13
A centipede, *Lithobius* spp., that is a common predator in turf and surrounding habitats.

been modified into a stout, jaw-like fangs located just below the mouth. Most centipedes are beneficial predators, though some are large enough to bite humans if restrained or handled.

The *Diplopoda* or millipedes (Figure 3.14) are terrestrial and also have head and trunk regions with a single pair of antennae. Each trunk segment has

FIGURE 3.14
Several types of millipedes can be found in turfgrass, including this polydesmid type.

been joined with another, forming diplosegments, so that it appears that each visible segment has two pairs of legs. Millipedes are considered beneficial scavengers, though many feed on living plant tissues, especially of seedlings.

The *Symphyla* or garden centipedes are terrestrial and have a head and trunk with a single pair of antennae but no eyes or body pigmentation. They look like tiny centipedes but they do not have fang legs and most of the species feed on root hairs of plants. Occasionally, they reach large numbers in the soil and their feeding may cause stunting of the plants.

The *Insecta* is the largest group of animals and the members are characterized by having three distinct body regions (head, thorax, and abdomen), a single pair of antennae, and three pairs of legs, and adults usually possess two pairs of wings. They respire through a set of air tubes, trachea, that run throughout the body. There are numerous orders of insects that contain members that attack or are associated with the turfgrass environment. Probably of greater importance is whether the insect belongs to one of two categories according to life cycle or metamorphosis. The more primitive insects have a gradual (simple or incomplete) life cycle while the vast majority of species have a complete (complex) life cycle.

Pest Life Cycles

All pests of turf, whether weeds, diseases, or insects have life cycles. By understanding these life cycles, the turf manager can approach management by emphasizing controls targeted to the most vulnerable stage of a pest.

Insects can be divided into two general groups with similar life cycles—gradual and complete metamorphosis. Those insects with a gradual metamorphosis have immatures and adults that look much alike and generally live and feed in the same habitat. Chinch bug nymphs and adults have the same general body shape and feed on plant juices with their piercing, sucking mouthparts. On the other hand, insects with a complete metamorphosis have immatures that are very different in form than the adult and these two stages usually feed on different foods and live in different habitats. Cutworm larvae live in the turf and have chewing mouthparts while the adults fly to flowering plants and have sucking mouthparts. These adults return to the turf to lay eggs.

Insects with gradual life cycles are often easier to control because all stages are present and the only decision to be made is whether the population is large enough to cause problems. However, pests with complete life cycles are usually vulnerable to management during some critical phase of their cycle. Understanding when these periods occur and sampling just prior to the event, turf managers can maximize the efficacy of their control options.

Insect Metamorphosis

Insects with a *gradual metamorphosis* (Figure 3.15) have three life stages: egg, nymphs, and adult. Eggs hatch into the first nymph (called the first instar), which can add body mass but must shed its exoskeleton to continue growth. The second nymph (called the second instar) continues the process. As the nymphs molt and mature, they develop external wing pads (if they have wings as adults), and eventually molt into the adult stage (last or adult instar). Most insects with a gradual metamorphosis have egg, nymphs, and adults located in the same habitat and feeding in the same manner. Though the nymphs may have different coloration than the adult, they have the same general body shape and form.

Insects with a *complete metamorphosis* (Figure 3.16) have four life stages: egg, larva, pupa, and adult. The eggs hatch into the first larva (called the first instar), which can add body mass but must shed the exoskeleton to continue to grow. The second larva (called the second instar) continues the process, usually for a predetermined number of times. Most cutworm and sod webworm larvae have five to six instars while white grubs have only three instars. The last larval instar then molts into a transformation stage, the pupa. Within the pupal exoskeleton, most of the larval body is rearranged

FIGURE 3.15
Hairy chinch bug, *Blissus leucopterus hirtus*, life stages: (left to right) egg, first, second, third, fourth, fifth instar nymphs, full-winged adult, and short-winged adult.

FIGURE 3.16
A masked chafer, *Cyclocephala*, life stages: (left to right) egg, first, second, third instar larvae, pupa, and adult.

and reproductive organs are developed. The pupa usually displays the eventual adult appendages of wings and legs. The winged adult stage emerges from the pupal case, expands the wing pads, and is ready to continue the life cycle. Insects with complete life cycles usually have larvae that live in different habitats than the adult. White grubs feed in the soil while the adult beetles often feed on the leaves of trees and shrubs. Cutworms feed on the leaves and stems of turf while the adults must obtain sustenance from the nectar of flowers. Pests with complete life cycles can cause special problems for turf management because the eggs and pupae are usually resistant to controls. Likewise, if the larvae or adults are susceptible to controls, they are usually in different habitats.

The larvae of insects with a complete metamorphosis are usually given special names. Scarab beetle larvae are called white grubs; moth and butterfly larvae are caterpillars; most fly larvae are maggots; and, larvae of bees, wasps, and ants may be grub-like or caterpillar-like.

Mite Life Cycles

Plant-feeding mites associated with turf can be treated as if they have a simple metamorphosis (Figure 3.17). All mites have four stages: the egg, larva (an immature with only three pairs of legs), two nymphal instars (protonymph and deutonymph), and the adult. Usually, all stages are found on or near the host. The adults of some mites migrate out of their normal habitat to lay eggs in protected areas. The eggs of most mites are resistant to chemical controls, which make mites difficult to manage if the eggs enter periods of dormancy. The eggs of the winter grain mite go dormant for the entire summer while the eggs of the Banks grass mite may be either summer or winter dormant.

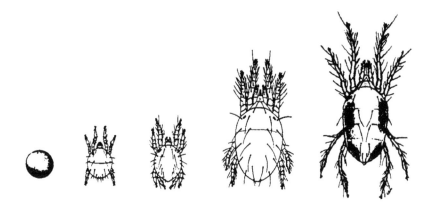

FIGURE 3.17
Illustration of twospotted spider mite, *Tetranychus urticae*, life stages: (left to right) egg, larva, protonymph or nymph I, deutonymph or nymph II, and adult. (USDA.)

Ticks are parasitic mites with considerably more complicated life cycles. Their larvae (called seed ticks), nymphal instars, and adults may require different hosts and can travel considerable distances while feeding on their hosts.

Zones of Activity (Turf, a Unique Habitat)

Most grasses used as turf are perennials that have great ability to recover from environmental stress and pest damage. Many species of grasses react to environmental stresses by going dormant during winter cold or summer drought. When managers try to force grasses to maintain color and growth during these normal dormancy periods, additional stresses can be caused and pests often take advantage of these situations. Golf course managers cut turf so short that it cannot develop deep and extensive root systems. This stressed turf may be less tolerant of white grub or mole cricket feeding on the roots.

Grasses come in infinite varieties with a wide range of attributes. Wise turf managers must learn how to select the best species and varieties with resistance and tolerance to the environmental stresses and pests located in the area. Renovation with resistant, tolerant cultivars should be the first line of defense against insects and mites attacking turfs.

Most turf can tolerate low to moderate levels of pest activity. By simply watering or fertilizing, much insect and mite damage can be masked. On the other hand, overreliance on watering and fertilizing can result in disaster if the irrigation is suddenly cut off or the pest populations build beyond the turf's ability to regrow.

Most turfgrasses produce a layer of dead and living organic matter—thatch. This thatch serves as a natural insulating layer for mediating soil temperatures and maintaining moistures. However, if this thatch layer becomes too thick, many insects and mites also take advantage of this mediated habitat. Thatch is very high in organic matter that can serve as a substrate for adsorption (sticking to a surface) of pesticides. Thatch and the organic matter in the underlying soil can support fungal growth. When these fungi dry out, they can form a water-repellent (hydrophobic) zone that results in what is called "localized dry spot." Therefore, if a thatch layer is too thick or hydrophobic, insecticides applied for grub, billbug, or mole cricket control will never reach their targets.

Turf constantly renews its roots, stolons, shoots, and leaves. This renewal, along with the activity of earthworms and other invertebrates, continues to increase the organic matter content of the soil immediately below the thatch layer. Recent studies have indicated that turfgrass stands are net positive carbon sequestration habitats. This organic matter increase is considered a major benefit of turf culture and can help explain why turfgrass also serves as a living filter to reduce groundwater contamination. On the other hand, this increase may also help build up levels of "organic matter feeders" such as white grubs and mole crickets.

Managed turfgrass is an incredibly diverse habitat, biologically! While most would dispute this statement because of the limited number of grass species used in the normal stand of turf, recent studies are finding that 20,000–50,000 arthropods per square meter are typical of established turfgrass stands! Fortunately, most of these arthropods are small and engaged in the processes of recycling the dead organic matter produced by turf. Reliant on the recycling guild are numerous predators and parasites that feed on these detritivores. Of course, there are thousands of species of bacteria, fungi, nematodes, and other invertebrates also associated with a turfgrass stand. Even where regular applications of insecticides, fungicides, herbicides, and fertilizers are made, most populations of arthropods in turfgrass remain unaffected and those that are suppressed tend to rebound within weeks of an application.

Monitoring

Pest monitoring is simply using those techniques and tools that allow the turf manager to determine *when* and *if* control action is needed.

Many turfgrass managers apply pesticides for control of anticipated turfgrass insect and mite pests through regularly scheduled "programs." These applications may be called "preventive-," "round-," or "calendar-" timed applications. On the other hand, some insect pests are not expected and "reactive" treatments are made after some damage has been detected. These can be called "curative" treatments. If made early enough, curative treatment allows the manager time to scout and determine if a pest is reaching damaging levels. However, if not done well, significant damage may occur and the manager may have to "rescue" the turf. In these cases, extra fertilizer, irrigation, and even reseeding may be needed. Obviously, avoiding rescue treatments is desired.

Preventive applications have been frowned upon by many entomologists. This is due to the fact that most early synthetic insecticides could easily kill insect or mite pests *after* they were discovered. However, many turf managers were using preventive applications to prohibit or eliminate a perceived, potential pest problem. Terms like "grub proofing," "pest proofing," and "guaranteed insect elimination" have been used. In many cases, the pesticide was not needed and the continual use of the same pesticide can cause pest resistance or accelerated degradation to occur. Unwarranted insecticide applications can also disrupt the beneficial impacts of predators, parasites, and pathogens that can help suppress pestiferous insects.

Occasionally, preventive pesticide applications are warranted. Where pests are certain to occur (because of previous experience or predictive models indicate that a major pest outbreak will occur), prevention is an acceptable, perhaps preferred, strategy. Some newer insecticide chemistry has modes of action that work better when used as preventives, especially when the young insects are targets. This is not different than using a preemergent herbicide to prevent annual grassy weeds. Most fungicides prevent fungal infections, not cure them. Insecticides with sustained residual ability or those with insect growth regulator (IGR) action are often more effective when used as preventives.

Curative applications of insecticides are still useful. Many insect and mite pest populations vary considerably from year to year and if these pests are easy to detect, making a control application that knocks them out or reduces their damage to tolerable levels is perfectly acceptable. Turf-infesting caterpillars are usually easy to knock back whenever you discover them, as are chinch bugs. But, billbugs can be very difficult to "cure."

Tools and Strategies for Timing of Controls

Useful strategies for timing of controls are available and should be used to reduce the problems associated with regular and repeated pesticide applications. These alternatives include active monitoring and sampling of pest populations, using weather-mediated models and pest mapping.

Hands and knees method is still the most commonly used method of sampling pests in turfgrass! This is more than just a cursory visual inspection of turfgrass made as you drive by! Chinch bug and billbug damage can look like summer drought stress or the beginning of a disease. Making an assumption that you are viewing stress rather than pest damage can be a disaster! Experienced turf managers are diligent in checking everything out. Separate the turf canopy and see if there are any insects or disease present; use a soil probe to check the soil moisture and thatch thickness; inspect the root–thatch zone for likely issues. In general, these visual inspections are not quantitative, that is, yields a number of pests per unit of turfgrass area. Mere presence or absence of a pest is important to know, but does not help make critical decisions about whether to use a control tactic or not.

Quantitative monitoring and sampling of pest populations is at the heart of all IPM programs. Before proper controls can be applied, one needs to know if a pest is present and if the population or potential population will cause significant damage to the turf. In addition to the traditional *visual inspection* of the turf, several trapping and sampling tools are useful for monitoring and quantifying turf insects and mites. The most common ones are as follows:

- *Pitfall traps* are simply cups or cans sunk into the turf to capture crawling insects such as billbug adults (Figure 3.18). Obviously, a pitfall trap is not appropriate where children may twist an ankle,

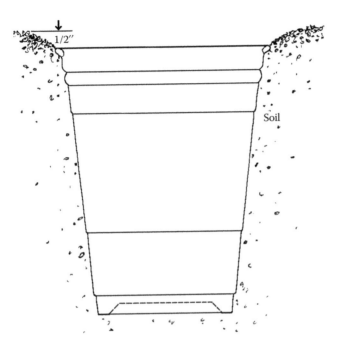

1/2″

Soil

FIGURE 3.18
A small pitfall trap can be constructed by using a 16–20 oz. plastic cup placed into a cup-changer hole.

but they can be used next to flower beds or under a tree to check for insect activity. Some insects are best monitored using linear pitfall traps. These are simply plastic pipes with a groove cut down one side, buried into the ground and attached to a collecting container of some sort. Linear pitfall traps are commonly used to assess mole cricket activity, billbug adults, and annual bluegrass weevil adults.

- *Light traps* that use UV–black lights are very attractive to sod web-worm, cutworm, June beetle, and masked chafer adults (Figure 3.19). Light trap sampling can assist you in determining whether insects are early or late according to normal "calendars."
- *Pheromone traps* contain the sex and/or attractant chemicals used by sod webworms, cutworms, and Japanese beetles (Figure 3.20). These can be used, like light traps, to determine insect activity periods.
- *Cup changer samples* merely use a cup changer to pull plugs in a transect or grid across the turf to sample for billbug larvae or white grubs (Figure 3.21). You can quickly determine if a potential prob-lem exists when the majority of samples contain a billbug or white grub larva. The standard 4.25-inch-diameter cup changer is almost

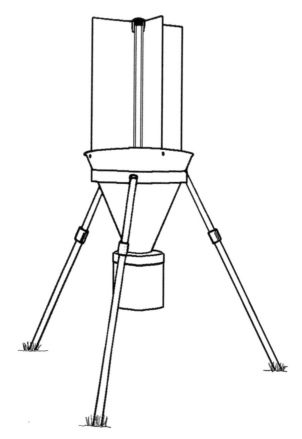

FIGURE 3.19
A typical light trap. (Ill. Coop. Ext. Serv.)

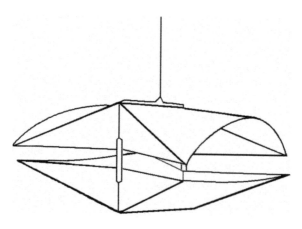

FIGURE 3.20
A cardboard "wing trap" type of pheromone trap.

FIGURE 3.21
Using a standard golf course cup changer to monitor white grubs.

exactly 1/10-ft^2. Of course, using a heavy knife to cut through the turf on three sides of a square foot area and lifting back the back to expose the insects below, or even a shovel are acceptable sampling tools.

- *Disclosing solution* (soap flush) uses one tablespoon of liquid dishwashing detergent per gallon of water for sod webworm and cutworm monitoring. Simply spread two gallons of the mix over a square yard of turf and the caterpillars will quickly come to the surface (Figure 3.22). This soap solution is also useful in disclosing mole crickets, black turfgrass ataenius adults, annual bluegrass weevil adults, and other insects. Adult mole crickets and crane fly larvae often need two flushes to force them to the surface.

- *Damage (grid) ratings* use a wood, metal, or plastic frame with cord or wire strung between the sides to form square grids (Figure 3.23). The frame is dropped onto the turf and the presence or the absence of pest activity is recorded for each grid, or a rating is used for each grid and the average is taken.

FIGURE 3.22
Application of a detergent solution will coax up a wide variety of turf-infesting pests.

Weather-mediated predictive (degree-day) models are developed by monitoring weather parameters and comparing these to insect or mite activity. Though these models help determine better timing of controls, they still do not answer the question of whether the pests are present in sufficient numbers to cause damage or warrant controls. Models have been developed and published for chinch bugs, bluegrass billbugs, masked chafers, Japanese beetles, and sod webworms. Most of these models have not achieved widespread usage because the models are too complex or use base temperatures not normally used. Some areas are also using *plant phenology* models. Instead of using weather measurements, plant phenology models use the visible displays of plant development (e.g., first leaf emergence, first flowering, and petal drop) correlated to pest development. A common example would be the arrival of black turfgrass ataenius adults when Vanhoutte spirea is in full bloom.

Pest mapping is simply good record keeping. Since insect and mite pests generally require specific habitats to build up damaging populations, turfgrass

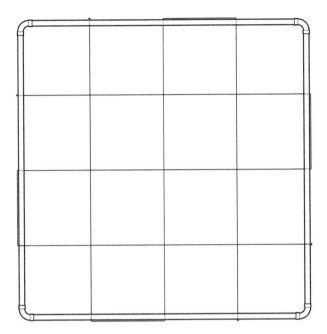

FIGURE 3.23
A PVC pipe frame strung into grids for sampling turf insects and their damage.

areas that have had problems in the recent past are the most likely places that will be attacked again. In short, if a damaging pest population occurred last year in an area, the probability is much higher that the same thing will occur again in the future. Keeping a useful record of pest occurrence is pest mapping.

Pest mapping in lawn care can be easily accomplished on a customer or neighborhood basis. Technicians making visits to several lawns in a neighborhood often describe them as grub, billbug, or chinch bug lawns. The turf and soil conditions in various neighborhoods seem to encourage certain pests and exclude others. Relatively new neighborhoods often seem to have very similar conditions from lawn to lawn. The developer likely purchased the sod or seed from a single source. Therefore, most of the lawns will have the same species or blends of grasses. If the "contractor's blend" was used, most of the lawns will have common perennial ryegrass and fine fescues predominating. These will be very susceptible to sod webworms, chinch bugs, and billbugs. If pure Kentucky bluegrass sod was used, billbugs and chinch bugs can be major pests, almost immediately. Lawns that are in the 4–6 years age period often have sufficient organic matter and thatch to support white grub outbreaks. By "mapping" neighborhoods with common characteristics, the wise lawn care manager can route better timing of controls into those neighborhoods that need them. In short, there is no sense in applying chinch bug insecticides in a billbug neighborhood!

Likewise, golf course superintendents commonly admit that certain putting greens or fairways have cutworms or white grubs while others never have problems. Mapping and treating these high-risk areas is wiser than "going wall to wall" with an application. Sports field managers often see different pest complexes on their fields. This can result from the different levels of management, stress, and ages of the fields. Lights at night can also attract certain pests into areas where they are more likely to lay eggs.

To accomplish pest mapping, staff need to be trained to take the time to record what, where, and when they see pests! For lawn care technicians, obtain a large area map of the operating areas and "code" neighborhoods where specific pests have been detected. This can be as simple as having color-coded pins that are used to mark where pests were found. Soon, you will see clusters of the same color in specific neighborhoods. Turf facility and golf course managers can download area maps from the Internet. Such maps can clearly show sports fields and fairways. Reducing these maps so that they will fit onto single pages of a notebook makes a good record book. Simply use colored pencils or pens to mark and date the occurrence of different pests.

Sampling, monitoring, and record keeping take time! However, such efforts often save time in the future and help catch problems before major damage occurs. A lawn care technician will need extra time to sample a lawn for billbugs, chinch bugs, or white grubs. If the results of these efforts are recorded, a neighborhood map of pest activity will save time next year. Some golf managers have found that using a sampling crew (two to three people who are trained to use proper sampling techniques) to scout the turf for grubs can reduce the need for treatments 50–75%! A crew of three can cover an 18-hole golf course in 1–2 days.

Selecting Appropriate Controls

Pest Management versus Pest Eradication

Managing insects and mites that attack turfgrass has generally relied on the use of pesticides. Whether this is good or bad is beyond the scope of this discussion, but we must ask whether alternative controls are available and appropriate. Before we can consider the alternatives, we should review our current concept of insect and mite pest management. Pest "management" as opposed to "eradication" implies that some pests will always be around. It is the goal of pest management to keep the pest populations suppressed at a level where damage is not overly evident. In field crops, this has generally been termed as an economic threshold level. In turfgrass management, the aesthetic threshold level (the population of a pest that causes noticeable, unacceptable, visual damage) is the term to be used.

Integrated Pest Management

IPM is the selection, integration, and implementation of pest control (biological, chemical, and/or cultural) based on predicted economic, ecological, and sociological consequences. In other words, when a pest control action is used, we must consider the direct costs of products and their application, as well as potential changes to the ecosystem and attitudes of humans. Using the IPM approach, several important concepts must be adopted:

1. No *single* pest control method will be successful, long term. All of the control options—biological, chemical, and cultural—must be used.
2. *Monitoring* (sampling and record keeping) of the pest is constantly needed to evaluate the status of a pest populations (e.g., not present, present but not causing aesthetic damage, and present and causing aesthetic damage).
3. *Mere presence* of a pest is not a reason to justify control action.
4. IPM is a *decision-making process*, not a predetermined program to be followed through the season.

There has been considerable misunderstanding about the IPM approach, IPM control options, and the underlying concepts. IPM is often called a "program," though it is really a method or way of approaching pest management. IPM is not a biological control or "organic" program, though biological controls are useful and organic materials can be used. It is not the "goal" of IPM to reduce or eliminate pesticide use, though pesticide usage is often dramatically reduced through pest monitoring and increased usage of biological and cultural controls. IPM is neither the easiest nor the least expensive management technique. Turf managers using IPM methods need the knowledge and training to make decisions, because IPM is a decision-making process. The initial costs of investing time in monitoring and biological/cultural controls are eventually returned as improved turf and using fewer pesticides.

Monitoring in IPM

Monitoring techniques and tools have been discussed above but the position of monitoring in IPM needs to be reemphasized. Monitoring pest activity and population levels is the key to successful IPM. Unfortunately, most feel that monitoring must be a complicated and time-consuming process where someone must constantly watch each and every turf area. This is simply not true. In most cases, monitoring can be done through careful record keeping. Knowing which lawns (or neighborhoods), sports fields, or golf course holes have experienced insect or mite damage in previous years can greatly lessen the day-to-day need for direct monitoring! Wet soil areas on southern golf courses are where mole crickets always show up first! A lawn damaged by Japanese beetle grubs last year is at a high probability of repeating the problem this year. Age of the turf is also important. Recently established turf rarely has sufficient organic matter (thatch and upper soil) to support white grubs, but high fertility on such turf may encourage outbreaks of cutworms and armyworms. A quick look at the turf under your care is always better than assuming that there are no pests present!

Turfgrass and associated ornamental plants seemingly present an impossible task for monitoring and record keeping! If each plant within these systems has a few to a couple of dozen pests, we would be dealing with hundreds of pests rather than the half dozen found in a field crop. To overcome this bewildering issue of complex environments, we often use the concepts of *key plants* and *key pests*.

- *Key plants* are trees, shrubs, flowers, and turfgrasses that are known to have perennial pest problems. They are "pest prone." As an example, Kentucky bluegrass is often attacked by billbugs, chinch bugs, sod webworms, and white grubs, while endophyte-enhanced, turf-type tall fescues are resistant to billbugs, chinch bugs, and sod webworms. European white-barked birch is readily attacked and killed by borers, but oak trees are relatively resistant, even though commonly infested by other insects.

- *Key pests* are those that cause significant damage or may kill trees, shrubs, flowers, and turfgrasses. These pests often have special times (windows of vulnerability) when they are susceptible to controls. Clover mites may discolor turf and sod webworms can dine on leaf blades, but they rarely kill turf. On the other hand, billbugs, white grubs, and mole crickets often kill turfgrass plants by destroying roots and eating crowns. In these cases, clover mites and sod webworms are nuisance or aesthetic pests, but the others are "key pests" that can kill the turfgrass. On ornamental trees, aphids or galls rarely cause significant dieback or death but a single borer burrowing around the cambium area has the potential to kill the plant!

Control Options

As mentioned above, IPM uses three general control options—biological, chemical, and cultural controls. These are our control tools and we must understand the benefits and limitations of each option. Since we are dealing with turf culture, many of the pest problems can be traced back to the direct result of poor turf maintenance. In other words, turf placed in urban habitats, pushed to perform in stressful conditions (e.g., mowed too closely and allowed to develop thick thatch layers), or not suitably adapted (e.g., fine fescue in the sun) are the ones most likely to be severely attacked by pests. Overfertilizing (builds up thatch layers), poor mowing practices, and irrigation can assist insect and mite populations to build up to damaging levels. Today, we cover these "turf health care" components under cultural controls.

Cultural Controls

The cultural control option should be our *first* consideration in turfgrass IPM. Cultural controls in field crops have generally included sanitation, crop rotation, tillage, host plant resistance/tolerance, mechanical/physical destruction, and quarantine. If we look at these techniques, we may wonder how they relate to turf culture.

1. *Sanitation* helps remove inoculums or hiding areas of pests. In turf, thatch removal and management are similar operations. Where turf seed was produced, and in some southern areas where warm-season grasses go dormant for the winter, surface burning was a common practice to reduce diseases and surface insect survival. Today, burning is considered to be an air-polluting practice and is discouraged.

2. *Crop rotation* is generally used in field crops (e.g., corn rotated with soybean) and should be considered for turf. Though we have little scientific knowledge on how a turf stand changes over time, anecdotal evidence suggests that original blends of various grass species are not maintained over time. Likewise, many turf areas were developed without the benefit of modern, resistance factors. These older areas should be renovated and replaced when possible. Therefore, renovation is the equivalent of crop rotation. A less obtrusive way to "rotate" is to interseed with improved grass species or cultivars. We have evidence that slit-seeding perennial ryegrass or turf-type tall fescues into Kentucky bluegrass can introduce endophytes into the system, which makes the stand resistant to billbug, chinch bug, and sod webworm attack. Most turf agronomists now advise against using species blends (using two or more species of grasses in the seed mix), but recommend the use of cultivar blends (using two to three cultivars of the same species of grass). The cultivars are selected for their genetic diversity so that the entire stand has several genomes that can better react to weather extremes as well as disease and insect attack. In short, it is recommended that turfgrass managers ask themselves, "should I just start over?" The up-front costs renovating (crop rotation) often pays for itself in the long run through future cost savings by reducing fertilizer, water, and pesticide expenses.

3. *Tillage* in field crops exposes resting pests and breaks up the soil for better air and water movement. In turf, core aeration, verticutting, and top dressing are similar processes.

4. *Host resistance* uses plants that are less susceptible to pest attack (tolerance) or produce actual toxins (antibiosis) that kill or stop pest growth. This tactic can be one of the most important in turf culture. In fact, most insects and diseases that are persistent turf problems

can be greatly reduced with the use of resistant turf cultivars. For people concerned with the overuse of pesticides, this is the major option to be considered. Constantly check with seed suppliers to see what has been developed in the area of turf resistance (to insects and diseases). Golf courses and sports fields often totally renovate with new cultivar blends of grasses, while overseeding and interseeding (slit seeding) can improve stands in lawns.

5. *Mechanical/physical* techniques are as simple as crushing the pest underfoot to using large industrial vacuum sweepers to suck up pests. In turf, we need to constantly remind ourselves that simple crushing of pests can be an effective management tool! Research has indicated that core aeration can reduce grub populations by half. Adapting tools and equipment to fully use this technique are still in need of development. When armyworm egg masses are found on golf course hole markers, significant reductions in armyworm attack can be achieved!

6. *Quarantine* is a legal method of restricting movement of contaminated plant material. Unfortunately, this technique is rarely effective even though we know that many pest problems arrive on infested plant material. Therefore, we should pay special attention to avoid use of sod that may be infested with billbugs or chinch bugs.

7. *Turf health care (= good turf management)* is one of the simple but commonly ignored methods of pest management. In other words, a "healthy" plant can generally fend for itself against insects, mites, and diseases. Therefore, one of the most important control alternatives that we can use is tending to the proper needs of turf—soil, location, water, fertilizer, and mowing.

Biological Controls

Biological control is using *parasites, predators,* and *pathogens* (diseases) to control pests. We have to realize that in urban landscapes and turf-covered areas, there is a multitude of beneficial insects and mites that can prey on pests. In many cases, these naturally occurring beneficials will do a good job of controlling the pests if we do not disturb the system too much. As stated above, we may disrupt this system by overusing pesticides that kill the beneficials better than the pests. Where new turf pests are introduced, especially from other continents, one strategy is to find biological controls from this native habitat and introduce them where the pest has newly established itself. This has been an especially effective, long-term strategy. Occasionally, we may need to increase the biological controls because the local population may be slow to react to the pest or may have been lowered because of other factors. The classical way to use the biological control option is through introductions, conservation, and augmentation.

- *Introductions* of exotic parasites, predators, or disease are made when foreign pests become established. This is an attempt to create some of the checks and balances found where these pests are naturally controlled. Occasionally, a foreign biological control is found that may better control a native pest! Introductions have to be done carefully so as to not cause secondary problems (like the Asian multicolored lady beetle that annoys home owners).

- *Conservation* is using control materials, usually pesticides, in a manner to have the least adverse affect on predators, parasites, and diseases already present. Spraying pesticide on a flowering plant for aphids may also kill a nectar-seeking wasp that controls white grubs. The aphids could be sprayed with insecticidal soap, which would have almost no affect on the wasps. Conservation can also be the providing of habitat or food needed by biological controls. Planting flowers around turf areas to provide nectar and pollen for predators and parasites can be the key gaining biological control of a pest.

- *Augmentation* is usually the rearing and release of biological control agents. Predators, parasites, or pathogens that are easy to rear are good candidates for the augmentative tactic. Unfortunately, several biological controls are available, but they do not fit the definition of a "good" biological control. If these are used, money is generally wasted!

A "good" biological control is a poor term to use. "Useful" biological control is more appropriate. Useful predators, parasites, and pathogens have the following characteristics:

1. *High reproductive potential* allows a biological control to keep up with the pest population. Pests are often "good" at being pests because they reproduce rapidly.

2. *Good mobility* of a biological control means that it will be able to search out the pests and be able to move to new locations when a pest has been controlled.

3. *Host-specific* biological controls will have reduced chances of adversely affecting nontarget organisms. Praying mantids are fun biological controls, but they eat caterpillars, grasshoppers, and honey bees equally.

4. *Persistent* biological controls will remain when pest populations become low. They will tend to carry over from one season to the next.

5. *Easily reared or encouraged* biological controls are generally less expensive to use and economically competitive with other control costs.

6. *Tolerance of other controls* is a requirement of all our control tactics. To fit into a true IPM approach, biocontrols need to be tolerant of cultural and chemical controls, and vice versa.

Unfortunately, we often think that we have to actively introduce predators, parasites, and pathogens in our landscapes. Since most of these animals already exist, we merely have to be able to recognize them, avoid using poorly targeted insecticides, and provide food and habitat (most parasites and predators need nectar and water).

Important *predators* commonly found in turf are

1. *Lady beetles* are commonly sold as adults and are useful control agents in ornamental plants and crops, if properly handled (Figure 3.24). Lady beetles found in turf often indicate that aphids or mealybugs are present. These species are often not the same ones sold on the commercial market.

2. *Green lacewings* are also found in turf where aphids, mealybugs, or mites are active (Figure 3.25). The predaceous larvae feed on these pests and can be effective in reducing populations. Eggs can be purchased and sprinkled over the area but are of dubious benefit in turf.

3. *Ground beetles* and *rove beetles* are some of the most active and common predators present in most soil/turf habitats (Figure 3.26). Both the adults and larvae feed on a wide variety of pests. Unfortunately, most of these beetles are highly intolerant of pesticides. Target pesticide applications only to the places where pests are reaching damaging levels to conserve these beetles.

4. *Bigeyed bugs* are very common gray to black bugs that are commonly confused with chinch bugs (Figure 3.27). The adults and nymphs prefer to frequent open or sparse turf areas where they suck dry any small insect unfortunate enough to wander by. Bigeyed bugs are common predators of chinch bugs.

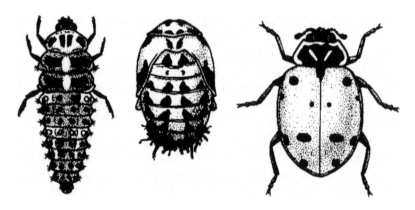

FIGURE 3.24
Diagram of the convergent lady beetle, *Hippodamia convergens*: (left to right) larva, pupa, and adult. (USDA.)

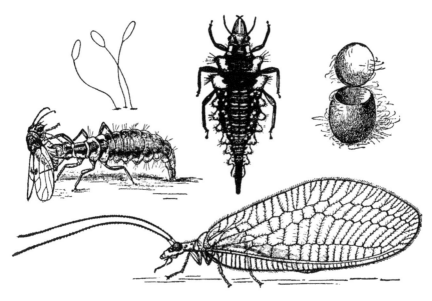

FIGURE 3.25
Diagram of the green lacewing, *Chrysoperla oculata*, stalked eggs, larva, opened cocoon, larva eating psyllid, and adult. (USDA.)

FIGURE 3.26
Most ground beetle adults and larvae are predators of other insects.

FIGURE 3.27
A common bigeyed bug, *Geocoris*, found in turfgrass.

5. *Spiders* are common predators that live and use turfgrass habitats
   (Figure 3.5). Some make webs and others are active stalkers that do
   not make webs to trap prey. Larger spiders, like the wolf spiders,
   are alarming to turfgrass users, but these are essentially harmless to
   humans. Others, like the bowl-and-doily spiders make small sheets
   of webbing on top of the turf canopy. In the morning, these can be
   covered with dew and may be mistaken for fungal mycelium or silk
   from caterpillars! In the fall months, we occasionally see turf covered
   with fine webs (called gossamer webs). These are usually produced
   by recently hatched spiderlings that are "ballooning" (catching a
   ride on the wind by using a strand of silk) to new areas.

Parasites are insects (often called parasitoids) with larvae that usually feed
on the inside of their host, thereby killing or sterilizing it. Some common
parasites that frequent turf habitats are

1. *Chalcid wasps* are usually very small to microscopic wasps that
   lay their eggs in the eggs, larvae, pupae, or adults of other insects.
   Though not well studied or confirmed in turf insect pests, species of
   these wasps have been recovered from sod webworms, cutworms,
   billbugs, and mealybugs. In fact, an introduced species for control of
   the rhodesgrass mealybug has been credited with keeping this pest

under control in much of Texas through Florida. Because of their small size, these parasites go largely unnoticed.

2. *Tiphiid* and *scoliid wasps* are medium-sized to large insects that attack white grubs (Figure 3.28). When abundant, they cause alarm because of people's fear of being stung. Native species often keep masked chafer and green June beetle grub populations in check and there are imported species that attack the Japanese beetle grub. They are usually hairy wasps that are black or they may have bright yellow or orange markings. Adults may hover over the turf during egg-laying periods.

3. *Sphecid wasps* (solitary wasps) are medium-sized to large insects that actively hunt their prey, paralyze it with a sting, and then lay an egg on the hapless victim. The most famous of these wasps in turfgrass is the cicada killer. This wasp digs holes in the soil, often along golf course sand traps or in thin turf, and buries cicadas in chambers where its larvae feed and complete their development. These are considered nuisance pests in turfgrass. On the other hand, the mole cricket wasp enters mole cricket burrows in search of mole crickets. If found, the wasp will paralyze the mole cricket long enough to attach an egg. The wasp larva then feeds on the mole cricket body contents until everything is eaten except for the exoskeleton!

4. *Ichneumonid* and *brachonid wasps* are small- to medium-sized wasps that commonly attack caterpillars in turf. The larvae usually emerge from the dying host and spin small white or yellow cocoons. Small

FIGURE 3.28
Tiphia wasp larva parasitizing a white grub.

masses of these cocoons are occasionally found in the sparse turf left behind after cutworm or armyworm attacks.

5. *Tachiniid flies* are medium-sized, hairy flies that attack a variety of insects. In turf, species are known that attack May/June beetles, various caterpillars, and mole crickets. The red-eyed fly attacking mole crickets was introduced from South America.

Pathogens are simply a variety of disease-causing organisms that kill insects. They are usually bacteria, virus, fungi, and protozoa. Insect pathogens are nearly ideal in that they are very host specific. They are also very noninfective to vertebrates. Examples are

1. *Bacteria*—have been the easiest of the pathogens to utilize because they can often be cultured *in vitro* (in artificial culture) and form spores fairly resistant to adverse environments. Examples are

   a. *Bacillus thuringiensis* (Bt)—has several strains that produce toxins lethal to various insect groups (and are thus technically a chemical control). The most common types are

      i. Bt *kurstaki* affects only young, leaf-feeding caterpillars. Some strains are effective against armyworms and tropical sod webworms. Black cutworms and many of the standard sod webworms appear to be unaffected by these Bt strains, especially if they have reached detectable size in the turf.

      ii. Bt *israelensis* (= BTI) affects aquatic fly larvae such as mosquitoes and black flies. These strains do not seem to be active against the turf-infesting flies such as crane flies, Australian sod fly, and march flies. However, the BTI strains are useful around golf courses that have water hazards where mosquitoes may breed.

      iii. Bt *tenebrionis* (= *san diego*) affects certain leaf-feeding beetles such as the elm leaf beetle. These strains do not seem to be active against white grubs.

      iv. Bt *japonensis* strain "buibui" is very active on various white grub species, especially Japanese beetles and masked chafers.

   b. *Paenibacillus* (= *Bacillus*) *popilliae* (= white grub milky disease) has one strain available that kills Japanese beetle grubs (Figure 3.29). Several strains have been identified that kill other species of grubs but these strains are not commercially available. Milky diseases seem to be "weak" pathogens that typically kill 30–50% of a grub population. This can be significant in moderate grub infestations but probably inadequate in heavy infestations.

   c. *Serratia* sp. (= bacterial honey disease) has been recovered from various grubs but the only commercial preparations are currently

FIGURE 3.29
Two Japanese beetle grubs: normal one (left) and one infected with milky disease (right).

used in management of the New Zealand grass grub. These are not available in the United States.

2. *Fungi* have been identified that attack a variety of turf-infesting insects, but they are difficult to utilize because the spores are easily dried out or need high moisture and/or water to germinate. Examples are

   a. *Beauveria* spp.—have been identified infecting a wide variety of insects, including bugs and beetles (Figure 3.30). A commercial strain is available in Europe for Colorado potato beetle control and numerous companies have tried to develop commercial preparations for chinch bugs. Problems with formulating and maintaining viability have kept commercial preparations from being overly successful in the marketplace.

   b. *Metarhizium* spp.—have been identified infecting numerous soil insects, including white grubs (Figure 3.31). No commercial strains are available in the United States but several strains have been successfully used in New Zealand, Japan, and Europe.

3. *Viruses* are common pathogens of insects but are one of the most difficult to use because they require living insects to grow. While insect tissue culture had promise to increase viral disease use, this has not happened. Only one commercial product, nuclearpolyhedrosis virus (NPV) for gypsy moths, has been commercialized. Recently, a virus attacking black cutworms has been identified and patented, but no commercial products are available.

FIGURE 3.30
Beauveria-infected Japanese beetle grubs.

FIGURE 3.31
Metarhizium-infected green June beetle grubs.

4. *Entomopathogenic nematodes* are a group of tiny parasitic roundworms that carry a bacterium lethal to insects (Figure 3.32). Once
   a juvenile nematode gains entry into an insect, it regurgitates the
   bacterium that paralyzes the insect. The nematode then feeds on
   the reproducing bacteria. Commercial products contain the infective juvenile stage (J3) of various species. Each species and strain
   of nematode seems to be most active against rather narrow groups

FIGURE 3.32
A masked chafer grub that was attacked by the insect-parasitic nematode, *Heterorhabditis* spp.

of insects. These infective juveniles can be applied through conventional spray systems, but since they are living organisms, they need to be irrigated into the turf/thatch/soil before the application dries. When using insect parasitic nematodes, it is advised that you contact commercial suppliers before you need the product. This will give the supplier time to rear up sufficient nematodes for the optimal time of application, thereby avoiding having the nematodes losing vitality through storage. The most commonly mentioned species are

a. *Steinernema carpocapsae* has several strains good at attacking insects that live in the upper soil or on the soil surface. Numerous commercial products are available.

b. *Steinernema riobravos* is a moderately good strain used for mole cricket control as well as other active insects.

c. *Steinernema scapterisci* was imported from South America for control of mole crickets. It can be obtained and this species is often released by governmental agencies for control of mole crickets in pastures and large commercial agricultural lands.

d. *Heterorhabditis* spp. are better at attacking insects that live deeper in the soil. This group can also bore through the insect cuticle.

Chemical Controls

Chemical control has been considered to be the most successful pest management tool in turf maintenance. Unfortunately, we have overused and misused this option so that many turf users cast a wary eye to its use. Chemical control to most people means pesticides, though other chemicals such as attractants and pheromones are increasingly important in our IPM process. Even if pesticides are our principal weapon, we need to understand that not all pesticides are created equal. In IPM, we want to use the ideal pesticide—a material that only kills the target pest and has no other effect. Unfortunately, we do not have these "silver bullets." Most of the modern pesticides currently in use have short residual life spans (this reduces accumulation in the environment), are more selective (this reduces the chance of killing nontarget animals), and are used at lower rates (this reduces the total chemical "load" used). Because of these characteristics, we need to be able to better target our applications to control pests.

Another general public misconception about pesticides is that "natural" pesticides are better than "synthetic" pesticides! IPM principles do not make this distinction. Using pesticides in IPM is evaluated on economic, ecological, and sociological impacts together. In other words, there is a natural botanical insecticide (e.g., nicotine sulfate with an LD_{50} of 55, plus being a mutagen and carcinogen) that is much more toxic and has more adverse environmental effects than a very effective, synthetic insecticide (e.g., chlorantraniliprole with an $LD_{50} > 5000$ and is practically nontoxic to birds, fish, earthworms, and mammals). Because of these issues, chemical controls used in an IPM setting should be selected on their total attributes, not just their origins!

By knowing that we do not have ideal pesticides, we must use great caution to limit their adverse effects. Generally, this means that we should only *target applications* to those areas that need them—*not cover sprays*. General cover sprays (spraying everything whether needed or not) tend to cause several problems. Cover sprays tend to tip the balance of control in favor of the pest. As incredible as this seems, cover sprays usually kill beneficial insects and mites (predators and parasites) better than they kill pests! Since pests usually have good reproductive ability, they "rebound" faster than their natural controls. This causes what we call *pest resurgence* and *secondary pest outbreak*. Cover sprays tend to cause development of resistance. Pests and potential pests often develop resistance to pesticides when they are under constant pressure from a specific pesticide. In other words, a few insects on a plant may not be causing significant damage, but if we constantly spray these insects, we are forcing them to develop resistance. Then, when they reach damaging levels, our pesticide is no longer effective. A more recently identified problem with general cover sprays of pesticides has been identified to be *enhanced* (accelerated) *degradation*. Since most of our current pesticides are organic compounds (i.e., containing carbon, hydrogen, and oxygen), microbes are able to use the chemicals as foods or nutrients. Generally, these microbes are beneficial in aiding in the removal of these pesticides from the environment. However, when constantly "fed" through general cover sprays, these microbes "learn" to "eat" these pesticides more rapidly than normal. In summary, if we are going to use the chemical control option, we need to *use target sprays only when needed.*

To use the chemical control option to best manage turf-attacking insects and mites, the turf manager must have knowledge of the major chemical groups and of the specific problems associated with using these compounds in the turf environment.

Insecticide Groups: Chemical Categories and Modes of Action

The chemical control option of IPM contains traditional pesticides as well as repellents, attractants (commonly called pheromones), and desiccants.

A. *Pesticides* are chemicals that directly or indirectly kill the target pest. Pesticides that kill insects are called insecticides, while those that target mites and ticks are called miticides. There are over 40 recognized insecticide chemical categories and 28 distinctive modes of action (how they kill the insect). This is beyond the scope of this chapter, but you are encouraged to learn about these in other books or through websites. In this section, we will cover some of the broad categories of insecticides and miticides.

   1. *Inorganics* are pesticides without carbon. They can be natural earth minerals or man-made compounds.

      a. Boric acid is used for cockroach and ant control inside houses or buildings, but it is not registered for turfgrass use.

      b. Diatomaceous earth is the glass-like remains of single-celled organisms, diatoms, that scratch insect cuticle or puncture gut cells. This material acts mainly as a desiccant, and is rarely useful in turf unless combined with an insecticide. This is probably because of the humid microclimate located within the turf canopy.

      c. Elemental sulfur is an ancient control for insects and mites. Because of the large quantities needed for control, no products are currently available for usage against turf insects or mites.

      d. Heavy metal salts (e.g., mercury, lead, and arsenates) have been used extensively in the past, and until recently, were available for fungal and weed pest problems. These products are now generally considered too dangerous to use because of adverse effects on nontarget animals, as well as accumulation in the environment.

   2. *Oils* are petroleum- or plant-based hydrocarbon chains that have insecticidal/miticidal activity. Toxic action appears to be accomplished by causing cellular membrane disruption, thereby killing cells that are contacted.

a. Petroleum oils are highly refined mineral oils used on dormant trees and shrubs (dormant oils) or applied to actively growing plants (summer, verdant, or horticultural oils). Petroleum oils have not been developed for turf usage because of phytotoxicity and lack of exposed insects on leaf surfaces.

b. Citrus oils (i.e., D-limonene) have been shown to have insecticidal properties at low dosages. These are usually combined with other insecticides and the products are most effective as contact pesticides.

3. *Fatty acid salts* or *soaps* are generally man-made and consist of fatty acid chains that have been rendered with strong ionic salts, usually containing potassium or sodium, that form molecules that are hydrophilic (attracted to water) on one end and lipophilic (attracted to fats or oils) on the other end. Since cell walls are a lipoprotein matrix, soaps tend to tear cell membranes apart. Fatty acid chains containing 6–10 carbons have insecticidal/miticidal properties. Soaps with longer carbon chains can be photytoxic and are available as contact herbicides. Soaps are useful for control of soft-bodied insects such as aphids, mealybugs, caterpillars, and mites. Because soaps can be phytotoxic, use only registered products on the plants that are listed.

4. *Microbial toxins* are molecules produced by bacteria, fungi, protozoa, and similar microbes. Toxins like the Bt endotoxin (from bacteria) are relatively low in toxicity to mammals, while botulism toxin (another bacterium) is one of the most toxic molecules known to science. The toxins may be used by extracting the microbes cultivated in culture or by using the whole organism. In either case, the chemicals produced by microbes are used rather than expecting an infection by the microbes.

a. *Bacillus thuringiensis* (Bt) is a bacterium species that has numerous subspecies and serotypes (genetically distinct). Some of these produce a protein crystal that attaches to the gut lining of insects or nematodes, destroys some cells, and allows gut contents to leak into the body cavity. The strains of several of the Bts are discussed under biological controls.

b. Avermectins (includes abamectin, and avermectin isomers) are molecules isolated from the fermentation products of *Streptomyces avermitilis*. These products are used primarily for insect and mite control in trees, shrubs, and field crops.

c. Spinosyns are molecules isolated from *Saccharopolyspora spinosa*, a soil-dwelling fungus-like organism. The spinosyns have mite and insect activity and apparently interfere with nerve function. Products containing these molecules are available for turfgrass management.

5. *Botanicals* are plant extracts, usually alkaloid chemicals, that have insecticidal properties. Many people believe that since these are "natural" products, they are safer than other pesticides. Many of these chemicals can have strikingly adverse effects on mammals and other nontarget animals. Many cause severe allergic reactions (e.g., pyrethrin and sabadilla), have high acute toxicity (nicotine), or are even confirmed carcinogens (nicotine).

   a. Pyrethrin is derived from a specific species of chrysanthemum originally grown in Iran. The natural product (ground-up flower petals) is an irritant to insect nervous systems and can cause quick knockdown, but many insects recover. To combat this phenomenon, pyrethrin is usually often mixed with a synergist, such as piperonyl butoxide (PBO), to produce better kill of insects. Some people are very allergic to the material, which can cause intense itching. Pyrethrin-containing products are available for turf, but little efficacy data are available.

   b. Azadirachtins (active isomers in neem tree/seed extract) are extracted from an Asian tree grown primarily in India and nearby countries. Neem extract has been used generally as a cleaning agent, antimicrobial, and breath freshener. The isomers interfere with insect cuticle formation and are considered to be a type of botanical IGR. Products are available for turfgrass usage and products that contained specified amounts of azadirachtins produce consistent results. Crude neem extracts often contain variable amounts of azadirachtins.

   c. Rotenone (= Cube, Derris), sabadilla, ryania, and nicotine are other plant-based insecticides that have been taken off of the market because of toxicity and nontarget issues and lack of patent protection. All can be harmful to humans and other animals if not used carefully.

6. *Synthetic organics* are man-made compounds containing carbon, oxygen, and hydrogen, and are usually synthesized from petroleum-based compounds. This is the group most people refer to when they mention pesticide. Because of the diversity and number of materials in this group, no major attempt will be made to cover all the products and compounds in this category. Currently, there are nearly 40 chemical categories that represent almost 30 different modes of action! Understanding modes of action and other attributes of these pesticides (e.g., length of residual action,

water solubility, and binding capacity) is extremely important in achieving the best level of control when using these tools.

a. Organochlorines (= chlorinated hydrocarbons) are characterized by having extremely long residual life spans and they often had a broad spectrum of insects that were controlled. Unfortunately, these same attributes caused restrictions on their use due to residues ending up in the food chain, in nontarget environments, and in nontarget organisms. Chlordane and dieldrin were commonly used in the 1960s for grub control. A single application controlled grubs for years!

b. Organophosphates generally had shorter environmental life spans, but their neurotoxic effects were commonly seen in nontarget organisms, including humans. Products in this category ranged from highly toxic (category 1, $LD_{50} < 50$) to low toxicity (category 3, $LD_{50} = 501–5000$), and most were banned from urban landscape use in the 1990s. Chlorpyrifos, diazinon, isofenphos, and acephate were commonly used in turf.

c. Carbamates also had shorter environmental effects and their mode of action is nearly identical to the organophosphates. Again, most of these were banned from urban landscape use in the 1990s. Carbaryl and bendiocarb were used extensively in turf.

d. Pyrethroids are synthetics that act like the botanical insecticide, pyrethrin. Pyrethroids are generally in categories 2 (moderate toxicity, $LD_{50} = 51–500$) through 3, but they affect both insect and nontarget animal nervous systems. While pyrethroids are currently widely used in turfgrass management, it is likely that these will also be banned from urban landscape use in the near future. Nearly a dozen pyrethroids have been used in turf. Nearly all end in -thrin (e.g., permethrin, bifenthrin, cyfluthrin, and deltamethrin).

e. Neonicotinoid insecticides came to the market when organophosphates and carbamates were phased out. This group has relatively long activity and has a considerable differential in toxicity to insects compared to other animals. This group basically blocks normal neural functions in insects and affected insects just sit or do not behave normally (feeding, drinking, and escaping predators). Another distinctive attribute of this category is that most molecules are systemic in action, that is, they are taken up by plants. Pesticide in this category range from medium toxicity to practically nontoxic ($LD_{50} > 5000$). Currently, there are four major neonicotinoids used in turf insect management: imidacloprid, thiamethoxam, clothianidin, and dinotefuran.

f. Fiproles or phenylpyrazoles have a unique mode of blocking neural action and this category is represented by one compound, fipronil. This molecule can be used at very low rates and it has systemic as well as long residual effects. It appears to be most active against ants but is also used for control of mole crickets.

g. Anthanilic diamide is the most recent insecticide category to be used in turf insect management. This is represented by chlorantraniliprole. This molecule affects the way that calcium is used in insect muscles, so it only affects insects and related arthropods. Because of the low nontarget toxicity and relatively benign environmental footprint, this is currently considered to be a true, practically nontoxic, pesticide.

h. IGRs represent several chemical categories and each one affects some type of insect growth system (e.g., reproduction, molting, and cuticle formation). In the past, these were synthetic chemicals, but azadirachtin (neem extract) is a botanical insecticide that interferes with insect cuticle formation. Since IGRs act by interfering with or enhancing natural insect hormones, they usually have relatively low vertebrate toxicity and are low in toxicity to other invertebrates. IGRs work rather slowly and mortality takes time. In turf, IGRs have been used to target susceptible immature stages, early in their development.

B. *Repellents* are compounds, both natural and synthetic, that cause a pest to stop feeding and/or move away. Most repellents are used as products applied to the skin or clothing to repel biting flies, ticks, and so on. Azadirachtin (neem) products appear to repel some insect feeding on plants. Various types of pepper, onion, garlic, and herbal extracts have been used to repel insects, but none of these products have demonstrated usefulness in turf.

C. *Attractants* and *pheromones* are compounds that attract a pest that "thinks" that the compound is food or another of the species (aggregation or sex pheromones). Most of the compounds in this group have not been used effectively to reduce turfgrass pests, but attractants and/or pheromones are used in traps to monitor pest activity.

Using Pesticides to Manage Insects and Mites in Turf

Using insecticides and miticides appears to be a relative simple process, but many problems can be encountered, especially in the unique turf environment. Besides targeting the correct "window of opportunity," these pesticides have to deal with thatch and soil organic matter, volatilization, degradation from ultraviolet light or microbes, spray hydrolysis, tank-mix incompatibilities, and pest resistance.

Influence of Thatch

Most turf varieties tend to form a layer of dead leaves and stems held together by grass roots and stolons—thatch. Though a thin layer of thatch seems to be beneficial to the growth of the turf, thick layers can cause the turf to grow above ground and can contribute to disease and insect problems. This layer of living and dead organic material can severely restrict the movement of pesticides. In recent studies on the movement of insecticides through this layer, over 95% of all insecticide types get bound to this layer. Therefore, if the target pest is white grubs or mole crickets, very little insecticide will reach the target if thatch is in the way.

Studies that have used pre- and postapplication irrigation, wetting agents, and core aeration as aids for moving the insecticide through the thatch layer have met with mixed or no improvement in efficacy. Turf managers should attempt to keep thatch to a minimum. Thatch can also be compacted with traffic and partially decomposed thatch with a dry fungal layering can be totally impervious to water-based insecticide applications.

Volatilization

Many insecticides and miticides are liquids at room temperatures. These compounds can volatilize (evaporate) and volatilization generally increases with increased temperature. Emulsible concentrates, when applied to turf foliage and not irrigated shortly after application, may have considerable product loss. Flowable and granular formulations generally reduce these problems. In any case, irrigation as soon as possible after a pesticide application will help move more product into the thatch/soil zone. On the other hand, pesticides applied for foliar feeders such as sod webworms or cutworms should be left on the turf foliage. Ideally, these products should be applied as late in the day as possible, after the turf canopy has cooled and air movement is reduced.

Ultraviolet Light Degradation

Exposure of many chemical compounds to ultraviolet radiation causes changes in the chemical bonds. This may break the compound into inactive molecules or cause it to combine with other chemicals, which also inactivate the pesticide. Pyrethroids, IGRs, microbial pesticides, and botanical insecticides are often highly susceptible to UV degradation. Manufacturers may add UV blockers to their formulations but it is wise to water in applications of susceptible products before significant exposure to direct sunlight occurs. Again, if these products are being targeted for control of foliar pests, apply them in the late afternoon so as to avoid direct exposure.

Accelerated Microbial Degradation

Numerous microbes, usually bacteria and single-celled fungi, are active in the thatch and soil where they help recycle complex organic compounds. These microbes break down complex compounds into components, which can be used as energy supplying food or building blocks of growth. When these microbes are constantly challenged with the same organic compound (a pesticide), especially in a moist, highly organic environment (thatch/soil), the microbes often "learn" how to break down the organic compound and use the constituent parts.

At present, certain pesticides appear to be very susceptible to accelerated microbial degradation while others seem to be fairly resistant. In any case, accelerated degradation is always a threat and wise and judicious use of insecticides should be exercised.

Degradation management usually includes using the pesticide no more than once in a season, alternating the pesticide class with other chemical classes, and applying only the recommended amount at the best time.

Chemical Hydrolysis

Hydrolysis is the general term used in chemistry to indicate that hydrogen atoms are being added to a compound. The addition of hydrogen usually breaks double bonds in a molecule, replaces other atoms, or generally alters the chemical structure. This alteration often causes the chemical pesticide to be much lower in toxicity or completely nontoxic. Most pesticides begin hydrolysis when subjected to extremes of alkalinity (i.e., sweet or high pH) or acidity (i.e., sour or low pH). Alkaline hydrolysis in a tank-mix is one of the most common reasons for pesticide deactivation. Though most pesticides that are extremely susceptible to alkaline hydrolysis have buffers added to their formulations, periodic measurement of a tank-mix is a wise effort. Most tank-mixes should be kept near normal (pH = 7) or slightly acidic (pH < 7).

Generally, if you are using highly alkaline water, mix the pesticide with the water, agitate for a few minutes, and take a pH reading. Meters for measuring pH are relatively inexpensive or pH indicator papers can be used. If the tank-mix remains near pH 8 or above, the addition of a commercial buffer or acidifier may be in order. Most pesticide labels now contain information on when a pH adjustment is needed. If other chemicals have been added to the tank, especially fertilizers, shifts in the pH can occur as agitation and warming of the tank-mix progresses during the day. If the tank-mix is not to be used completely shortly after mixing, take periodic pH readings to see if the mix needs to be adjusted.

Tank-Mixing

To save time, insecticides are often mixed with fertilizers, fungicides, herbicides, or wetting agents. Occasionally, pesticides are not compatible with some of these other materials and either deactivation of the pesticide occurs or some kind of undesirable settling or gelling can occur. Since most pesticide companies no longer provide mixing compatibility charts, it is wise to perform a formulation compatibility test before mixing a large quantity. In a quart container, mix a test batch of the materials to be combined. Stir and shake well and let set for 10–15 min. If a precipitate settles out (especially if wettable powders or flowables were not used), which is difficult to suspend, do not use the mix. Likewise, if a gel or greasy mass forms on the edge of the container, the mix is probably incompatible. Unfortunately, mixing compatibility tests do not indicate whether any pesticide deactivation has occurred. If you do not obtain expected results from a pesticide mixed with other materials, suspect chemical incompatibility.

Pest Resistance

Pest resistance to pesticides is a well-documented phenomenon. When enough of a population is challenged by a pesticide, those few animals that survive often have a genetic factor that allowed them to survive. These survivors pass on this factor to their offspring and the next generation has more individuals carrying the resistance factor. If this is allowed to continue, eventually most of a pest population carries one or more resistance factors to the pesticide and they are no longer susceptible to the compound.

Resistance seems to occur where pesticides remain in the environment for a long time (thus the pest is constantly under pressure to change) or a large proportion of the population is continually exposed (all the crop, or turf, is sprayed with the same material several times in a season). Obviously, good methods of managing this problem are to use pesticides with relative short residual periods, apply a pesticide only when needed, apply the pesticide only to the area needing pest reduction, and alternating products.

At present, since the loss of the long residual organochlorine pesticides, few turf insects have been found with resistance to other classes of pesticides. However, certain pests such as the southern chinch bug and green bug are notorious for having resistance to organophosphate and carbamate insecticides.

Insecticide/Miticide Affects on Nontarget Animals

Though herbicides, fungicides, insecticides, and miticides are all pesticides used to manage pests, insecticides and miticides have the unique position of acting on the nervous systems or metabolic pathways in animals. Because of their mode of action, insecticides and miticides may have dramatic effects on nontarget animals. When used according to label instructions, most insecticides and miticides will have only temporary and minimal adverse effects on nontarget animals. However, turf managers need to be constantly aware of the potential for problems.

Most beneficial insects found in the turf environment, especially ground beetles and rove beetles, are very sensitive to insecticides. Fortunately, if insecticides are not applied over the entire area (a complete golf course, including the roughs), these highly mobile predators rapidly recolonize the treated area. However, if periodic, repeat applications are made, these predators can eventually lose their effectiveness. This is another reason why general insecticide cover applications should be replaced with targeted applications to those areas needing management.

Several pesticides, especially carbamates, appear to have adverse effects on earthworm populations. Earthworms are significant actors in the environment's decomposition of thatch. Current research indicates that earthworm populations can recover rather rapidly if applications of such pesticides are restricted to once per season.

Several insecticides have warnings concerning bird and fish toxicities. For birds, the greatest concern is for herbivorous waterfowl such as geese and some ducks. Turf managers need to survey areas to be treated to see if lakes, ponds, or streams are nearby or waterfowl are known to forage in the area. Of greatest danger are granular insecticides that may be picked up as the waterfowl works the edge of a waterway, and surface and liquid applied insecticides that are allowed to remain on the turf foliage.

Fish and other aquatic inhabitants are additional problems for turf managers trying to deal with pests in turf near waterways. Most contamination of ponds, lakes, and streams comes from contaminated surface water. The two most common causes are applications of insecticides to areas surrounding water followed by too much irrigation or sudden rain storms, or applications of pesticides to water-saturated soils followed by additional irrigation or sudden rain. Insecticides that require posttreatment irrigation for safety or efficacy reasons should be applied to turf with underlying soils, which can absorb the required irrigation. If the soils are saturated, wait a few days until conditions improve. Though light rain can assist in the movement of insecticides into the thatch/soil zone, sudden summer downpours can be very unpredictable. It is often wiser to wait until the potential threat is gone before making the pesticide application.

Equipment for Making Insecticide/Miticide Applications

For insecticides and miticides to have their maximum efficacy, they must be applied at the right time and should be applied evenly at the correct rate. Therefore, selection of the appropriate application equipment by the turf manager is very important for consistent results. Most insecticides are applied as liquid sprays or dry, granular materials.

Liquid sprays are applied through large area applicators (usually spray booms) applying at 20–40 gallons of spray per acre. Boom sprayers leave most of their application on the turf foliage. When emulsible concentrates (EC) and soluble powders are used, most of the insecticide will remain on the turf foliage. Therefore, a liquid-applied EC targeted for white grubs should be irrigated in immediately after application, and preferably before the application is allowed to dry. Wettable powders and liquid or dry flowable formulation are somewhat less adversely affected by drying before irrigation occurs. On the other hand, if surface or upper thatch dwelling pests are the targets, no irrigation should be applied.

Granular products are generally applied using drop or broadcast spreaders. Granulars need to be applied uniformly over the area and both drop and broadcast spreaders can cause problems in getting a uniform application. The drop spreader requires that each granule be calibrated for its individual size and weight. Broadcast applicators (rotary, flaying arm, etc.) require granules of rather uniform size and density. Otherwise, larger and heavier granules are liable to travel further than smaller and lighter granules. Each broadcast applicator produces a unique pattern with each granular product. Careful calibration is needed to ensure a uniform pattern.

Subsurface applicators are being used to inject or slit-insert both liquid and granular insecticides into the soil/thatch zone. In the southern states, this has appeared to improve activity against mole crickets. However, applications for white grub management have been less consistent or even dramatically less in control than conventional surface applications. Subsurface applicators should reduce the chances of leaving insecticide residues on the turf surface and should place more insecticide in the zone where the pest may make contact. Considerably more research is needed to bring this technology to its full usefulness.

Leaf- and Stem-Infesting Insect and Mite Pests

This category includes those arthropods that feed on the upper leaves and stems of turfgrass plants. Many of these pests often hide in the thatch, others remain exposed on the leaf surfaces, and the rest hide in the spaces beneath leaf sheaths and between nodes. Those insects with chewing mouthparts eat entire leaves and stems. These are usually the larvae of various moths and butterflies. The rest of the pests have rasping or sucking mouthparts and include the mites, thrips, aphids, and mealybugs.

Leaf-chewing pests leave behind ragged edges on the leaves, sunken spots in the turf and, in the case of severe infestations, they eat all the green material down to the brown thatch. Pests with sucking mouthparts tend to discolor the turf, leaving it yellowed, rusted, or blanched white in color.

Bermudagrass Mite

Species: *Eriophyes cynodoniensis* Sayed (Phylum Arthropoda, Class Arachnida, Order Acarina, Family Eriophyidae) (Figures 3.33 and 3.34).

Distribution: In North America, this pest is found in all the states where bermudagrass is grown. It has also spread worldwide.

Hosts: Bermudagrass is the only known host.

Damage symptoms: Damage is first noticed when bermudagrass does not have vigorous growth in the spring and is often yellowed. The turf appears stunted, and close inspection reveals that the stem length between nodes is greatly reduced. Leaves and buds become bushy,

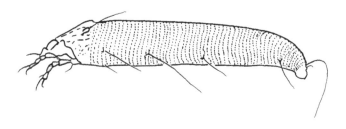

FIGURE 3.33
Diagram of the bermudagrass mite, *Eriophyes cynodoniensis*. (Ariz. Coop. Ext.)

FIGURE 3.34
Witchesbrooming of bermudagrass caused by the bermudagrass mites.

forming a rosette or tuft that is called "witchesbrooming." Heavy infestations produce clumps that eventually turn brown and die.

Description of stages: This mite has stages typical of the eriophyid mite group. These are extremely small, pear-shaped mites with worm-like, soft bodies.

Eggs: Round, translucent white eggs are about 0.002 inch (0.06 mm) in diameter.

Nymphs: The first nymph has the tapered shape of the adult but is almost clear. These mites have only two pairs of short legs, rather than the four pairs of legs found on most mites. The abdomen has minute rings, which look like segments. The second nymph is about 0.005 inch (0.12 mm) long, and more white in color.

Adults: Only females are known. These look like the nymphs and are only 0.006 inch (0.2 mm) long when fully grown, but have a creamy white color.

Life cycle and habits: Because of their small size, these mites are very difficult to study, and little is understood about their life cycles and habits. Most eriophyids lay less than a dozen eggs during their adult span and these usually hatch in 2–3 days. At 75°F (23.9°C), it is estimated that adulthood is reached in 7–10 days, and eggs are laid for 2–5 days. Thus, a cycle can be completed in 1.5–2 weeks. This short time period allows for a rapid buildup of a population during summer temperatures. The bermudagrass mite seems to be quite tolerant of high temperatures, having moderate mortality at 120°F (48.9°C). Cold temperatures tend to stop development, though survival during the winter can take place where bermudagrass remains green at the soil surface. This mite can apparently spread by being blown or carried on the bodies of other insects. However, the most common method of spreading is by transportation of infested turf. The mites cannot survive on bermudagrass seed.

Control options: Eriophyid mites usually are not killed by normal miticides, but are killed by many insecticides. Controls are probably warranted when 4–8 witchesbroomed tufts per square foot are encountered in an area.

Sampling

A 3×4 ft (0.9×1.2 m) plastic rectangle should be strung on 1 ft^2 grid (12 ft^2 total). The sampling hoop consists of four sections of 1-inch PVC pipe. Two sides are 3-ft segments and the others are 4-ft segments. The segments are joined with right angle connecters and glued. Holes are drilled through the sides at 1, 2, and 3 (on the 4-ft segments only) ft from each corner. A monofilament string is then threaded, in a grid pattern, through the holes. This hoop is tossed onto the turf and 10 of the square foot grids are rated for mite activity. Each tuft of turf with a witchesbroom is counted. Fairways and roughs should be sampled every 50 yards, and four hoop samples should be taken for each green and apron. Tees should have two hoop samples. When the presence of the mite activity is noted, samples should be taken every month. If the activity is increasing and exceeds four to eight tufts per ft^2, chemical controls are probably warranted. Below four tufts per ft^2, cultural controls should be used.

Option 1: Cultural control—use resistant varieties: Common bermudagrass is often attacked, but improved varieties such as Tifgreen (238) and Tifway (419) have shown considerable resistance.

Option 2: Cultural control—turf maintenance: The bermudagrass mite does not do well in short turf. However, mowing too short may scalp the turf. Good fertilization and water will help reduce stress and mask mite populations. Mite attacks are seldom damaging during wet periods.

Option 3: Chemical control—soft pesticides: Though not specifically registered for this pest, some of the insecticidal soaps can be used on turf. Industry reports indicate that these soaps, when used with sufficient water to thoroughly wet the turfgrass blades and stems, are effective in controlling turf attacking mites.

Option 4: Chemical control—traditional pesticides: Proper identification is needed because water stress can look like early mite damage. A microscope with at least 30× magnification will be needed to adequately see the mites. Short residual pesticides may have to be reapplied in 7–10 days to kill mites hatching from eggs. At present, there are no miticides or insecticides registered in the United States for control of this mite. In the past, diazinon, fluvalinate, and chlorpyrifos have been effective. The only true miticides that have activity against eriophyid mites are avermectin and spiromesifen.

Clover Mite

Species: *Bryobia praetiosa* Koch, with several biotypes recognized (Phylum Arthropoda, Class Arachnida, Order Acari, Family Tetranychidae) (Figure 3.35).

Distribution: A cosmopolitan species found in North and South America, Europe, Asia, Africa, and Australia.

Hosts: This pest attacks a wide variety of plants, including several turfgrasses such as Kentucky bluegrass and ryegrasses.

Damage symptoms: These mites rasp the surface of grass blades and leave a silvery appearance to the upper surface. The major problem with these mites is their nuisance activities. They tend to migrate into houses during population flushes in the spring and fall. Though they do not transmit any disease, nor do they bite, they leave a red stain when crushed, which is difficult to remove.

FIGURE 3.35
Clover mites, cast skins, and eggs located on the side of an irrigation control box.

Description of stages: This mite has egg, larval, protonymphal, deutonymphal, and adult stages.

Eggs: The small round eggs are shiny red-orange colored and about 0.005 inch (0.12 mm) in diameter.

Larvae: The larvae have only three pairs of legs and are reddish in color.

Nymphs: The nymphs develop another pair of legs and have the typical adult form of a slightly depressed, oval body with elongate front legs. Under high magnification, the body is covered with tiny fingerprint-like ridges.

Adults: Only females are known, and these are reddish- to chestnut-brown in color. They have the front legs about twice the length of the other legs and are about 0.016 inch (0.4 mm) long.

Life cycle and habits: In the cool-season turfgrass zones, this pest overwinters in the egg stage but adults can sometimes be found in protected areas. Clover mites also oversummer in the egg stage, being active during the cool of the spring and fall. In the southern zones, this pest oversummers in the egg stage but adults and nymphs are more common during the entire winter months. In the spring, when the temperatures rise above freezing, the eggs hatch, or in milder climates, the adults become active and begin to lay eggs. This means that both eggs and adults may be present during the spring. The spring eggs hatch in a week at 28–48°F (–2.2–8.9°C), while overwintered eggs will hatch in 12–18 h. Overwintering adults continue to lay eggs until mid-April, while new spring adults lay eggs that do not hatch until the following fall. The spring generation takes about a month to mature and is active from snow melt to mid-June. The oversummering eggs hatch in September and the mites mature in 25–35 days. These mites may lay additional eggs, which hatch at this time, or the eggs may delay hatch until the following spring. The mites are strongly attracted to warm surfaces during cool weather and will climb up the sides of buildings and trees. On buildings, they may enter doors or windows and become a nuisance inside.

Control options: This is a sporadic pest that has nuisance populations during favorable years when long cool springs or falls help build up populations. Generally, little turf damage occurs, but home invasions may become unacceptable.

Option 1: Cultural control—reduce oviposition sites: Since this mite prefers to lay eggs away from the turf and on the sides of buildings and on the trunks of trees, creating wide mulch areas or making a border of small stone or gravel will reduce oviposition.

Option 2: Chemical control—barrier treatments: Since the mites may enter houses, and populations often build up in the shady sides of buildings, treat the parameter turf of buildings with an appropriate miticide.

Option 3: Chemical control—general turf sprays: Monitor mite populations during spring or fall when cool temperatures have lingered. If the mite populations are building, general cover sprays of a miticide will reduce the population and prevent house invasion. Use selective miticides that do not harm predatory insects and do not delay treatments in the spring, as the mites will lay summer eggs that are not susceptible to the miticides.

Banks Grass Mite

Species: *Oligonychus pratensis* (Banks) (Phylum Arthropoda, Class Arachnida, Order Acarina, Family Tetranychidae) (Figure 3.36).

Distribution: Originally described from the Pacific Northwest, but now known to occur from Washington state to Florida and south. Also known from Hawaii, Puerto Rico, Central America, Mexico, and Africa.

Hosts: Commonly attacks Kentucky bluegrass in Washington state, Oregon, and Colorado, but switches to bermudagrass and St. Augustinegrass in the southern states.

Damage symptoms: Lightly infested plants have small yellow speckles along the grass blades. As damage progresses, the leaves become more straw colored and they eventually wither and die. This often happens in hot, dry spells and the damage may be mistaken for summer dormancy. This mite overwinters in all stages, and in warm winters, large numbers of mites may cause significant damage by the following spring. This mite produces considerable webbing at

FIGURE 3.36
Banks grass mites and damage on St. Augustinegrass leaf blade.

the bases of turf tillers and this may be easily seen in the morning dew. In St. Augustinegrass, the mite egg shells and webbing are sometimes mistaken for molds or dust. Concentrations of the mites on the tips of southern grasses cause general yellowing and dieback similar to heat or drought scorch.

Description of stages: This is a true spider mite that has egg, larval, protonymphal, deutonymphal, and adult stages.

Eggs: The eggs are spherical and about 0.005 inch (0.125 mm) in diameter. They are first pearly white but change to a light straw-yellow color before hatching.

Larvae: The larvae are oval in shape and have only three pairs of legs. When newly hatched, the larvae have red eye spots and salmon-colored bodies. After feeding, the gut contents cause the body to become light green. The front pair of legs remains light orange in color.

Nymphs: Both the protonymphs and deutonymphs have four pairs of legs and continue to be bright green as they feed.

Adults: The adults are quite sexually dimorphic: the females are broadly oval and about 0.016 inch (0.40–0.45 mm) long, while the males have a strongly tapered abdomen and are only about 0.013 inch (0.33 mm) long. During the spring to fall feeding periods, the adults are bright green with light orange legs, but during the winter, the green fades and the mites take on a bright orange-salmon color.

Life cycle and habits: This spider mite can be active any time during the year when temperatures are high enough. Usually, mature, mated females overwinter at the base of grass plants and in the soil, though a few males and nymphs may also be present. These overwintering females lose their green color and become orange-salmon colored. As spring arrives, the surviving mites begin to feed on emerging grass and gradually the mites turn greenish. After the normal color is established, the females begin to lay eggs in their copious webbing or on the plants. The females lay 50–70 eggs during their life span, though females laying over 100 eggs are known. The eggs take 4–25 days to hatch, depending upon the temperature. If turf temperatures are above 70°F (21.1°C), the larval stage takes only 2 days, the protonymph takes 1 day, and the deutonymph takes 2 days to mature. This means that during hot weather, eggs can hatch in 4 days and development may be completed in 5 days. During cool spring and fall temperatures, complete development of the immatures may take 25–37 days. This means that six to nine overlapping generations may occur in a season. As soon as the adult females emerge from the deutonymph stage, males begin copulation. If no males are present, the females have the ability to lay unfertilized eggs, which develop only into males. These males, in turn, can mate with the female, their mother, and she can then produce female offspring. Only mated females can produce female mites. During hot, dry summer weather, immature mites and sometimes adults migrate to the center of dormant grass clumps and rest until the grass returns to active growth following rains.

Control options: Since this mite is a true spider mite, general miticides are needed for control; most insecticides are not effective.

Option 1: Cultural control—watering the turf: This mite requires warm, dry conditions for optimum survival and reproduction. Irrigation at regular intervals seems to greatly reduce mite populations.

Option 2: Chemical control—spray scheduling: Since the eggs may take 2–3 weeks to hatch in cooler weather, reapplications of miticides based on their residual activity periods may be necessary. Set up a reapplication schedule that will catch any immatures that hatched since the last application. Also keep in mind that you want to control these immatures before they reach the adult stage and additional eggs are laid. Most miticides have activity against this spider mite.

Winter Grain Mite

Species: *Penthaleus major* (Duges) (Phylum Arthropoda, Class Arachnida, Order Acari, Family Eupodidae) (Figure 3.37).

Distribution: Common in small, grain-producing states west of the Mississippi river. Recorded attacking bluegrass and fescue in the

FIGURE 3.37
Winter grain mites have distinctive orange-red legs and a dorsal anus.

northeast. Also found in Australia, China, Europe, New Zealand, South Africa, South America, and Taiwan.

Hosts: Cool-season grasses such as bluegrass, ryegrass, and fescue; occasionally bentgrass.

Damage symptoms: Damage is often misidentified as winter kill or snow mold. Turf leaves first appear silvered on the tips with more severe damage producing scorching and browning. Damage often occurs under snow, and brown patches appear when the snow melts.

Description of stages: Typical mite life cycle except the eggs oversummer.

Eggs: Freshly laid eggs have a glistening reddish-orange color. These are glued to the bases of grass plants on the roots or on pieces of thatch. After drying for a day, the eggs become wrinkled and more straw colored.

Larvae: As with all mites, the larvae have only three pairs of legs. Larvae are reddish-orange just after hatching but turn dark brown to black as they feed. The mouthparts and legs remain reddish-orange.

Nymphs: Two nymphal instars are found. When the larva molts into the nymphal stage, a fourth pair of legs is gained. The nymphs look like the adults, olive-black with reddish-orange legs, but are smaller. The nymphs also generally have a more tapered abdomen.

Adults: The adults are relatively large for mites, up to 3/64 inch (1 mm) long. They are the only turf-inhabiting mites with olive-black bodies, redorange legs and mouthparts, a pair of white eye spots, and a dorsal anus. Only females are found, though males have been reported.

Life cycle and habits: The most distinctive feature about the winter grain mites' life cycle is the oversummering eggs and winter mite activity. In the northern United States, the mites appear to hatch in mid to late October when soil surface temperatures are approaching 50°F (10°C). The larvae feed by rasping the surface of grass blades and sucking up the cell contents. Within a few days, the larva molts into the nymphal stage, which feeds in the same manner for a week or two. The mites tend to hide during daylight and can be found clustered on the crown of grass plants, in the hatch, and at the soil surface during warm bright winter days. Apparently, snow cover does not inhibit feeding and may actually afford protection. Females can live up to 5 weeks, during which they may lay 30–65 eggs. Eggs laid from November through March usually hatch that winter, but eggs laid from March onward usually oversummer to hatch the following fall. It appears that two overlapping generations may occur during the winter, with peak populations being found in late December and late February. Winter grain mites often produce a droplet of liquid from the anus if disturbed. This may be a defensive action, though undisturbed, feeding individuals also produce droplets. This mite is also very susceptible to desiccation, often becoming inactive and rapidly shriveling if moved to a dry spot. This mite also seems to be reactive to carbamate insecticides. Carbamates seem to kill natural mite predators as well as stimulate reproduction of some mites.

Control options: Damage by this pest is almost impossible to predict. When turf has been treated in summer with a carbamate, such as in cutworm or webworm control, winter grain mites often attack during the winter.

Option 1: Cultural control—mask damage: Since this pest rarely kills the turf, the normal damage is spring silvering of the turf. This can be rapidly masked by applying a dormant fall fertilization or applying a light spring fertilizer application to encourage rapid recovery of the turf.

Option 2: Chemical control—avoid regular use of carbamates: Alternate or use other types of pesticides to control summer turfgrass pests. This may reduce the amount of destruction of natural mite predators.

Option 3: Chemical control—early applications of miticides: Since it is hard to apply pesticides to turf under snow or in cold temperatures, treatment of the turf in the fall has been suggested. This has not worked well in the past, though new mite ovicides (egg poison) are being developed, which may prove useful in the future.

Option 4: Chemical control—spring applications of pesticides: Most organophosphate insecticides will satisfactorily kill this pest in the spring
when temperatures allow spraying. Miticides used to control spider mites may not kill this pest, but some insecticides do! Where
allowed, chlorpyrifos has been effective.

Greenbug

Species: *Schizaphis graminum* (Rondani) with several biotypes (Phylum
Arthropoda, Class Insecta, Order Homoptera, Family Aphidae)
(Figures 3.38 and 3.39).

Distribution: Reported to have damaged turfgrass from Kansas to New
York and south into Kentucky and Maryland; worldwide pest of
cereal grains in Europe, Africa, and North America.

Hosts: Different biotypes prefer wheat, sorghum, oats, and over 60 members
of the grass family; prefers Kentucky bluegrass but will survive
and reproduce on Chewings fescue and tall fescue.

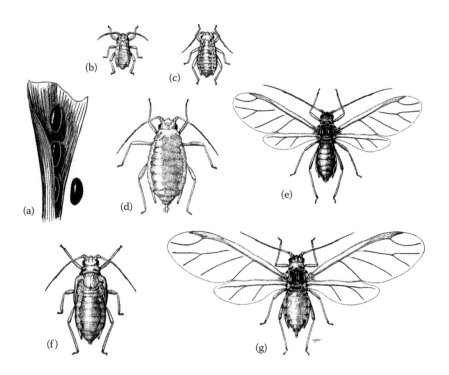

FIGURE 3.38

Greenbug, *Schizaphis graminum*, life stages: (a) eggs on leaf blade; (b) first instar nymph;
(c) second instar nymph; (d) wingless live-birthing female; (e) winged male; (f) "pupa" of
live-birthing female; and (g) winged live-birthing female. (USDA.)

FIGURE 3.39
Greenbugs on grass blade.

Damage symptoms: Individuals suck plant juices, which rob the plant of nutrients and water. However, this aphid has a toxic agent in its saliva that causes the leaf tissue around the feeding site to turn yellow and then burnt-orange. Injury usually begins in shaded areas under trees or next to buildings, often along the north and east sides. This damage may then spread toward sunny areas with the general turf color changing from green to yellow to brown. Damage begins in June and may continue until a killing frost occurs. This pest seems to appreciate hot dry conditions, as rainy periods reduce population outbreaks. Severe damage is most evident in November but may also show up from late July through September.

Description of stages: Though aphids technically undergo gradual metamorphosis, they have complicated life cycles in which the females give live birth to nymphs asexually (ovoviviparous, parthenogenesis) until late fall. At this time, sexual forms are produced and overwintering eggs are laid. Several other species of aphids may be found on turfgrasses, but most of these are dark or not green in color.

Eggs: The eggs are elongate oval, about 1/32 inch (0.8 mm) long, and are green when first attached to grass blades. They turn a shiny black in a couple of days.

Nymphs: These look like adults except they are smaller. The pear-shaped body is light green and usually has a darker green stripe down the back. The tips of the legs, antennae, and cornicles (the tail-pipe-like structures on the abdomen) are black. Nymphs destined to become winged forms have obvious wing pads in the last instar and are called "pupae."

Adults: Adults are about 5/64 inch (2 mm) long and have the same green color of the nymphs, as well as black markings. Winged forms usually appear whenever crowding occurs, often after considerable damage has taken place. Winged adults are usually darker green and have the wing veins marked with black.

Life cycle and habits: Greenbugs were suspected to migrate into turfgrass from small grain, agricultural fields in the south and west. However, eggs of some of the biotypes have been found that can overwinter in the northern climates, and nymphs have been found in northern turfgrasses too early in the spring to have been blown in. Also, certain lawns will be infested year after year, while adjacent lawns are not attacked. All aphids that hatch in the spring are females that reproduce asexually (parthenogenesis) by giving birth to first instar nymphs (ovoviviparity). In cool weather, these nymphs may take 2 weeks or more to mature but as summer temperatures rise, only 7–10 days are needed. Thus, populations can literally explode in a few weeks as mature females may produce two to three nymphs per day. Greenbugs seem to prefer the shade and build up populations in the shade of trees, buildings, and fences. Though greenbugs do best in the shade, they reproduce most rapidly in warm, dry weather. As daylight periods decrease in the fall, nymphs are produced, which grow into winged sexual forms that can fly to new areas. Apparently, these forms seek out suitable places to lay overwintering eggs. Greenbugs produce less honeydew than most of their relatives, but enough is often present to attract ants, bees, and other sugar-seeking insects.

Control options: Many natural controls normally keep this pest in check, but these may be missing in turfgrass habitats. Some populations of this pest may be resistant to organophosphate and/or pyrethroid insecticides.

Option 1: Natural control—allow natural control agents to attack greenbugs: Lady beetles, lacewings, and parasitic wasps often seek out the greenbugs and effectively reduce populations if preventive insecticide sprays used to control chinch bugs, sod webworms, and cutworms have not been overused.

Option 2: Cultural control—use resistant turfgrasses: Perennial ryegrass, zoysiagrass, and bermudagrass are not attacked. Research has shown that endophyte-enhanced turfgrasses are generally resistant to greenbugs. Kentucky bluegrass clones that have been identified as being resistant to two greenbug biotypes are Kenblue 46, A-34 (GB-3), Wabash 21, Delta 20, and Cougar 6. Others are being identified, so contact seed source companies for new releases. If a lawn has been so severely damaged by greenbugs as to require renovation, consider using a resistant turf variety.

Option 3: Chemical control: Greenbugs are relatively easy to control with contact and systemic insecticides applied to active populations. However, some biotypes appear to have some resistance to organophosphate and/or pyrethroid insecticides. If an application of an insecticide has not controlled a population, select another pesticide with a different mode of action.

Sampling

Greenbugs are most commonly detected by looking for the yellow-orange discoloration of turf around trees and next to buildings. By looking closely at this turf, the aphids can be seen lined up along the upper surface of leaf blades. Color-blind individuals may do better by taking an insect sweep net and using it in the area. The darker aphids readily show up next to the white cloth of the net. Sweep netting will also help detect the presence of lady beetles and lacewings.

Insecticides and Application

Because greenbugs may be resistant to several of the insecticide groups, select an insecticide with a different mode of action if a treatment fails to control the insects. Greenbug populations are usually very localized in turfgrass, so an application of a neonicotinoids or other aphid-active insecticides only need to be applied where the aphids are active.

Sod Webworms (= Lawn Moths): Introduction

Species: Many species of sod webworms attack turfgrasses in North America. The most common type, several species that used to belong to the large genus *Crambus* (family Crambidae), rests on grass blades during the day and characteristically rolls the forewings tube-like around the body (Figure 3.40). The imported tropical sod webworms are more devastating to southern turfgrasses, hold the forewings roof-like over the body, and belong to the family Pyralidae. Usually two or three species may be causing damage in any given area, and species complexes vary across the continent. (Phylum Arthropoda, Class Insecta, Order Lepidoptera, Family Crambidae.)

Distribution: Different species are common across North America. Species that seem to prefer cool-season grasses are bluegrass sod webworm, *Parapediasia teterrella* (Zincken); larger sod webworm, *Pediasia trisecta* (Walker); the western sod webworm, *Pediasia bonifatellus*

FIGURE 3.40
Striped sod webworm, *Fissicrambus mutabilis,* in its typical resting pose.

(Hulst); striped sod webworm, *Fissicrambus mutabilis* (Clemens); the elegant sod webworm, *Microcrambus elegans* (Clemens); and the cranberry girdler, *Chrysoteuchia topiaria* (Zeller). Some of these also occur in the warm-season zones. Additional species of crambids are more common in warm-season turfgrasses but the imported tropical sod webworm, *Herpetogramma phaeopteralis* Guerne, is the principal pest.

Hosts: All species of turfgrasses are attacked.

Damage symptoms: The crambid types generally construct tunnels in the soil and thatch, lining them with silk. From these hiding places, they cut down individual blades of grass. This eventually gives a sparse and ragged appearance to the turf. Extensive infestations may lead to irregular brown patches of turf, especially in dry periods. On the short-cut surfaces of golf greens and tees, sod webworm larvae make silk-lined tunnels across the soil surface, just below the mow line (Figure 3.41). This activity produces distinctive trails or marks on the surface. These trails do not interfere with ball roll, but foraging birds can peck the surface in search of the underlying larvae. The tropical sod webworm will web foliage and

FIGURE 3.41
Sod webworm burrow mark on the surface of a bentgrass putting green.

often feeds along the tips and edges of grass blades. Large populations can greatly thin and web over turf. Birds are commonly seen
feeding where sod webworm populations are high.

Life cycles: These pests have complete life cycles with eggs, larval, pupal,
and adult stages. Species in the northern areas have one to three
generations per year, while southern species are inactive only during cold weather.

Identification of species: Over 30 species of sod webworms have been identified in North America. However, only about half of these species
commonly occur in turfgrass areas and even fewer species reach
pest status. The adults are fairly easy to identify to species by
using wing color patterns and male genitalia. The larvae are quite
difficult to identify to species, and an expert should be consulted if
larval identification is needed. The crambid types lay ribbed eggs
by dropping them into the turf, and the tropical sod webworms
attach flat scale-like eggs to blades of grass.

Control options: Most sod webworms are easy to control, though they may
be difficult to reach within their silken tunnels. Sampling can be
done by using a disclosing solution over a square yard of turf and
counting the number of emerging larvae. Adult activity can be
monitored using an insect net or light trap.

Option 1: Cultural control—use fertilizer and water: Damage can often be outgrown if water is continually available and fertility levels are
appropriate for regrowth. Considerable damage may occur if

irrigation is not possible during periods of drought, or close mowing is used.

Option 2: Biological control: Natural parasites are known, but ground beetles
and rove beetles are major predators of eggs and smaller larvae.
Fungal and viral diseases have also been identified, but these usually do not provide consistent control. The insect parasitic nematodes, *Steinernema* spp., seem to provide adequate control of this
group when used at 1×10^9 juveniles per acre. Nematode efficacy
can be improved by applying them in the early morning or late
afternoon when sunlight is at a minimum, the thatch has been
thoroughly moistened, and irrigation occurs immediately after
application (before the spray droplets dry).

Option 3: Cultural control—use resistant turfgrass varieties: Resistance against
the tropical sod webworm has been observed in bermudagrass
and zoysiagrass selections, and resistance to crambids has been
demonstrated in bluegrass cultivars. Perennial ryegrasses, tall
fescues, and fine fescues with fungal endophytes are also highly
resistant to sod webworm attacks.

Option 4: Biobased control—microbial toxins, BT: Several strains of the bacterium *Bacillus thuringiensis* have been shown to control sod webworm larvae. Bt products are most effective against young larvae.

Option 5: Chemical control—use contact and/or stomach pesticides: Most sod webworms are easily controlled if the pesticides are ingested or penetration of the webbing tunnels is achieved. Since the larvae feed
shortly after dark, the best control is achieved by spraying in the
late afternoon and not irrigating after the application (to leave the
pesticide residues on the leaf blades). Some species have multiple
generations in a season and this may require additional treatments
to control new larval populations produced by adults that have
flown in from untreated areas.

Sampling

Sod webworm larvae (Figure 3.42) are sometimes difficult to confirm in
turf. They may build webbed tunnels through the thatch or into holes in
the ground. In the short-cut turf of golf course greens and tees, the larvae produce horizontal, silk-lined burrows just under the mow line. These
will appear as short, brown streaks on the surface. Visual inspection often
reveals larger, sawdust-like fecal pellets (=frass) with silk webbing. Green
frass indicates recent or current activity. A soap disclosing solution should
use two gallons sprinkled over a 1 yd^2 area to force any caterpillars to the
surface. Generally, 5–10 larvae per ft^2 may warrant control. However, on golf
greens and tees, this threshold is much lower due to potential bird feeding!
Bird feeding may indicate sod webworms, but is not a confirmation of their
presence.

FIGURE 3.42
A nearly full-grown sod webworm larva exposed in the thatch of lawn turf. Notice the distinctive spots and green frass material.

Insecticides and Application

Most of the pyrethroid insecticides registered for turf use are very effective against sod webworms. These have generally replaced the organophosphate and carbamate insecticides due to use restrictions. Other insecticides used for white grub and mole cricket control are often not very effective against caterpillars. Liquid applications that are not irrigated in have performed better than granular applications of the same insecticides.

Bluegrass Webworm

Species: *Parapediasia teterrella* (Zincken) (Phylum Arthropoda, Class Insecta, Order Lepidoptera, Family Crambidae) (Figure 3.43).

Distribution: Found in the eastern half of North America.

Hosts: Prefers Kentucky bluegrass but also feeds on ryegrass, fine and tall fescues (without endophytes), as well as weed grasses such as crabgrass and orchardgrass.

Damage symptoms: Closely mown turf shows symptoms more rapidly than poorly maintained turf. Individual larvae can cause small depressed pot marks of brown grass. As the larvae grow, these spots may enlarge into individual areas of several inches in diameter. Heavier infestations have the enlarging dead patches touching to form irregular patterns of sparse turf. Poorly maintained turf may have a general ragged appearance with many scattered dead stems. Birds often make probing holes in the areas of

FIGURE 3.43
Bluegrass webworm adult, *Parapediasia teterrella*.

dead turf. Many infestations in golf course fairways and roughs as well as in home lawns are passed off as being summer dormancy.

Description of stages: The stages are typical of moth complete life cycles.

Eggs: Cylindrical, with ends bluntly rounded and with 16–18 longitudinal ridges. These 0.02 × 0.012 inch (0.5 × 0.3 mm) diameter eggs also have smaller cross ridges between the longitudinal ones. The eggs are pure white when laid and gradually change to a deep straw-yellow until they hatch.

Larvae: Freshly hatched larvae are 3/64–5/64 inch (1–2 mm) long, have blackish brown head capsules and a translucent yellowish body. After feeding, the body becomes greenish from food in the gut. The body spots are barely visible, though minute hairs are conspicuous. As the larva molts and grows through five (male) to six (female) instars, the head capsule turns brownish-yellow tinged with green, the body ground color becomes straw yellow, and the segmental spots are reddish brown. Mature larvae are 3/8–5/8 inch (9–16 mm) long.

Pupae: The amber-yellow pupa is nondistinctive and is 3/8 × 3/32 inch (8–10 × 2.5 mm).

Adults: The adults have a wing span of 9/16–13/16 inch (15–21 mm). The palps and head are white above, and grade to a light brown below. The most distinctive feature of the forewings is the seven spots along the tip. A curved brownish-orange line runs across the wing just inside the tip, and the veins are usually distinctly lighter in color. The hind wings are lighter than the forewing, and have a narrow brown line around the margin.

Life cycle and habits: Adults can be found during most of the summer months, but peaks in adult flights in June and August suggest that two generations are normal in the middle states. Larvae overwinter in silk-lined chambers placed in the soil or thick thatch. Mature larvae may feed briefly before pupating in mid-May. Partially grown larvae feed rapidly to mature, and pupate in late May and early June. The pupa may take 5–15 days to mature, depending on temperatures. The adults emerge from the pupa after dark, expand their wings, and usually mate in the middle of the night. Some couples may remain together after sunrise but most finish mating by daylight. Those who have not mated in the night of emergence usually accomplish this task the following night. Males often die within a day and adult females live only 5–7 days. Dry conditions may contribute to shorter life spans. Females begin laying eggs the night after mating by hovering over the turf. They rarely fly higher than 2 ft. Maximum egg laying occurs about an hour after sunset and may continue for a couple of hours. A female may lay 200 eggs before expiring. The eggs have no adhesive and tend to work into the thatch. The eggs can hatch in 5–6 days at 70°F (21.1°C) or above, but take longer at lower temperatures. The larvae hatch by breaking open the end of the egg and soon spin some webbing along a leaf blade. Here, the larvae eat the surface tissues, and after a molt or two drop to the ground to form a larger tube-like silken tunnel. This tunnel has pieces of thatch attached and often has piles of green fecal pellets near its opening. Older larvae usually feed at night and come to the opening of their tunnel to clip off blades of grass or entire stems. The larvae usually take about 40–45 days to mature during the summer. The second-generation larvae, maturing in the fall, dig deeper into the thatch or soil to overwinter in a silk-lined chamber. Extra molts may occur in the spring if larvae do not obtain enough food during the fall feeding period.

Control options: See: Sod Webworms: Introduction.

Larger Sod Webworm

Species: *Pediasia trisecta* (Walker) (Phylum Arthropoda, Class Insecta, Order Lepidoptera, Family Crambidae) (Figures 3.44).

FIGURE 3.44
Larger sod webworm adult, *Pediasia trisecta*.

Distribution: Common in the northern half of North America. Most common in the bluegrass and tall fescue-growing regions.

Hosts: Seems to prefer Kentucky bluegrass but may attack ryegrass, and fine and tall fescues.

Damage symptoms: See: Bluegrass Webworm.

Description of stages: The stages are typical of the moth complete life cycles.

Eggs: Elongate oval, with ends rounded and with 14–19 longitudinal ridges. The 0.02 × 0.012 inch (0.5 × 0.3 mm) eggs also have faint cross striations between the long ridges. The eggs are light yellow when laid, and turn brownish-yellow before hatching.

Larvae: Newly hatched larvae are 1/16 inch (1–2 mm) long and have a reddish-brown head capsule marked with black. The body is pale yellow but turns reddish, especially toward the posterior. As the larva grows through seven or more instars, the head capsule becomes brownish-yellow with darker markings, and the general body color becomes yellowish with green food contents. The body spots are distinctly chocolate-brown. Mature larvae are 1.0 inch (24–28 mm) long.

Pupae: The light brown pupa is nondistinctive and about 7/16 × 1/8 inch (11 × 3 mm).

Adult: The adults have a wing span of 7/8–1⅜ inch (21–35 mm). The head, palps, and thorax are straw colored and speckled with brown tipped scales. The forewing color is rather nondistinctive and varies considerably from almost solid cream to grayish with light-colored veins. The tip has silvery gray scales interrupted with white scales where the veins end. Usually, three small black spots are present at the tip.

Life cycle and habits: This webworm appears to have two to three generations per year, with peak adult flights in mid-June, late July, and mid-September. However, some adults may be found from mid-May to early October. In the Pacific Northwest, this species is listed as being univoltine (one generation per year). The larvae overwinter in silk-lined chambers dug into the ground beside turf roots. Feeding resumes for a short time in the spring and pupation occurs in May. Within 10–20 days, the adults emerge. The adults emerge at night, with males appearing early after dark and females emerging a couple of hours later. Most emergences occur from 9:00 PM to 1:00 AM and by 1:00 AM, males are swarming about the turf in search of receptive females. Swarming and mating continues until dawn. Mated females begin laying eggs the following night, beginning half an hour after sunset and continuing for a couple of hours. Females lay an average of 180 eggs, ranging from 80 to 220, over 4–6 nights. Eggs dropped by the females hatch in 5–8 days, depending on the temperature. The young larvae crawl about until finding a suitable grass blade upon which to build a first home and begin feeding. A loose net-like web is spun in the fold of the blade and feeding is done by eating the tissue in a trough between two veins. After feeding, the larva builds a tighter web over the feeding area and attaches its fecal pellets. After molting three times, the larva can no longer nest in the blade's groove and soon drops to the base of the plant. Here, it forms a new webbing tunnel running along the ground and incorporating soil and thatch. The larva feeds at night by cutting off blades of grass and dragging them into the tunnel. While feeding on a fresh grass blade at one end of the tunnel, green sawdust-like fecal pellets (frass) are deposited at the free end of the tunnel. When this tunnel becomes too fouled with frass, a new tunnel is often constructed in another direction. Several of these abandoned tunnels may be constructed before the larva matures. At maturity, the larva forms a cocoon of silk spun near its feeding tubes. The cocoon is elongate oval and is covered with soil and thatch pieces. The cocoon is lined inside with grayish silk and an opening exists at one end. Larvae not maturing early enough to emerge in the fall overwinter in cells constructed in the soil or thatch. These cells are roughly spheres of silk in which the tightly coiled larvae rest. In summer, the larvae take 30–50 days to

mature, depending on temperatures. The pupal stage then develops for another 10–20 days before the adult emerges.
Control options: See: Sod Webworms: Introduction.

Western Lawn Moth

Species: *Tehama bonifatella* (Hulst) (Phylum Arthropoda, Class Insecta, Order
    Lepidoptera, Family Crambidae).
Distribution: Commonly found in the Rocky Mountain plateau west to the
    Pacific Coast. It appears to be most common from Washington
    through California.
Hosts: Prefers Kentucky bluegrass, perennial ryegrass, fine fescue, and bent-
    grass. It may move over to bermudagrass in southern climates.
Damage symptoms: See: Bluegrass Webworm.
Description of stages: The stages are typical of moth complete life cycles.
    Eggs: Slightly barrel shaped, with about 35 longitudinal rows of
    depressed plates, approximately 1/16 × 7/128 inch (1.58 × 1.35 mm).
    The eggs are yellow when laid and gradually change to a light
    orange-purple at hatch.
    Larvae: Freshly hatched larvae are 3/64 inch (1.2 mm) long and
    have brown head capsules and a translucent yellowish body. After
    feeding, the body becomes greenish from food in the gut. The
    body spots become distinctly dark by the fourth or fifth instar. As
    the larva molts and grows through seven or more instars, the head
    capsule turns dark brown with darker mottling. Mature larvae are
    about 5/8 inch (15 mm) long.
    Pupae: The yellow, turning to brown, pupa is nondistinctive and
    is 5/16 × 3/32 inch (8 × 2.5 mm).
    Adults: The adults have a wing span of 11/16–15/16 inch
    (17–23 mm). The palps and head are white above and grade to buff
    below. The most distinctive feature of the buff-colored forewings
    is the three elongate spots running along the middle. The forewing
    ends with a chevron mark, followed with a series of black dots. The
    hind wings are darker than the forewing and have a silvery cast.
Life cycle and habits: Adults can be found during most of the summer
    months but peaks in adult flights indicate that four to five broods
    occur each summer. The second and third broods, in July and
    August, appear to cause the most significant damage. Late fall lar-
    vae overwinter silk-lined chambers placed in the soil or in thick
    thatch. Mature larvae may feed briefly before pupating in mid-May.
    Partially grown larvae feed rapidly to mature, and pupate in mid-
    to late April. The pupa may take 5–15 days to mature, depending
    on the temperature. The adults emerge from the pupae after dark,
    expand their wings, and mate during the first night. Males often
    die within a day or two while adult females live only 5–7 days. Dry,

hot conditions may contribute to shorter life spans. Females begin laying eggs the night after emerging. They drop eggs into the turf canopy while hovering between 12 and 24 inches. Maximum egg laying occurs about an hour after sunset and may continue for a couple of hours. A female may lay 100–300 eggs before expiring. The eggs have no adhesive and tend to work into the thatch. The eggs can hatch in 5–10 days, depending on the temperature. Newly hatched larvae eat the surface tissues of leaves, and after a molt or two, drop to the ground to form a larger, tube-like, silken tunnel. This silken tunnel may have plant debris attached and often has piles of green fecal pellets near its opening. Older larvae prefer to feed at dusk or dawn. The larvae usually take about 30–40 days to mature during the summer. A complete generation takes about 45 days, and three to four generations can occur over the summer. The second and third generations appear to be the largest in numbers. These larvae are feeding in June and late July to early August.

Control options: See: Sod Webworms: Introduction.

Tropical Sod Webworm

Species: *Herpetogramma phaeopteralis* (Guenee) (Phylum Arthropoda, Class Insecta, Order Lepidoptera, Family Pyralidae) (Figures 3.45 and 3.46).

FIGURE 3.45
Tropical sod webworm adult, *Herpetogramma phaeopteralis.*

FIGURE 3.46
Tropical sod webworm larva exposed and frass in St. Augustinegrass.

Distribution: Found worldwide in the tropical zones. Attacks turfgrasses in Florida and the Gulf states, where it may not freeze during the winter.

Hosts: Known to attack St. Augustinegrass, bermudagrass, and centipedegrass. May also attack zoysiagrass and bahiagrass.

Damage symptoms: Outbreaks seem to appear overnight when the larger larvae begin to forage on the aboveground turf leaves. Early symptoms are ragged edges on the leaves, followed by stripping of the foliage much like armyworm damage.

Description of stages: The stages are typical of the moth complete life cycles.

Eggs: Unlike the barrel-shaped eggs of crambids, these eggs are flat and scale-like. The eggs are oval, 0.02 × 0.028 inch (0.55 × 0.7 mm), and clustered in small masses. They appear translucent white, and become brownish-red before hatching.

Larvae: Look very much like crambid larvae, but can be distinguished by the arrangement of spots and head markings. The head capsule is dark, yellowish brown and the body is clear to cream color that is tinted with green by the food in the gut. Mature larvae are 5/8–3/4 inch (15–20 mm) long.

Pupae: The reddish-brown pupa is nondistinctive and about 3/8 × 3/32 inch (9.0 × 2.6 mm).

Adults: The adults are very different from crambids because they do not roll the forewings around the body. Instead, the wings are held roof-like over the body, and the general body shape looks much like a swept-wing jet. Adults may be dingy brown to straw colored and have irregular darker marks. The wing span is about 13/16 inch (20 mm).

Life cycle and habits: This webworm has continuous generations during the year and only slows down its cycle during cooler temperatures. However, outbreaks seem to occur most commonly during the hotter summer months following rainy spells. Females lay eggs on grass blades or overhanging vegetation of ornamentals during the early night and these eggs generally hatch in 6–10 days when temperatures are in the 70s°F (21.1°C). Young larvae usually spin webbing in the V-shaped depression of a blade of grass and feed by removing the tissues between veins. The larvae continue feeding in this manner for the first three or four instars and move to new blades of grass if needed. These early instars take from 13–18 days, depending on the temperature. At this stage of development, the larvae begin to feed on the edges of the grass blades and leave a ragged edge. However, this is still rather hard to detect in healthy turf. The last instars, 7 and 8, feed for 10–12 days by eating entire leaves from the grass plants. This is the time when they begin to be noticed, as the larvae seem to work like a living mowing machine radiating from the area where the eggs were deposited. In fact, the last instar larvae can eat over 10 times the amount of grass per day compared to the amount eaten by the fourth instar larvae. The larger larvae are active only at night and hide below the grass surface during the day. By parting the turf at the edge of the damage margin, the resting larvae can be found curled up at the soil surface. The larvae also tend to leave trails of silk when they move from one grass blade to another and this is easily seen in the morning if dew is present. At maturity, the larvae spin a loose bag of silk in which they pupate. This bag usually incorporates bits of grass and debris on the surface, rendering it hard to detect. Within this cocoon, the pupa rests for 7–14 days, depending on the temperature. The adult moths emerge at night and do not fly during the day unless disturbed. The adults mate in the evening and seem to require a nectar meal to survive for any time, usually no more than 2 weeks. The adults are strongly attracted to lights but do not become active at low temperatures. In fact, this insect is very sensitive to low temperatures. Larvae that mature in 25 days at 78°F (25.6°C) take up to 50 days to mature at 72°F (22.2°C). Likewise, pupae take twice as long to emerge at the lower temperature. This pest seems to overwinter with difficulty where temperatures get to freezing during the winter. Populations seem to decline when soil temperatures drop below 60°F (15.6°C), and this may account for the absence of any major damage until late summer populations build up.

Control options: Also see: Sod Webworms: Introduction.

Option 1: Cultural control—use resistant turfgrass varieties: "Common,"
   PI-289922, Fb-119, and "Ormond" varieties of bermudagrass
   (*Cynodon dactylon*) had much less foliar damage than "Tifway" and
   "Tifgreen" (*C.* × *magenissi*). Other varieties should be checked for
   resistance. Resistance has also been demonstrated in various zoysiagrass cultivars.
Option 2: Biobased control—use BT microbial t\oxins: the use of *Bacillus thuringiensis* has had mixed results against this pest, and repeat applications may be needed to prohibit damage from reinvading larvae.
   Some of the Bt var. *spodoptera* strains have had better activity
   against this pest.

Grass Webworm

Species: *Herpetogramma licarsisalis* (Walker) (Phylum Arthropoda, Class
   Insecta, Order Lepidoptera, Family Crambidae).
Distribution: Suspected to have originated from Southeast Asia. It is
   currently found in those countries and adjoining islands and
   Australia. The pest was first found in Hawaii in 1967 on Oahu.
   It has since spread to the other islands of Hawaii-Hawaii, Kauai,
   Maui, and Molokai.
Hosts: Kikuyugrass, *Pennisitum claudestimum* Hochst ex Chior, seems to be
   the major preferred host, though 13 other grass hosts are known in
   Hawaii. Bermudagrass, sunturf bermuda, centipedegrass, and St.
   Augustinegrass are common turfgrasses attacked.
Damage symptoms: The larvae feed on leaves, stems, and crowns of turf-
   grasses. Initial feeding produces grass blades with a ragged edge, but
   continued feeding produces irregular brown patches. Considerable
   webbing and frass (fecal pellets) are usually very evident.
Description of stages: This insect's stages are typical of moths with complete life cycles.
   Eggs: The eggs are flat, oval, and scale-like. They measure
   0.02 × 0.35 inch (0.6 × 0.9 mm) and can be laid singly or in groups.
   They are usually attached to the upper surface of leaves, along the
   midrib. They start out being creamy white and change to dark
   orange before hatching.
   Larvae: The larvae are light green to brown in color and have the
   conspicuous rings of dark brown spots, typical of pyralid larvae.
   There are five instars, with the first being about 3/32 inch (2.3 mm)
   long and the mature larva about 13/16 inch (20 mm) long.
   Pupae: The light brown pupae turn dark brown just before the
   adults emerge. They are placed in a tightly woven silken case and
   are about 13/32 inch (10.5 mm) long.
   Adults: These moths are fawn to light brown in color, and the
   wings have faint darker spots and a zigzag line near the outer

margin. The body is about 3/8 inch (10 mm) long and the wing span is 15/16 inch (23–24 mm).

Life cycle and habits: Adults apparently emerge from their pupal cases at night, and mating occurs during the same night. The females need 3–6 days before they are ready to lay eggs. They usually feed on flower nectar and moisture on leaves. Once the preoviposition period is over, an average of 250 eggs are laid over 5–7 nights. The eggs take 4–6 days to hatch. The newly hatched larvae feed on leaf upper surfaces. They strip away strips of the upper tissues, leaving the lower epidermis intact. The larvae molt every 2 days, and the third instars begin consuming entire leaf margins. The last instar, the fifth, takes over 4 days to mature. The larvae hide in silken tunnels in the turf thatch during the day and emerge to feed at night. The larval development takes 10–14 days. Mature larvae construct a silken hibernaculum-like cocoon. This structure has frass and plant debris attached. The pupae take 6–7 days to mature before the adults emerge. A generation takes about 32 days to complete. The adults can be found in large numbers resting in tall grasses or in low shrubs and plants alongside turf.

Control strategies: See: Sod Webworms: Introduction. Grass webworm outbreaks can occur at any time during the season but are usually associated with rainy weather. Periodic sampling for the larvae should be performed if adults are noticed flying about. Remedial controls may be needed when 10–15 larvae per yd^2 are found.

Option 1: Cultural control—resistant turfgrasses: Egg-laying preference and larval feeding studies indicate that Sunturf bermudagrass, Tifdwarf bermuda, St. Augustinegrass, and centipedegrass are the most attractive to this pest. Common bermudagrass and Tifway bermuda are more resistant to attack.

Option 2: Biological control—conserve parasites: The tiny wasp, *Trichogramma semifumatum* (Perkins), attacks the grass webworm's eggs. Up to 96% of the eggs in an area may be attacked. Various other larval parasites have lesser effects. Delay or reduce applications of traditional insecticides to conserve these parasites.

Cutworms and Armyworms: Introduction

Species: Several species of thick-bodied, nonhairy caterpillars may attack turfgrasses. In the northern zones, the black cutworm, *Agrotis ipsilon* (Hufnagel), bronzed cutworm, *Nephelodes minians* Guenee, and armyworm, *Pseudaletia unipuncta* (Haworth), are common. In southern and transition zones, black cutworms, armyworm, as well as the granulate cutworm, *Felta subterranea* (Fabricius), fall armyworm, *Spodoptera frugiperda* (Smith), and yellowstriped armyworm, *Spodoptera ornithogalli* (Guenee), are commonly found attacking turfgrasses. Other species are occasionally found in turfgrasses, though often in lesser numbers (Phylum Arthropoda, Class Insecta, Order Lepidoptera, Family Noctuidae) (Figures 3.47 through 3.55).

Distribution: The above-mentioned cutworms and armyworm (often referred to as the common armyworm) are found all across North America, and some are worldwide in distribution. The *Spodoptera* armyworms are tropical and semitropical species that migrate from Central America and the Gulf states northward every summer.

Hosts: All species of turfgrasses may be attacked.

Damage symptoms: The true cutworms are semi-subterranean pests. They usually dig a burrow into the ground or thatch and emerge at night to chew off grass blades and shoots. This feeding damage often shows up as circular spots of dead grass or depressed spots, called "pock marks," that resemble ball marks on golf greens. The armyworm and fall armyworm feed on the grass blades, leaving

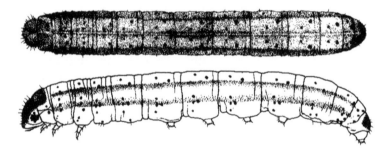

FIGURE 3.47
Diagram of black cutworm larva, *Agrotis ipsilon*. Dorsal view (upper) and lateral view (lower). (After Rept. Ill. State Entomol.)

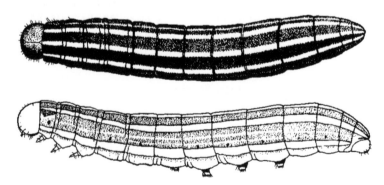

FIGURE 3.48
Diagram of bronzed cutworm larva, *Nephelodes minians*. Dorsal view (upper) and lateral view (lower). (After Rept. Ill. State Entomol.)

FIGURE 3.49
Black cutworm adult.

FIGURE 3.50
Black cutworm larva feeding on short-cut bentgrass from burrow at night.

FIGURE 3.51
Armyworm adult.

FIGURE 3.52
Armyworm larva.

FIGURE 3.53
Fall armyworm adult.

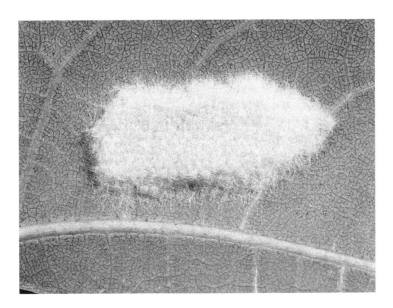

FIGURE 3.54
Fall armyworm egg mass on tree leaf.

FIGURE 3.55
Fall armyworm larva showing white, inverted Y-mark on head capsule.

a ragged, torn appearance to the turf. However, in short-cut turf, these pests can also leave pock marks. Heavy infestations can result in the turf being eaten down to the ground by a moving "army" of caterpillars.

Description of stages: Cutworms and armyworms are the larvae of moths and thus have complete life cycles with eggs, larvae (caterpillars), pupae, and adults.

Eggs: The eggs are usually round, 0.02–0.03 inch (0.5–0.75 mm) in diameter, flat on the lower surface, bluntly pointed at the top, and often with sculpturing, lines and ridges, on the surface. Cutworms often lay eggs singly or in small masses while armyworms tend to lay hundreds of eggs in a single mass, often covered with scales from their bodies.

Larvae: The larvae have generally hairless bodies except for a few bristles scattered over the body. Besides the three pairs of true legs, these caterpillars have five pairs of fleshy prolegs on the abdomen. Most cutworms have characteristic markings on the head and body that aid in species identification. Full-grown cutworm larvae are 1/4 inch (6.0 mm) wide and 1⅜–2.0 inches (35–50 mm) long. Most cutworms and armyworms coil into a spiral when disturbed.

Pupae: The pupae are brown, reddish-brown, or black and 1/2–7/8 inch (13–22 mm) long. The antennae, wingpads, and legs are firmly joined together, but the abdomen is free to twist around if the pupa is disturbed.

Adults: The adults are generally dull-colored moths with wing spans of 1⅜–1 7/8 inch (35–45 mm). At rest, the wings are folded flat over the abdomen.

Life cycle and habits: The armyworm and bronzed, variegated, and granulate cutworms appear to overwinter as larvae or pupae in the northern states. The black cutworm and *Spodoptera* armyworms overwinter only in the southern states in all stages of development. In the northern states, the nonmigrating caterpillars may have two to four generations per year, while in the southern states, three to seven generations may occur depending upon the length of the season. The bronzed cutworm has a single generation per year. Females mate and feed at night on flowers of trees, shrubs, and weeds. Mated females seek out crops or grasses and lay clusters of eggs on the leaf blades. A single female may lay 300–2000 eggs over several days. Under optimum conditions, these eggs hatch in 3–10 days and the young larvae begin to feed on the leaves. The cutworms excavate a hole into the ground or use existing holes, such as aerification holes, to hide during the day. From these retreats, the larvae venture forth at night to feed on plant material. Often, they will drag a leaf or stem back to their burrow to feed on during the daytime. The armyworms do not remain in a hiding place but may hide during the day under a leaf or in the thatch. Nearly mature armyworms do not even hide during the day but feed continuously. Most of these species take 20–40 days to complete their larval development. The pupae may be located in the cutworm retreat or in the case of the armyworms, in the thatch. The pupa takes about 2 weeks to mature. These developmental times may be greatly lengthened during the cooler parts of the season.

Control options: These pests can generally be controlled by using one of the contact or stomach pesticides. However, if populations are high in surrounding areas, such as in unmowed areas or nearby field crops, continual reinfestations may occur. Because of this, contact with local cooperative extension services or pest management consultants may be beneficial in determining local pest activity.

Sampling

Young cutworms and armyworms are difficult to locate, so the use of a disclosing solution is beneficial in determining population pressure. Use two gallons of soapy water applied to 1.0 yd^2 (4.5 L/m^2) area. Within 3–5 min, the caterpillars will come to the surface and can be easily counted. Remedial controls are needed if 5–10 larvae per yd^2 of fairway or lawn turf are found. Thresholds on golf putting greens and tees may only tolerate several "pock marks" per surface!

Option 1: Cultural control—use resistant turfgrasses: In transition and northern turf, using endophyte-enhanced fescues or perennial ryegrasses can eliminate armyworm problems, but black cutworms appear not to be affected by endophyte toxins. Several cultivars of bermudagrass are also resistant to yellowstriped and fall armyworm attack. If turf is to be renovated because of past cutworm or armyworm attack, consider using resistant turfs.

Option 2: Biobased control—use BT: The bacterium *Bacillus thuringiensis* has multiple strains that are commercially available and some strains adequately control cutworms and/or armyworms, especially if enough material is ingested by young larvae. Be sure to check with your supplier as some strains work well on certain cutworms, but these strains may have not armyworm activity. Likewise, there are specific, *Spodoptera*-active strains for use against the fall and yellowstriped armyworms. Use of this microbial toxin works best when the caterpillars are in the first to third instars. Do not wash the material off of the turf foliage after the application, and applications late in the day to avoid UV-light contact can improve efficacy.

Option 3: Chemical control—use poison baits: These caterpillars will often feed on grain baits, and these can be used when populations are causing reinfestations. Be sure to have the cutworms identified because some species do not accept grain baits. This technique is more commonly used in agricultural fields or pastures.

Option 4: Chemical control—sample populations and spraying: If a disclosing solution test has indicated considerable activity (5–10 or more larvae per yd^2), a pesticide application may be needed. On golf course putting greens and tees, only a few cutworms or their damage spots may be tolerated. The armyworms are easily controlled by contact/stomach poisons as long as the larvae are actively feeding. Some problems with cutworm control have been noted. Apparently, mixed age populations may be present, and short residual insecticides will not kill any larvae that hatch several days after the initial application. Observe the sizes of larvae flushed out and if small to mature larvae are present, reapplication may be necessary in a week to 10 days, depending on the residual action of the product used. For improved efficacy, apply pesticides in the afternoon or evening and do not irrigate after the application to maximize the amount of residues remaining on the turfgrass foliage. Since most of the control is obtained through the caterpillars ingesting the pesticide, liquid sprays are usually more effective than granular applications. Pyrethroids are the most commonly used curative insecticides, some organophosphates, and carbamates are also used on golf courses. The anthranilic diamides have provided long residual action against turfgrass caterpillars.

Black Cutworm

Species: *Agrotis ipsilon* (Hufnagel) is also called the greasy cutworm and the dark sword grass moth (Phylum Arthropoda, Class Insecta, Order Lepidoptera, Family Noctuidae) (Figures 3.47, 3.49, and 3.50).

Distribution: The black cutworm is found all across North America but is also found in Europe, Asia, and Africa. Though this pest has trouble spending the winter where soil temperatures may reach below 15°F, adults are strongly migratory and northern regions can become reinfested each summer.

Hosts: All species of turfgrasses, except for Kentucky bluegrass, may be attacked.

Damage symptoms: This is a true cutworm and has semi-subterranean habits. They usually dig a burrow into the thatch/soil or use existing cracks and crevices or aerification holes, from which they emerge at night to chew off grass blades and shoots. This feeding damage often shows up as circular spots of dead grass or depressed spots (often called "pock marks") that are similar to ball marks on golf greens.

Description of stages: Cutworms are the larvae of moths and thus have complete life cycles with eggs, larvae (caterpillars), pupae, and adults.

Eggs: The eggs are usually round, 0.02 inch (0.5–0.6 mm) in diameter, flat on the lower surface, and bluntly pointed at the top. When freshly laid, they are greenish but become tan to dark brown before hatching.

Larvae: The larvae have generally hairless bodies except for a few bristles scattered over the body. The general body color is gray to nearly black on the upper half, and slightly lighter gray below. Often, a pale and indistinct middorsal line is visible. Under a microscope, the body integument appears to have a pebble-like surface. Mature larvae can be 1³⁄₁₆–1¾ inch (30–45 mm) long and 1/4 inch (7.0 mm) wide.

Pupae: The pupae are brown, reddish-brown, or black and 1/2–7/8 inch (13–22 mm) long. The antennae, wingpads, and legs are firmly joined together, but the abdomen is free to twist around if the pupa is disturbed.

Adults: The adults are generally dark gray to black-colored moths with some brown markings. The forewing span is 1⅜–1¾ inch (35–45 mm) and a black, dagger-shaped mark appears at the outer edge.

Life cycle and habits: The black cutworm is not able to survive hard freezes during the winter, so larvae and pupae tend to overwinter south of Tennessee. Around the Gulf states and that latitude, adults may also be present almost any time during the year. Because of this, five to six generations can occur in the Gulf states each

season. At the level of the Carolinas across to Oklahoma and
Kansas, black cutworms likely have four to five generation per
season, and from Pennsylvania across to Iowa, two to three generations are commonly observed. In subtropical regions, this pest
has continuous, overlapping generations. Adults emerge at night
from pupae formed in the soil or thatch. The females usually
mate within the first two nights after emergence. Adults actively
feed on nectar from flowering plants at night. Each female has
the capacity to lay 1200–1600 eggs over 5–10 days. While it is
known that black cutworm females lay clusters of eggs on broadleaf weeds in field crops, in turfgrass, the females attach one to
two eggs onto the tips of grass blades. The eggs hatch in 3–6 days,
and the first instar larvae usually feed on the leaf surface. After
several days and molting, the larvae drop to the thatch and continue feeding without making burrows for the first three instars.
Young larvae tolerate each other, but when the larvae reach the
fourth and fifth instars, they become cannibalistic and they may
move 50–80 ft in one night. Larger larvae create a burrow in the
ground or horizontally within the thatch. They tend to occupy
these burrows for 3–5 days before moving to a new location. Most
feeding is done at night, but the larvae may extend their heads
from their burrows during the day to nibble on grass blades. On
golf courses, the larvae readily occupy aerification holes. Like
most cutworms, male larvae undergo five larval instars and the
females under six instars (molts). Under moderate temperatures
(in the 70s and 80s°F during the day), the larvae finish development in 20–40 days. The pupa is usually formed in the last cutworm burrow. The pupa takes about 2 weeks to mature. These
developmental times may be greatly lengthened during the
cooler parts of the season.

Black cutworm adults often migrate in large numbers in the
spring from Gulf states into northern regions. Flights often arrive
with spring storm fronts and adults have been collected in light
traps in late April and early May in Pennsylvania across to Iowa.
These adults lay eggs in field crops (mainly corn) and on turfgrass
that result in the first damage by the larvae in early June.

Control options: See: Cutworms and Armyworms: Introduction. Black
cutworms are generally controlled by using one of the contact
or stomach pesticides. However, if populations are high in surrounding areas, such as in field crops, continual reinfestations
may occur. Because of this, contact with local cooperative extension services or pest management consultants may be beneficial in
determining local pest activity. First and second instar cutworms
are difficult to locate, even when using a disclosing solution (soapy
water). However, if cutworms are suspected, a disclosing solution

is one of the easiest ways of determining population pressure. If a soap flush reveals 5–10 larvae per yd^2 on golf course fairways or high-cut turf of lawns and sports fields, remedial controls will be necessary. Only several cutworm spots on greens may require treatments due to the potential for ball roll disruption. Adults can be monitored using the black cutworm pheromone trap or black light traps. Larval evidence can be expected about 7–10 days after peak trap catch.

Option 1: Biobased control—microbial toxins: Some of the *Bacillus thuringiensis* var. *spodoptera* toxins have had good activity against the black cutworm's relatives, the armyworms. For maximum efficacy against the black cutworm, these pesticides need to be applied when the young larvae (first through third instars) are active. Make applications about 7 days after peak adult trap catches are observed.

Option 2: Biological control—entomopathogenic nematodes: The insect parasitic nematodes are generally effective against the larvae of this pest. Products containing *Steinernema carpocapsae* or other *Steinernema* species have performed best. The best efficacy is obtained by applying the nematodes in the late afternoon, just before sunset and lightly syringing the turf immediately after the application.

Option 3: Cultural control—turf mowing and aerification management: Since the eggs of this pest are attached to grass blade tips, clippings that are removed from golf course greens and tees should be disposed of far away from the green and tee surfaces. Ideally, these should be placed in a compost pile where decomposition occurs quickly. Aerification performed in spring months should take into account the stage of development of black cutworms. If medium-sized to large larvae are present, they will readily move onto tee and green surfaces that have been cored. This can also happen later in the season. Soap flushing of areas around greens and tees will help you determine the larval numbers and age. Treatment of surrounding areas will keep cutworms from immediately occupying aerification holes.

Bronzed Cutworm

Species: *Nephelodes minians* Guenee (Phylum Arthropoda, Class Insecta, Order Lepidoptera, Family Noctuidae) (Figure 3.48).

Distribution: Northern North America, primarily east of the Rocky Mountains. This is a true cool-season turf pest.

Hosts: Appears to feed on most of the cool-season grasses, including endophyte-enhanced ryegrasses and fescues.

Damage symptoms: This pest feeds primarily in the fall and early spring seasons, but there are examples of the larvae feeding under the cover of snow! Heavy feeding under the cover of snow results in turfgrass appearing to have been severely damaged by snow mold. When the snow melts, all the green stems and leaves are missing, revealing matted thatch. These areas may require reseeding. Light to moderate populations thin turf or cause localized brown areas.

Description of stages: The bronzed cutworm has a typical moth life cycle with egg, larval, pupal, and adult stages.

Eggs: The eggs are initially light green, round, about 3/32 inch (0.5 mm) in diameter, flat on the lower surface, bluntly pointed at the top, and with fine sculpturing ridges on the surface. They become dark green just before hatching.

Larvae: Newly hatched larvae are light green with five pale stripes down the bodies with three white stripes on the green pronotal shield (just behind the head). The first three instars are green, but the later instars take on a deep chocolate-brown color, often with a bronze sheen. These larger larvae also have the pronotal shield black in color. Mature larvae are often 1⅜–1¾ inch (35–45 mm) long and 3/8 inch (4.5 mm) wide.

Pupae: Pupae are typically dark brown, rounded on the head end, and pointed at the abdominal tip. They are about 3/4 inch (20 mm) long.

Adults: The moths range from a light bronze-brown to a dark chocolate-brown color. The wings are mottled with indistinct, darker patterns. Wing span ranges from 1⅛–1½ inch (30–38 mm).

Life cycle and habits: Adult moths fly from late August through September and they are commonly attracted to lights. Females attach small clusters of their green eggs to grass blades and weeds found within turf. About half of these eggs hatch in about 2 weeks but other eggs remain dormant until spring conditions. Larvae that hatch in the fall will feed on grass blades until the soil freezes. In moderate conditions, these fall-emerging larvae are often able to continue feeding under the cover of snow. Eggs that successfully overwinter appear to hatch in March into early April. The first instar larvae are green and first feed within the fold of a leaf blade, but soon they move into the thatch. The first three instars remain green and feed primarily on grass blades. The larger larvae hide during the day in the thatch but feed on grass leaf blades and stems at night. Overwintered larvae tend to finish their development and pupate within the soil under the turf by early June. March- and April-hatching larvae may not finish their development until the end of June. The pupae are formed in the soil under the turf and they remain dormant for 2–3 months.

Control options: See: Cutworms and Armyworms: Introduction. This pest often avoids insecticide treatments because of its cool-season activities. It is also difficult to predict when and where damaging populations will occur. Lawns, grounds, sports fields, or golf course roughs that have experienced bronzed cutworm damage should be soap flushed in early to mid-November to detect the presence of larvae. If found, treatments with quick-acting insecticides would be warranted.

Armyworm

Species: *Pseudaletia unipuncta* (Haworth) is often called the "common" armyworm to differentiate it from the Spodoptera species of armyworms. It has been listed in the past under the generic names *Leucania* and *Cirphis* (Phylum Arthropoda, Class Insecta, Order Lepidoptera, Family Noctuidae) (Figures 3.51 and 3.52).

Distribution: The armyworm is found throughout the United States, generally east of the Rocky Mountains. It is occasionally a turf pest in Utah and California.

Hosts: All grasses and grass crops may be attacked.

Damage symptoms: As their name implies, armyworms can reach large numbers that literally "march" across turf, eating every bit of green leaf and stem material! They often occur in large numbers, especially where turf is being grown near small grain crops, like corn and wheat fields. They will also build up large populations in grassy meadows and along rights-of-way for roads or utilities. When small, the larvae leave a ragged leaf margin, but as the larvae grow, they devour everything to the thatch/soil level. This moving front of feeding caterpillars can consume many square feet of turf in a night. Most people describe the event as: "My turf disappeared overnight!" The armyworm has been a pest of cereal and forage crops since the colonial times.

Description of stages: The armyworm has a typical moth life cycle with egg, larval, pupal, and adult stages.

Eggs: The eggs are round, shiny white with a slight greenish case, and have no surface ridges. They are about 0.02 inch (0.5 mm) in diameter, and are laid in several rows with up to 500 in a mass.

Larvae: Newly hatched larvae are light green and about 1¹⁄₁₆ inch (2 mm) long. As they grow, the larvae remain olive green to brown, and have a darker stripe along each side and a broad, darker stripe down the back. Mature larvae may be 1³⁄₁₆ inch (30 mm) long.

Pupae: The pupae are typical brown forms with the legs and wing pads showing through the pupal shell. They are about 9/16 inch (15 mm) long and can rotate the abdomen when disturbed.

Adults: The adults are light tan with darker shading. They have a distinct, small, light-colored spot on the midforewing, often with a tiny black dot in the middle. Adults are about 3/4 inch (20 mm) long and have a 1 3/16 inch (30 mm) wing span.

Life cycle and habits: The armyworm is able to withstand most winter temperatures in the larval and pupal stages located in the soil. It emerges as new adults in late March into April and the first generation can cause crop damage in May to mid-June. A second generation runs through June and July, with the third generation active in August to the first frost. The females lay clusters of up to 500 eggs in rows lined up on grass surfaces. The eggs are often covered by a fold in the leaf or by another leaf stuck to their surface. Each female is capable of laying several thousand eggs. Since adults are attracted to lights at night, masses of eggs can occur on or around these structures. When the larvae first emerge, they use a looping motion to move across leaf surfaces. They tend to stay together and feed on the same plants until everything is devoured. The larvae then move together to the next plant. On sunny days, the caterpillars tend to hide in spaces between leaves or on the undersurfaces of leaves. As the caterpillars near maturity, they do not hide and can move, in mass, across grassy areas, with each caterpillar devouring individual plants. Armyworm outbreaks most commonly occur after they have attacked small grain crops in their first generation. The adult moths that emerge seek more tender food, and turfgrasses are commonly selected. Mild winter temperatures combined with moderate, moist spring and early summer conditions favor armyworm outbreaks. The second or third generation is the one most likely to cause damage on turf.

Control options: See: Cutworms and Armyworms: Introduction. The true armyworm rarely kills turf, because the "hoard" tends to eat only the leaves and upper stems of turf. Since the crowns are not destroyed, the turf should soon recover with irrigation. However, defoliation in midsummer can result in crown desiccation or heat death, so periodic syringing until the grass blades are restored is recommended.

Fall Armyworm

Species: *Spodoptera frugiperda* (Smith) is probably the most important of the "Spodoptera" complex. However, a close relative, the yellow-striped armyworm, *Spodoptera ornithogalfi* (Guenee), is commonly found intermixed with the fall armyworm (Phylum Arthropoda, Class Insecta, Order Lepidoptera, Family Noctuidae) (Figures 3.53 through 3.55).

Distribution: This pest is thought to be semitropical in origin, probably from Mexico or Central America. It is a permanent resident of the Gulf states and is able to migrate northward during the spring and summer. Since it is not able to withstand freezing temperatures, the northern populations die each fall.

Hosts: This pest attacks several hundred grassy and broadleaf plants.

Damage symptoms: The younger larvae feed on the margins of leaves that produces a ragged look. As the larvae mature, they eat all the above-ground leaves and stems. The larvae do not seem to be overly gregarious and migratory as found in the armyworm, and feeding damage in turf can appear as progressive thinning of the turf. The adults of these species appear to be able to detect fertility levels and larval attacks on recently seeded or sprigged turf is common.

Description of stages: The fall armyworm has typical egg, larval, pupal, and adult stages found in all moths.

Eggs: The eggs are laid in masses of 100–250 in two to three layers. They are about 0.016 inch (0.4 mm) in diameter, begin as greenish gray, turn blackish brown at hatch, and have fine square reticulations on the surface. Each mass is covered with scales from the abdomen of the female.

Larvae: First instar larvae are about 1/16 inch (2 mm) long and are light green in color. As they grow, they obtain distinctive lines down the body, range from light tan to olive green in color, have distinctive black spots at the base of larger hairs, and reach 1³⁄₁₆ inch (30 mm) in length. The head has a diagnostic, white, inverted Y-mark.

Pupae: The pupae are droplet-shaped, with the narrow end at the abdomen. They are a dark chestnut-brown and are about ⅝ × ³⁄₁₆ inch (15 × 4.5 mm).

Adults: The male and female markings are fairly different. Both are generally shades of gray with white markings. The females can be almost entirely gray without distinctive marking. Both males and females have a characteristic drop-shaped lighter mark running from the middle of the forewing and trailing to the hind tip margin. They range from 11/16–1.0 inch (17–25 mm) long with wing spans of ¾–1³⁄₁₆ inch (20–30 mm).

Life cycle and habits: This pest is a constant threat to southern turf. It rarely damages transition or northern turf because it cannot overwinter where the turf may freeze. All stages are present, year round, in the areas of the Gulf states within 100 miles of the coast. With the arrival of warmer spring temperatures, adults may lay eggs on preferred grasses and small grain crops. However, the females seem to prefer to lay egg masses on the leaves of trees, shrubs, flowers, and other structures overhanging managed turf. Upon hatching, the larvae drop to the underlying turf and begin feeding. The eggs take 7–10 days to hatch in cool weather and can hatch in 2–3 days in the heat of July and August. The young larvae eat the egg shell before turning to plant foliage. They feed together until they reach the fourth or fifth instar. At this time, they may become cannibalistic and begin moving outward. Larger larvae feed during the day and if disturbed by a shadow or touching, they will fall from the plant and remain tightly coiled where they fall. They soon recover and return to feeding. The larvae can take as little as 12 days to mature in July and August and up to 28–30 days to mature in cooler weather. The larvae consume an average of 13 cm² of leaf tissue, of which 12 cm² is consumed in the last two instars! This is why they appear to suddenly strip the turf down within a couple of days. Upon maturation, the larvae move into the lower thatch area and form pupae within some loosely spun silk webbing. The pupa takes 9 days to mature in August and up to 20 days in cooler weather. The adults actively feed on nectar and overripe fruits. They can live for about 2 weeks, and the females lay three to five egg masses. The adults have strong migratory urges and commonly travel hundreds of miles on rapidly moving storm fronts. Though they commonly arrive in northern states by June and July, they rarely attack turf upon their first arrival. Their main northern hosts are corn and soybeans, but they often move to turf in August and September.

Control options: See Cutworms and Armyworms: Introduction. The fall and yellowstriped armyworms commonly reach plague proportions from July through October in the Gulf states. State agricultural agencies commonly release armyworm alerts when these great outbreaks are bound to occur. Turf managers should be looking for these notices because these are the times that turfgrasses will be under the greatest pressure.

Lawn Armyworm

Species: *Spodoptera mauritia* (Boisduval) (Phylum Arthropoda, Class Insecta, Order Lepidoptera, Family Noctuidae).

Distribution: This pest seems to be a native of the Oriental and Indo-Australian regions, but has been transported to many of the Pacific Islands, including Hawaii. It is not known in North America.

Hosts: Like many of the *Spodoptera* armyworms, this species feeds on a wide variety of grasses and sedges. It is especially fond of bermudagrass and zoysiagrass.

Damage symptoms: As their name implies, lawn armyworms can literally march across the turf eating most of the leaves and stems. Moderate populations produce a ragged appearance to the turf. Large populations produce a fairly well-defined line between green, undamaged turf and a completely eaten and brown area. This front can move forward by a foot per night.

Description of stages: Armyworms are moths with complete life cycles.

Eggs: The eggs are laid in masses attached to the leaves of trees, on buildings or other objects overhanging or near turf. The mass is usually elongate oval in outline and has five to seven layers of eggs. The females usually cover the mass with their abdominal hairs. The masses contain 600–700 eggs and each egg is about 0.02 inch (0.5 mm) in diameter.

Larvae: The first instars are about 3/64 inch (1.2 mm) long and are greenish after feeding. Later instars develop patterns of brown to black stripes and spots. Each body segment has a pair of jet black dashes next to the longitudinal yellow stripes. Mature larvae are 1⅜–1¾ inch (35–40 mm) long.

Pupae: Pupae are first light brown but soon turn a dark chestnut-brown. They are about 5/8 × 3/16 inch (16 × 4.5 mm).

Adults: Adults resemble the fall armyworms. The male is more distinctly marked than the female. They have a drop-shaped light spot about midway down the wing, which is followed with a dark black spot. The wing is generally mottled, dark gray. The females have a wing span of 1⁵⁄₃₂–1¾ inch (34–40 mm), and the males are slightly smaller.

Life cycle and habits: This tropical species has continuous generations throughout the season. The adult moths emerge at night and the females are mated within a day. The adults may feed on nectar and water from rain or dew. After a 4-day preoviposition period, the females begin to lay masses of eggs. Egg laying begins at dusk and is generally completed before midnight. Egg masses are normally attached to the foliage of trees and shrubs near turf. They are almost never laid on grass blades. Since the adults are attracted to lights, many egg masses are often laid on nearby buildings or trees. The eggs take 3 days to hatch and the young larvae drop to the turf. During the first five instars the larvae feed during both day and night. They seem to prefer to stay together, often with several feeding on the same plant. As they strip off the foliage, they move forward to new turf. The older larvae are easier to see and usually hide in the thatch and plant debris during the day. The seven to eight instars take nearly 28 days to complete development. The mature larvae burrow into the thatch and pupate. Pupation is completed in 11 days.

Control strategies: Also see: Cutworms and Armyworms: Introduction. Considerable effort has been made to import biological controls for this pest in Hawaii. Under normal conditions, these parasites and predators do an adequate job but occasionally they do not build up fast enough to prohibit turf damage. If turf is suspected to have a damaging population of lawn armyworms, sampling should be performed.

Controls are needed if 10–15 larvae per yd^2 of lawn turf are found with a soap disclosing flush.

Option 1: Biological control—conserve predators and parasites: Two tiny wasp egg parasites attack this pest (*Telenomus nawai* Ashmead and *Trichogramma minutum* Riley). The larval attacking wasp, *Apanteles marginiventris* (Cress.) seems to be the most important natural enemy of the lawn armyworm. By providing nectar-producing flowers and targeting insecticide sprays only to the areas needing them, these parasites can be conserved.

Option 2: Cultural control—reduce night lighting: Since the adults are attracted to lights at night and they oviposite nearby, place lights away from sensitive turf areas. Lights within the yellow range are also much less attractive to night-flying insects.

Other Turf-Infesting Caterpillars

There are several other turf-infesting caterpillars that are occasionally encountered damaging turfgrasses. Most of these can be managed as if they were cutworms or armyworms.

Striped Grassworms (= Grass Loopers)

Species: Several species in the genus *Mocus. Mocus latipes* (Guenee) is the most common species damaging turf, but *Mocus disseverans* (Walker) and *Mocus marcida* (Guenee) may be present (Phylum Arthropoda, Class Insecta, Order Lepidoptera, Family Geometridae) (Figure 3.56).

Distribution: *Mocus* spp. are natives of the New World ranging from northern Canada to Argentina east of the Rocky and Andes mountain ranges. Generally, *M. latipes* is a periodic pest from Texas across to Florida.

Hosts: Bahiagrass, bermudagrass, and St. Augustinegrass are preferred.

Damage symptoms: The larvae feed on grass blades and heavy populations can completely strip all the leaves so that only the stolons remain. Light infestations leave a ragged appearance to the turf.

Description of stages: These lepidopterous pests have complete life cycles with egg, larval, pupal, and adult stages. The following descriptions are based on *M. latipes*, but the others are very similar.

Eggs: The round eggs are about 0.024 inch (0.6–0.7 mm) in diameter and have longitudinal ridges with faint cross ridges. The eggs are first light green and become mottled with dark brown with age.

Larvae: The larvae are typical loopers, that is, inchworm-like. They have only three sets of prolegs on the abdomen. The first instar nymphs are about 3/16 inch (5 mm) long, and slightly striped with brown and cream colors. Subsequent instars are more strikingly striped from the head capsule to the tip of the abdomen. Mature larvae are 1.0–1⅜ inch (25–35 mm) long.

FIGURE 3.56
Striped grassworm larva, *Mocus latipes*.

Pupae: The pupa is chestnut-brown in color and is often covered with a fine white powder. The cocoon is spindle-shaped and often wrapped in loose silk with several grass leaves or clippings attached.

Adults: The adults are about the size of a cutworm, but the wings are broader. The wings are mottled gray and brown and usually have a darker border along the tip. The wing span is usually 1¾–2³⁄₁₆ inch (45–55 mm).

Life cycle and habits: Several generations occur per year, but the major pressure from this group of pests occurs in late summer and early fall. Females emerging in April mate and begin to lay eggs within 3 days. Each female may lay 300–400 eggs individually attached to grass blades. Most of the eggs are laid within a week. The eggs hatch in 3–4 days, and the first instar larvae actively "inchworm" around until a suitable grass blade is found. Here, the larvae spin some webbing over the V-shaped groove in the grass blade and begin to feed by stripping the epidermis from the top of the leaf. The larvae grow rapidly, molting every 2–3 days when temperatures are high. When the fourth instar is reached, the larvae begin to feed on the leaf margins, leaving ragged edges. The next several instars (up to 6 or 7) feed over 2–3 weeks and the larger larvae can completely strip the leaves from plants and often chew on young shoots. The mature larvae are about 5 mm and blend in well with the surrounding environment. The mature larvae, ready to pupate, seek out protected areas under the turf canopy and construct a spindle-shaped cocoon. This cocoon usually has pieces of grass clippings attached and is very difficult to find. The pupa matures in 8–10 days, and the new moths emerge in the late afternoon. At 80°F (26.7°C), the life cycle takes 30–33 days, and somewhat longer at temperatures down to 70°F (21.1°C). Development is greatly slowed at temperatures below 60°F (15.6°C).

Control options: See: Cutworms and Armyworms: Introduction. These insects are easily controlled by most of the methods used for surface-feeding caterpillars.

Fiery Skipper

Species: *Hylephila phyleus* (Drury). (Phylum Arthropoda, Class Insecta, Order Lepidoptera, Family Hesperiidae) (Figures 3.57 and 3.58). Several species of skippers can be found feeding on turfgrasses. Proper identification usually requires the help of an expert.

FIGURE 3.57
Fiery skipper adult.

FIGURE 3.58
Fiery skipper larva.

Distribution: This butterfly is found over most of North and South America. It is most abundant in the Gulf states, and was detected in Hawaii in 1970.

Hosts: The larvae seem to prefer bermudagrass, but St. Augustinegrass is commonly fed upon. Turf with crabgrass appears especially attractive.

Damage symptoms: The individual larvae feed in localized spots and cause isolated round spots of brown grass, 1–2 inches (2.5–5 cm) in diameter. If an extensive infestation occurs, the spots join together and larger irregular brown patches result.

Description of stages: This is a typical butterfly with a complete life cycle.

Eggs: The round eggs look like a small ball cut in half. They are about 0.03 × 0.02 inch (0.7 × 0.5 mm). When freshly laid, they are white, but change to a light blue-green in a day or two.

Larvae: Skipper larvae are unique in form. They have dark heads, a diagnostic constriction of the neck, and plump bodies covered with tiny bristles. The first instar larva is a pale greenish yellow and about 3/32 inch (2.5 mm) long. Mature larvae become yellow brown to gray brown, with an indistinct median longitudinal stripe. Mature fifth instar larvae are 1.0 inch (24–25 mm) long.

Pupae: The pupa is formed in the thatch or in loosely webbed-together turf debris. The pupa is 5/8 inch (15–18 mm) long, and changes from a light greenish-tan to an overall brown just before adult emergence.

Adults: The adults are robust yellow butterflies with orange and brown markings. The wings span about 1.0 inch (25 mm), with a body about 5/8 inch (16 mm) long.

Life cycle and habits: In tropical and subtropical regions, this pest is active all year. In these areas, three to five generations are normal. In the northern part of its range, only two to three generations are completed, with the adults being most numerous in mid- to late summer. Adult fiery skippers are strongly attracted to nectar-producing flowers such as lantana and honeysuckle. They will also visit other annual flowers. The males appear to be slightly territorial and often perch near flowers being visited by the females. They strongly pursue females or other males flying into the area. Newly emerged females appear to take 3–4 days before oviposition begins. During the middle of the day, females alight on turf and place eggs at random. The females rarely lay more than one egg per stop, and the eggs are usually cemented to the underside of a leaf blade. Egg laying continues for 4–6 days, with each female laying between 50 and 150 eggs. At 80°F (26.7°C) and above, eggs can hatch in 2–3 days. Up to 6 days may be needed at lower temperatures. The newly hatched larvae feed on the margins of turf grass blades, but soon drop into the turf canopy and spin a loose silk shelter. From within these silken shelters, the larvae emerge at night to remove and eat entire leaves. This produces small round damage spots in the turf. Once larval development is complete, they form their cocoon in the silken shelter. At 75°F (23.9°C), this pest took 48 days to complete development from egg to adult. Above 81°F (27.2°C), the cycle took only 23 days.

Control options: See: Cutworms and Armyworms: Introduction. This insect rarely produces enough numbers to damage large areas of turf. However, the scattered damage spots on golf course putting greens or tees can contribute to poor playing surfaces. There are only a couple of parasites known from this pest, and their influence on populations is not known. If suspicious spots are noted in the turf, sampling with a soap disclosing solution should be performed. If five to eight larvae per yd² of lawn turf or golf course fairway are found, treatments are warranted. Fewer numbers on putting greens may need attention.

Option 1: Biobased control—microbial toxins: Some of the strains of *Bacillus thuringiensis* (Bt) toxins are registered for sod webworms and armyworms in turf. These should also have activity against the fiery skipper larvae. For maximum efficacy against the fiery skipper larvae, these pesticides need to be applied when the young larvae (first and second instars) are active. Since the larvae are enclosed in silken tunnels and feed only at night, apply the material in the late afternoon or evening and do not irrigate until the next morning.

Stem- and Thatch-Infesting Insect and Mite Pests

This complex and poorly defined area is composed of mainly live stems and stolons, as well as organic material in various states of decay. Turf that is constantly irrigated and heavily fertilized may grow roots and crowns in this zone without having much contact with the soil. In this situation, true thatch/soil dwelling pests may invade this stem and thatch area. This zone provides high humidity, temperature mediation, and an abundant supply of living and dead organic matter that can be used for food.

Pests with sucking mouthparts seem to prefer this zone. They usually feed by removing plant fluids from the vascular system located in the stems, stolons, and lower leaf sheaths. Many of these sucking pests form salivary tubes or inject salivary materials that clog the vascular bundles. This results in discoloration, stunting, or death of the leaves or tip ends of stems. Of major concern is when these sucking pests attack the vegetative crowns of various grasses. When this happens, the entire plant may not survive. Sucking insects are often difficult to manage with traditional stomach insecticides that require ingestion of plant material containing residues. Since these pests do not chew and ingest plant matter, contact or systemic insecticides are more effective in management.

Pests with chewing mouthparts may enter the stems, stolons, and crowns, and are called borers during this phase. Others simply feed externally on the living and dead organic matter.

Pests located in the stem and thatch zone may escape detection until considerable damage has occurred. Turf managers must train themselves to thoroughly search this area, either visually or with other monitoring tools. Soap flushes and water flotation methods are often successful in disclosing pests located in this area.

Chinch Bugs

Several species of chinch bugs are considered important pests of turf. The *hairy chinch bug* is a pest of cool-season turfgrasses, primarily in northeastern North America. The *common chinch bug* is primarily a pest of small grain crops across North America and some suspect that this species occasionally attacks turfgrasses. Genetic studies are needed to separate hairy from common chinch bugs. In the warm-season turf zones, the *southern chinch bug* feeds on bermudagrass and zoysiagrass, but it is primarily a serious pest of St. Augustinegrass. This pest can be found from the Carolinas into southern California. In the Prairie states, where zoysiagrass and buffalograss are grown, the *western chinch bug* can become a pest. This pest is also called the buffalograss chinch bug.

Hairy Chinch Bug (and Common Chinch Bug)

Species: Common chinch bug, *Blissus leucopterus leucopterus* (Say); hairy chinch bug, *Blissus leucopterus hirtus* Montandon (Phylum Arthropoda, Class Insecta, Order Hemiptera, Family Blissidae) (Figures 3.15 and 3.59).

Distribution: The common chinch bug is normally found from South Dakota across to Virginia and south to a line running from mid-Texas across to mid-Georgia. The hairy chinch bug cohabits some of the northern range of the common chinch bug, but also extends throughout the northeastern states and into southeastern Canada.

Hosts: The hairy chinch bug prefers cool-season turfgrass species such as fine fescues, Kentucky bluegrass, perennial ryegrasses, and bentgrasss. The common chinch bug prefers grain crops but has been recorded attacking cool-season turfs.

Damage symptoms: Irregular patches of turf begin to yellow, turn brown, and die. Patches continue to become larger in spite of watering. Apparently, feeding by chinch bugs blocks the water- and food-conducting vessels of grass stems. By blocking the water, the leaves wither as in drought and the manufactured food does not get to the roots. Recently attacked plants often display purple margins and tips. Continued feeding results in plant death. Damage generally occurs during hot weather from June to September.

FIGURE 3.59
Hairy chinch bug damage to home lawns is often mistaken for disease or drought dormancy.

Description of stages: These pests are true bugs and have a gradual life cycle with egg, nymphal, and adult stages. All the species of *Blissus* are very similar in form, and an expert is needed to separate species and subspecies.

Eggs: The eggs are elongate bean-shaped, approximately 0.03 × 0.01 inch (0.84 × 0.25 mm), and are roundly pointed at one end and blunt at the other. The blunt end has several small tubercles visible through a dissecting microscope. The eggs are first white and change to bright orange just before hatching.

Nymphs: There are five nymphal instars that change considerably in color and markings. The first instar has a bright orange abdomen with a cream-colored stripe across it, a brown head and thorax, and is about 1/32 inch (0.9 mm) long. The second through fourth instars continue to have this same general color pattern, except that the orange color on the abdomen gradually changes to a purple-gray with two black spots. The fourth instar increases to more than 2 mm long. The fifth instar is very different because the wing pads are easily visible and the general color is now black. The abdomen is blueblack with some darker black spots, and the total body length is about 3 mm.

Adults: The adults are approximately 1/8 × 1/32 inch (3.5 × 0.75 mm). The males are usually slightly smaller than the females. The head, pronotum, and abdomen are gray-black in color and covered with fine hairs. The wings are white with a black spot, the corium, located in the middle front edge. The legs often have a dark burnt-orange tint. Individuals in a population, or in some cases, most of a local population, may have short wings, called brachypterous, that reach only halfway down the abdomen.

Life cycle and habits: In Ontario province, a single generation is common but in Ohio, Kentucky across to Maryland, two generations per season are typical. The hairy chinch bug adults overwinter in the thatch and bases of clumps of grass in the turf. The adults become active when the daytime temperatures reach 70°F (21.1°C). Females feed for a short period of time and mate when males are encountered. Eventually, the females begin to lay eggs by inserting them into the folds of grass blades, on the lower leaf sheaths, or in the thatch. This usually occurs in May from New York to Illinois and June in Canada. A single female may lay up to 200 eggs over 60–80 days. The eggs take about 20–30 days to hatch at temperatures below 70°F (21.1°C), but can hatch in as little as a week when above 80°F (26.7°C). The young nymphs begin to feed by inserting their mouthparts in grass stems, usually while under a leaf sheath. The nymphs grow slowly at the beginning of the season because of cool temperatures, but speed their development by July. Generally the first generation matures by mid-July. At this time, considerable numbers of adults and larger nymphs can be seen walking about on sidewalks or crawling up the sides of light-colored buildings. If a good hot, dry spring is available, turf injury by the first generation can be evident by June. Generally, major damage is visible in July and August when the spring-generation adults are feeding, and their second-generation nymphs are becoming active. During the hot summer months, the new females lay eggs rapidly, and the second generation of young may mature by the end of August or the first of September. The second-generation adults may lay a few eggs for a partial third generation if the season remains warm. However, most of these late nymphs do not mature before winter temperatures drop. When cool temperatures arrive, the mature chinch bugs seek out protected areas for hibernation. Clusters of adults are often found at the bases of bunch-type grasses, such as tall fescue and perennial ryegrass.

Control options: The common chinch bug is one the oldest known American insect pests. The first records of damage to crops are from the 1780s. Because chinch bugs are major crop pests, a tremendous number of control strategies have been developed. Only those useful in turfgrass management have been selected. Chinch bugs are relatively easy to control when they are detected early.

Option 1: Cultural control—watering the turf: This pest seems to prefer hot and dry conditions for optimum survival and reproduction, but needs actively growing turf to thrive. Early speculation suggested that regular irrigation during the spring and early summer may increase the incidence of pathogen spread, especially the lethal fungus, *Beauveria* spp. The adult chinch bugs can withstand water because of the protective hairs on the body, but the nymphs readily get wet and can drown. More recent studies have shown that *Beauveria* outbreaks are more common when the chinch bugs come under stress, such as when their host plants go into summer dormancy.

Option 2: Cultural control—use resistant turfgrasses: The hairy chinch bug seems to prefer Kentucky bluegrass, and endophyte-free perennial ryegrasses and fine fescues, especially if these are in the sun and have greater than 0.5 inch (13 mm) of thatch. Bentgrass is also attacked, but this turf is rarely used in lawns. Field studies have shown that perennial ryegrass and fescue cultivars that contain endophytes are highly resistant to chinch bugs. In fact, having between 35% and 40% of a stand of turf with endophytic stems will prohibit chinch bug outbreaks.

Option 3: Cultural control—recovery from damage: Turf with light to moderate damage will recover rather quickly if lightly fertilized and watered regularly. Heavily infested lawns may have significant plant mortality because of the toxic effect of chinch bug saliva, and reseeding will be necessary. In such cases, slit-seeding with endo-phytic grasses is recommended.

Option 4: Biological control: Several researchers have been trying to develop a usable formulation of the *Beauveria* fungus, but at present, no consistently effective product is available. An egg parasitic wasp is known, but this species rarely attacks more than 5% of a pop-ulation. Several predators, especially the bigeyed bugs, *Geocoris* spp., are noted to kill large numbers of chinch bugs. Bigeyed bugs are often mistaken for chinch bugs because of their similarity in size and shape (see: Nuisance Invertebrate, Insect and Mite Pests). Bigeyed bugs usually do not build up large populations until after considerable turf damage has occurred. Use of the insect parasitic nematodes, *Steinernema* spp. and *Heterorhabtitis* spp., have given inconsistent results when used against chinch bugs.

Option 5: Chemical control—use preventive applications: In turf areas where chinch bugs have been a perennial problem, early insecticide applications have been used to reduce the beginning spring popu-lation. This works well if applications are made in May after the adults have finished spring migrations and the young nymphs are just becoming active. It is recommended that preventive treat-ments be used only if sampling has determined that chinch bugs are present. In the past, chlorpyrifos or diazinon were used as pre-ventive insecticides, applied in early May. With the restrictions on the urban landscape use of these insecticides, the pyrethroids and neonicontinoids have been used as preventive treatments.

Option 6: Chemical control—curative treatments: Chinch bugs are rather easy to detect in turf, and targeted insecticide treatments can be applied to reduce populations that appear to be building to damaging lev-els. Field experience indicates that sprays tend to work faster and are more effective than the same insecticide applied in granular formulations.

Sampling

Several sampling schemes have been developed for assessing chinch bug populations in turf. The simplest method is to visually inspect the turf by spreading the canopy where brown turf grades into green. Chinch bug nymphs tend to hide in the deeper thatch, so dig deeper where thick thatch layers are encountered. The most accurate but time-consuming sampling method is to use the flotation technique. Push a metal cylinder (large coffee can with the top lid removed, and bottom with rim removed works well) into turf that is suspected of being infested; fill up the can with water and counting the number of adults and nymphs that float to the surface over a 10-min span. Populations of 25–30 per ft^2 (270–325 per m^2) warrant control, especially if these numbers are encountered in June through mid-August.

Southern Chinch Bug

Species: *Blissus insularis* Barber (Phylum Arthropoda, Class Insecta, Order Hemiptera, Family Blissidae) (Figures 3.60 and 3.61).

Distribution: This pest can be found from southern North Carolina to the Florida Keys and west to central Texas, Oklahoma, and Kansas. This pest has apparently been spread with sod to other countries. It is also found in Hawaii. Populations were suspected in Kansas and Oklahoma attacking zoysiagrass, but recent studies suggest that the western chinch bug is involved (this is often called the buffalograss chinch bug).

Hosts: This insect is a major pest of St. Augustinegrass, though it will occasionally attack centipedegrass, zoysiagrass, bahiagrass, and bermudagrass.

Damage symptoms: Usually irregular patches of St. Augustinegrass, zoysiagrass, or bermudagrass turn yellow and then brown. These patches are often circular or rounded and may expand or new patches may begin to form in the lawn. This can be mistaken for disease attack. As the southern chinch bug populations build up, they may kill extensive areas in a lawn and begin moving across

FIGURE 3.60

Southern chinch bug, *Blissus insularis*, nymph, and adult feeding on St. Augustinegrass.

FIGURE 3.61
St. Augustine grass damaged by southern chinch bugs.

sidewalks and driveways. Though they do not often remain on the surface of St. Augustinegrass, individuals frequently run to the surface, sun themselves briefly, and then return to the thatch zone.

Description of stages: These pests are true bugs and have a gradual life cycle with egg, nymphal, and adult stages. This species is difficult to separate from other *Blissus* spp., and an expert is generally needed if specific identification is critical. Generally, if this bug is in St. Augustinegrass, zoysiagrass, or bermudagrass in the southern states, it is most likely the southern chinch bug.

Eggs: The eggs are elongate oval, approximately 0.03 × 0.01 inch (0.75 × 0.23 mm), and are squarely cut off at the top. The blunt end has four small tubercles visible through a dissecting microscope. The eggs are first white but change to bright orange before hatching.

Nymphs: There are five nymphal instars that change considerably in color and markings as they develop. The first instars are bright orange with a cream-colored stripe running across the abdomen. The head and thorax become brownish with age. The second through fourth instars continue to have this same general color pattern, except that the orange color of the abdomen gradually changes to a dusky-gray with small black spots. The fourth instar increases to more than 3/32 inch (2 mm) long. The fifth instars are quite different because the wing pads have expanded and are easily visible. Occasionally, an additional instar, the sixth, will be formed, especially in cooler weather. The abdomen becomes blue-black with some darker black spots, and the total body length is about 1/8 inch (3 mm).

   Adults: The adults are approximately 1/8 × 1/32 inch (3.1 × 0.85 mm). The males are usually slightly smaller than the females. The head, pronotum, and abdomen are gray-black in color to dark chestnut-brown and covered with fine yellow to white hairs. The wings are white with a black spot, the corium, located in the middle front edge. The legs are chestnut-brown. Individuals in a population may have short, nonfunctional wings that reach only halfway down the abdomen.

Life cycle and habits: In the southern range of this pest, adults and a few individuals of all stages overwinter in St. Augustinegrass turf, especially in the thatch. In the northernmost part of the range, only adults overwinter. These insects may become active anytime the temperatures rise above 65°F (18.3°C), but reproduction generally does not occur until April or May. Females prefer to deposit their eggs by forcing them between the leaf sheath and stem. Each female may lay 45–100 eggs over several weeks. Some eggs are deposited in the thatch. The eggs hatch in 8–9 days at 83°F (28.3°C) and 24–25 days at 70°F (21.1°C). The young nymphs immediately begin feeding under the protection of the leaf sheath. Nymphs hatching elsewhere crawl into available spaces under leaf sheaths. Several nymphs are often congregated together. The nymphs go through five and occasionally six instars in 40–50 days during warm weather (above 80°F [26.7°C]). At cooler temperatures, below 70°F (21.1°C), the nymphs may take 2–3 months to mature. In most areas of the Gulf states, three to five overlapping generations occur each season. The first couple of generations are fairly well defined because of the initial start from the winter season. The first major adult peak usually occurs in June, and the second peak is in August. Subsequent peaks may be in October and December. The greatest amount of damage occurs during the dry season in the summer.

Control options: The southern chinch bug is very difficult to control because of management practices in St. Augustinegrass and bermudagrass lawns—frequent fertilization, irrigation, and use of one or two cultivars in any area. Populations in Florida into Texas are known to be resistant to organophosphate, carbamate, and pyrethroid insecticides, and some populations have also overcome the resistance characteristics of some St. Augustinegrass cultivars.

Option 1: Cultural control—watering the turf: See Hairy Chinch Bug—Option 1. Where regular irrigation is discontinued or irrigations heads get thrown off target, drought-stressed grasses are often more prone to successful southern chinch bug attack.

Option 2: Biological control: See: Hairy Chinch Bug—Option 4.

Option 3: Cultural control—modify agronomic management: Generally, St. Augustinegrass and bermudagrass can be overfertilized with quick-release nitrogen products, and this leads to rapid growth and immense buildup of fluffy thatch. In fact, many such turf areas have the roots and crowns growing in their own thatch layer, with minimal contact with the soil! This environment is very beneficial to southern chinch bug survival. Overly thatched lawns should be dethatched, verticut, or top dressed to improve the general conditions. Using more slow-release fertilizers and ensuring even irrigation coverage can greatly reduce populations of this pest.

Option 4: Cultural control—use resistant turfgrasses: Common St. Augustinegrass is highly susceptible to the southern chinch bug, while many of the zoysiagrasses and bermudagrasses are fairly resistant. The St. Augustinegrass variety Floratam has been the standard in Florida for resistance, but in recent years, this chinch bug has developed the ability to attack this cultivar in Florida and Texas. New varieties have been produced and are under development. It is suggested that you contact your local state turfgrass specialist or local cooperative extension office to inquire about improved cultivars for use in your area.

Option 5: Chemical control—use preventive/curative treatments: Since the southern chinch bug can be active over most of the year, there is usually no time when just overwintered adults are present. However, cool winter temperatures can greatly slow population growth and development, so early spring applications could be considered to be preventive treatments. Applications can be made any time when the chinch bugs are found, but insecticide applications without considering irrigation, fertilization, and grass resistance will like fail over the long haul!

Sampling

See Hairy Chinch Bug.

Insecticides and Application

Localized populations of southern chinch bugs are known to be resistant to organophosphate, carbamate, and/or pyrethroid insecticides. Most scientists believe that this is the result of making multiple applications of insecticides within the same season without consideration of rotation of chemistry and modes of action. If an application of an insecticide does not result in satisfactory control (within 7–10 days), do not make another application of the same insecticide! Select another insecticide with a different mode of action.

In some areas, combinations of pyrethroid plus neonicotinoid products have resulted in satisfactory control when the individual insecticide types did not yield control.

Twolined Spittlebug

Species: *Prosapia bicincta* (Say) (Phylum Arthropoda, Class Insecta, Order Hemiptera, Family Cercopidae) (Figures 3.62 and 3.63).

Distribution: Native of North and Central America, most common in Gulf states into Central America. Found as far north as Maryland across to Kansas.

Hosts: Attacks common in southern turfgrasses such as bermudagrass, St. Augustinegrass, bahiagrass, and centipedegrass. Adults feed on herbaceous perennials and have been noted to damage holly.

Damage symptoms: Large numbers of spittlemasses are unsightly and cause concerns by messing shoes and bare feet. Nymphs cause patches of turf to yellow, and the adults cause the leaves to turn brown and die. This gives the appearance of sparse, blighted turf. The adults may also cause damage to flowers or some ornamentals, especially Burford holly.

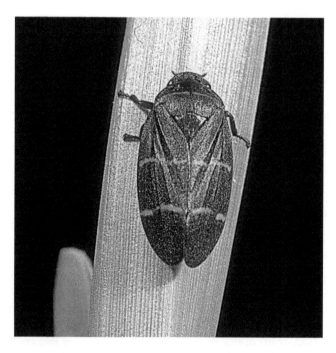

FIGURE 3.62
Twolined spittlebug adult.

FIGURE 3.63
Twolined spittlebug spittlemass in turf showing nymph.

Description of stages: This pest has stages typical of a gradual life cycle.

Eggs: The eggs are elongate oval and taper to a rounded point at one end. The 0.04 × 0.01 inch (1.09 × 0.27 mm) eggs are first pale orange and change to an orange-red in a few days.

Nymphs: Freshly hatched nymphs are pale yellow with a small orange spot on each side of the abdomen, and are about 5/64 inch (2 mm) long. The nymphs molt four times, during which time the orange spots enlarge to cover the entire abdomen. The final, fifth instar nymph has well-developed wing pads with two transverse orange bands, and is about 5/16 inch (8 mm) long. The nymphs are covered by the spittlemass.

Adults: The boat-shaped adults are about 3/8 inch (10 mm) long, and have a dark-brown to black color with two distinct reddish orange bands on the wings. The adults have orange-red bodies, do not produce spittlemasses, and can jump and fly short distances.

Life cycle and habits: The eggs overwinter in the turf and are usually deposited at the base of grass plants. Sometimes the eggs are inserted between a lower leaf sheath and stem, but more commonly they are inserted into the surrounding thatch. The eggs may be laid singly or in small groups. Though eggs are the normal overwintering stage, occasional adults may be found during the winter months in Florida. Eggs hatch in early spring when the turf is coming out of dormancy. The young nymphs must find a suitable feeding site within an hour or two or they die. After probing several places, the nymphs finally insert their mouthparts into suitable turfgrass tissue and they produce a spittlemass. This is actually an excretion from the anus. As feeding and growth continues, the nymphs may move about and sometimes several nymphs may share a single spittlemass. If too much moisture is present in the turf, the nymphs may move to the tips of the grass blades to form spittlemasses. Depending upon temperatures, the nymphs take 34–60 days to mature. The first peak adult emergences occur in late May and early June in Florida to July in South Carolina. Depending on temperatures and moisture, adults may be continuously present until October, or a second peak population of adults may be found in late August through September. Newly emerged adult females release a sex pheromone that attracts interested males. After mating, the females begin ovipositing in about a week. The females lay an average of 50 eggs over a 2-week period. During summer temperatures, the eggs take 14–23 days to hatch, but eggs laid in September or October do not hatch until the next spring. Nymphal feeding may only cause yellowing in the turf, but the adults cause a toxic reaction in which the affected stem turns brown and dies. Adults may also cause brown spots on flowers and ornamentals because of feeding. The adults fly for a period shortly after dark and are attracted to lights.

Control options: This pest rarely damages well-maintained turfgrasses (except centipedegrass) but is a nuisance because of the spittlemasses.

Option 1: Cultural control—restrict irrigation: Since the nymphs cannot survive dry conditions, especially when small, reducing irrigation in the spring and early summer may reduce populations. Using those turf management practices that reduce thatch buildup also makes it difficult for young nymphs to survive.

Option 2: Natural control: The fungus, *Entomophthora grylli*, attacks the adults, usually late in August through September. This fungus is probably encouraged by moist, warm conditions, so irrigation on warm evenings may help in its spread.

Option 3: Chemical control—curative insecticide applications: If spittlemasses are a nuisance problem or actual damage is occurring, applications of contact or systemic insecticides are generally effective. A reapplication may be needed in midsummer as adults can migrate in from surrounding turf. Ornamental plants, especially hollies, may also need protection if large numbers of adults are active.

Rhodesgrass Mealybug (= Rhodesgrass Scale)

Species: *Antonina graminis* (Maskell) (Phylum Arthropoda, Class Insecta, Order Hemiptera, Family Pseudococcidae) (Figure 3.64).

FIGURE 3.64
Rhodesgrass mealybugs attached to bermudagrass stem and showing waxy excrement tubes.

Distribution: Recorded from South Carolina to southern California; found worldwide in tropical and subtropical regions of Africa, Australia, Central America, India, Japan, Pacific Islands, and South China.

Hosts: Attacks over 70 species of grasses, including rhodesgrass, bermudagrass, and St. Augustinegrass.

Damage symptoms: This mealybug does not commonly kill turf unless stressful conditions are found. High-cut bermudagrass during drought is more prone to severe damage. This pest produces considerable honeydew, and ants or bees may frequent turf that is heavily infested.

Description of stages: Though this pest is a type of mealybug, it acts like true scales by being immobile once settled.

Eggs: Eggs are elongate oval and cream colored. The eggs are contained inside the remains of the female and her waxy cover.

Crawlers: These are the first instar nymphs, which are the only mobile forms. The crawlers are flat, oval, cream-colored insects with a median stripe tinged with purple. Short legs, six-segmented antennae, and two waxy tail filaments are evident.

Sessile nymphs: The crawlers settle on the grass crown or at nodes, insert their mouthparts, and begin secreting a waxy coat. After the first molt, the new sessile nymph takes on a sac-like form without legs. Only the thread-like mouthpart and anal filament emerge from the body. A second molt occurs while the waxy cover grows larger and the anal excretory tube elongates.

Adults: Only females are known and these reproduce asexually. The adult body is also sac-like, broadly oval, dark purplish brown, and 1/16–1/8 inch (1.5–3 mm) long. The fluffy waxy covering turns yellow with age, and openings at the anterior and posterior ends expose parts of the body. A very long, 1/8–3/8 inch (3–10 mm), waxy, tubular filament arises from the anus through which honeydew is excreted.

Life cycle and habits: This pest continues its life cycle year-round but is slowed by winter temperatures. Reproduction is considerably reduced during the winter, but activity increases in the spring as the grass begins rapid growth. Peak populations are reached by July, and populations are again reduced through July and August as summer stress is brought about by dryness. In September and October, the populations again resurge until peaking in early November. During the spring, females lay an average of 150 eggs over 50 days. As the crawlers hatch, they remain under the waxy cover of the female for several hours before emerging. These crawlers tend to first walk to the tops of plants but eventually settle down by wedging themselves beneath a leaf sheath, usually at a node. Here, the mouthparts are inserted and the excretory tube and waxy cover are started. In about 10 days, the first molt takes place and the walking legs disappear and the antennae are much smaller. The excretory tube continues to elongate and the waxy cover becomes thicker and larger. Two more molts take place under the waxy cover and maturity is reached in 25–30 days. During the summer, a generation averages 2 months but in winter, this may take 3.5–4 months to complete. This scale apparently travels by transportation of sod or grass cuttings. The crawlers may climb onto the legs of animals and "hitch a ride." High temperatures, especially near 100°F (37.8°C), reduce scale development and may actually kill individuals. Exposure of the scale to 28°F (–2.2°C) for 24 h is fatal. Thus, winter cold limits northern movement and survival of this pest. This scale increases turf mortality during periods of drought.

Control options: This pest rarely kills turf unless it is poorly managed. Do not cut the turf too short.

Option 1: Cultural control—water and fertilize: Frequent irrigation, fertilization, and mowing no shorter than 2 inches help prevent damage by this pest as actively growing turf appears to outgrow the insect populations.

Option 2: Biological control—conserve parasites: Two parasitic wasps have been successful in controlling rhodesgrass mealybug populations. A Hawaiian parasite, *Anagyrus antoninae* Timberlake, and an Indian parasite, *Dusmetia sangwani* Rao, have been established in the southern United States. Apparently, *Dusmetia* seems better adapted and can survive lower temperatures than *Anagyrus*. Both these parasites, once established, should be conserved by avoiding insecticide applications to areas known to have rhodesgrass mealybugs and the parasites. These areas will serve as harborage and a reservoir for the parasites.

Option 3: Chemical control—use of contact insecticides: Many contact insecticides, like the pyrethroids, do not effectively reach the protected stages, especially the eggs and resting crawlers. Repeated applications may be necessary if this strategy is used.

Option 4: Chemical control—use of systemic insecticides: Systemic insecticides are more effective for killing feeding mealybugs. Some eggs may escape harm if hatching occurs after the insecticide's residues disappear. Thus, a second application may be necessary for killing freshly settled crawlers. Most of the neonicotinoids insecticides have activity against mealybugs, though this species of scale is rarely on labels.

Bermudagrass Scale

Species: *Odonaspis ruthae* Kotinsky (Phylum Arthropoda, Class Insecta, Order Hemiptera, Family Diaspididae) (Figure 3.65).

FIGURE 3.65
Bermudagrass scales feeding on bermudagrass stem.

Distribution: This scale is found worldwide in tropical and subtropical regions. In the United States, this scale attacks bermudagrass from California to Florida. It is also known in Hawaii.

Hosts: This scale is most frequently reported on bermudagrass, though it has been found on centipedegrass, bahiagrass, St. Augustinegrass, and tall fescue.

Damage symptoms: Bermudagrass first appears to grow slowly and has a yellow color. This often looks like drought stress. Heavy infestations may dramatically thin and kill patches of bermudagrass. This type of damage is more evident during periods of hot, dry weather. Where bermudagrass enters winter dormancy, this scale can cause a delay in the spring green-up. Scale damage can be confused with nematode attack, spring dead spot and delayed spring green-up syndrome.

Description of stages: This is a typical armored scale that lays eggs and has winged male stages.

Eggs: The elongate oval eggs are pink to light burgundy colored and are located within the female shell.

Crawlers: The eggs hatch into a first instar nymph called a crawler. This stage is pink and the body is very flat and oval. They have short legs, antennae, and eye spots.

Settled nymphs: Settled crawlers begin to produce an oval, waxy test (shell), which is first straw yellow and then covered with white waxy secretions.

Adults: Adult females have shells, or tests, which are egg-shaped to oval in outline and 3/64–1/16 inch (1.0–1.75 mm) long. Often, there is a straw yellow area, the exuvium, to the side of the middle of the waxy test. The actual female body, inside the test, is oval and pinkish in color. Male scale tests are about one-half the size of females. The males emerge from their test and are small gnat-like insects with one pair of wings. Their yellowish-pink bodies are about 0.02 inch (0.5 mm) long and have two to three long white waxy threads arising from the tip of the abdomen.

Life cycle and habits: Little is known about the actual time periods needed for the development of this scale. Most studies have attempted to assess development by counting the numbers of different stages at various times of the year. From these studies, it appears that the bermudagrass scale may have two overlapping generations per year in the southern states. In Georgia and Florida, eggs are laid and crawlers are most active in the spring rainy season. Settled crawlers and adults can be found during much of the season, though little growth occurs during winter dormancy or during summer drought periods. It is suspected that this scale has continuous generations in warmer climates where the bermudagrass does not go dormant.

The eggs are laid and retained inside the female scale test. As the eggs hatch over several weeks, the tiny crawlers move along the stolons and lower grass stems. They prefer to settle under old leaf sheaths at the bases of crowns, but they may be found anywhere on the stolons and lower stems. Often, large numbers of settled crawlers and young adults can be found on stems and stolons in the soil or thatch layer. Rarely are the scales exposed on upper plant parts. As the settled scales grow, they begin to cover the body with loose waxy filaments, which eventually give way to the formation of a solid, waxy shell-like test. At maturity, the scales often extend slightly from under the old leaf sheaths that originally hid the body. Populations can be so large at nodes and crowns that the scales seem to be stacked on top of each other.

Control options: At present, no biological controls are known for this pest. Undoubtedly, small predators take their toll on the crawlers, but this has not been confirmed. Insecticide treatments have not been very successful in controlling this pest and none are currently registered for this purpose. Since eggs and crawlers may be present over extended periods, timed applications for crawlers are difficult to make. Damage assessment and control thresholds are not developed for this pest. However, when 30–40% of the turf shows yellowing from this pest, control procedures are probably warranted.

Sampling

When yellowed turf is encountered, the scale should be sampled for by digging out several affected stolons with attached above-ground stems. Inspect the nodes and bases of the stems for the oval, white scales. If the scale is confirmed, regular assessment of yellowing need to be done.

Option 1: Cultural control—water and fertilize turf: Bermudagrass that is well fertilized and watered can generally outgrow this pest. However, damage can begin to appear if irrigation discontinued in the summer.

Option 2: Chemical control—traditional insecticides: No insecticides are specifically registered for this pest, but some of the neonicotinoids used in turf also have excellent scale control capabilities in ornamental plants. One specifically good at controlling armored scales should be evaluated.

Billbugs: Introduction

Species: Billbugs are merely weevils, that is, beetles with a beak-like snout that also have the thorax and head almost as long as the wing covers. Billbug larvae look something like white grubs but lack legs and have a more pointed abdomen. Most of the billbug species, of which there are more than 60 species, breed in grasses (Graminae) and sedges (Cyperaceae). Fortunately, only four species have been recorded as common pests of North American turfgrasses and several others are considered occasional residents of turfgrass. The bluegrass billbug, *Sphenophorus parvulus* Gyllenhal, and hunting billbug, *Sphenophorus venatus vestitus* Chittenden, are the most common pests in cool-season turf and warm-season turf, respectively. The Phoenician (=Phoenix) billbug, *Sphoeniciensis phoeniciensis* Chittenden, occurs in warm-season turf in California and Arizona, while the Denver or Rocky Mountain billbug, *Sphoeniciensis cicatristriatus* Fahraeus, is a common pest in cool-season turf from northern New Mexico to Idaho. The lesser billbug, *Sphoeniciensis minimus*, is often found alongside the bluegrass billbug; the unequal billbug, *Sphoeniciensis inequalis*, can be found in both warm- and cool-season turf, and *Sphoeniciensis cosefrons* is occasionally found in considerable numbers in bermudagrass and zoysiagrass turf (Phylum Arthropoda, Class Insecta, Order Coleoptera, Family Curculionidae) (Figures 3.66 through 3.69).

Distribution: The different species of billbugs can be found wherever their favored hosts are cultivated. In fact, some species, like the hunting billbug, has been found around the world and other countries have billbugs that can damage turf.

Hosts: The bluegrass and Denver billbugs prefer Kentucky bluegrass, endophyte-free perennial ryegrass, and endophyte-free tall fescue, while the hunting billbug and Phoenician billbug are considered to be pests of bermudagrass and zoysiagrass. More recently, hunting

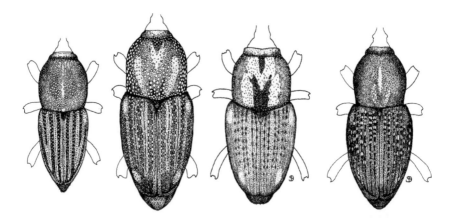

FIGURE 3.66
Illustrations of common turf-infesting billbug thorax and wing covers: (left to right) bluegrass billbug, *Sphenophorus parvulus*; hunting billbug, *S. venatus vestitus*; Phoenician billbug, *S. phoeniciensis*; and Denver billbug, *S. cicatristriatus*.

FIGURE 3.67
Bluegrass billbug adult and mature larva.

FIGURE 3.68
Bluegrass stems filled with sawdust-like billbug larval frass.

FIGURE 3.69
Hunting billbug adult walking across a golf green.

billbugs have also been found in the southern range of cool-season turf cultivation.

Damage symptoms: Billbug larvae bore into turfgrass stems. This kills the stems, causing them to turn straw colored as they die. This activity alone is not significant. However, larger larvae drop out of the stems and they can feed on the grass crowns, killing the entire plant. Larvae may also feed on roots and stolons. It appears that the hunting billbug larva may feed outside stems more than the other billbug species. Infestations begin to appear as scattered dead stems, and this can be mistaken for dieback of old seed-head stems. Eventually, 2–3 inch (5.0–7.5 cm) diameter spots die out; this is often misidentified as dollar spot disease. Heavy infestations may kill extensive turf areas, especially when drought conditions occur.

Life cycles: All the billbugs except for the hunting appear to have a single generation per year, though the bluegrass billbug will often attempt a second generation. The hunting billbug may breed continuously in southern states. Most billbugs spend the winter as adults, though a significant number of hunting and Rocky Mountain billbugs overwinter as larvae. Overwintered adults have a spring feeding period that is followed by egg laying, and larvae develop over the summer.

Identification of species: The larvae and pupae are extremely difficult to identify, so try to find adult specimens. The adults are best identified by the patterns of pits and furrows found on the pronotum and wing covers. A 10× hand lens will be needed. The four common species can be identified using the patterns of pits and furrows on the back (Figure 3.66).

Control options: Billbugs appear to be perennial pests once they have become established in an area. Often, whole neighborhoods with the same turfgrass type are attacked. Adult billbugs are notorious walkers. Because of this migratory habit, turf areas treated one year will often become reinfested from surrounding, untreated turf.

Option 1: Cultural control—use resistant turfgrass: Several cultivars of Kentucky bluegrass, bermudagrass, and zoysiagrass have demonstrated resistance or tolerance to billbug attack. Perennial ryegrasses, fine fescues, and tall fescues with endophytes are highly resistant to billbug attack. Contact seed suppliers for cultivars suited for your region. Billbugs generally do not seem to severely attack southern grasses such as centipedegrass and St. Augustinegrass.

Option 2: Cultural control—water and fertilizer management: Billbugs usually insert their eggs into grass stems, so they are less susceptible to dry soil conditions. Low to moderate billbug larval infestations can often be masked by using water and fertilizers in the summer that keeps the turf actively growing. Adequate water may reduce plant stresses due to root and crown damage. This is not a reliable method and should be used only if spring adult controls were inadequate.

Option 3: Biological control: Few natural predators or parasites have been identified attacking billbugs. However, adults are occasionally attacked by the fungus, *Beauveria*, and larvae have been found infected by *Beauveria* and *Metarhizium* fungi. Keeping the turf environment moist in the spring migration period may help spread this fungus. The parasitic nematode, *Steinernema carpocapsae*, has been used successfully to control billbug larvae and adults when applied at 1.0 billion infective juveniles per acre.

Option 4: Chemical control—spring adult preventive insecticide applications: An excellent strategy for control of most billbug species has been early applications of pesticides targeted to kill the adults emerging from hibernation sites. For the bluegrass billbug, this time is when the soil surface (at the l-inch [2.5 cm] depth) reaches 67–69°F (19.4–20.6°C). Pitfall traps can also be used to monitor spring migration for timing of pesticide use. Denver, hunting, and Phoenician billbug adults may move later in the spring.

Option 5: Chemical control—larval preventive-curative controls: Currently, most of the neonicotinoid insecticides appear to have systemic action that affects young billbug larvae burrowing in the stems. Together with the thatch/soil residues, such applications also seem to kill larger billbug larvae as they emerge from stems to feed on crowns, roots, and stems in the soil–thatch zone. Larval activity may be monitored by taking soil samples.

Bluegrass Billbug

Species: *Sphenophorus parvulus* Gyllenhal (Phylum Arthropoda, Class Insecta, Order Coleoptera, Family Curculionidae) (Figures 3.67 and 3.68).

Distribution: This pest is most common in northern North America but may be found in southern states. It prefers cool-season turf-growing regions.

Hosts: Kentucky bluegrass is the preferred host, but this pest has been known to infest perennial ryegrass, red fescue, and tall fescue.

Damage symptoms: Light infestations in well-kept turf results in small dead spots that look like the turf disease, dollar spot. Sometimes the damage looks like irregular mottling, browning, in the turf. Moderate infestations or chronic infestations result in sparse turf with some irregular dead patches. Heavy infestations result in complete destruction of the turf, usually in August. These infestations are sometimes confused with greenbug attacks. Billbug-damaged turf turns a whitish straw color rather than yellow. Soil under damaged turf is solid, not spongy as in white grub attacks. Turf can be pulled out to reveal stems that have been hollowed out and are packed with sawdust-like fecal material, frass.

Description of stages: Billbugs have complete life cycles with a single generation per year and rarely a partial second generation.

Eggs: The pearly-white, kidney-shaped eggs are inserted into a small hole chewed in grass stems and are approximately 1/16 × 1/48 inch (1.6 × 0.6 mm).

Larvae: The typical weevil grubs are a robust C-shape with a slightly tapering abdomen. These larvae have no discernible legs, and have light yellowish brown head capsules after hatching. The head capsules gradually darken as the mature larvae reach about 1/4 inch (6 mm) long.

Pupae: The pupae are 5/16 inch (7–8 mm) long, and are first creamy-white. These change to a reddish brown just before the adult emergence. The pupae have the conspicuous billbug snout and pronotum.

Adults: The adults are 5/16 inch (7–8 mm) long with a black body. When freshly emerged, the adults are reddish brown. Fresh adults have a light brown coating over the body that can be rubbed off to reveal the shiny black shell. The pronotum is covered with small uniform punctures and the wing covers have distinct longitudinal furrows with pits. Occasionally, the pronotal punctures fade along the top midline of the pronotum. Any other pattern indicates that another species may be present. If in doubt, this species should be confirmed by a specialist as several species of billbugs may be present in the turf.

Life cycle and habits: In most of its range, this pest overwinters in the adult stage. Adults have been found overwintering in thatch, cracks, and crevices in the soil, worm holes, and in leaf litter near turf. The hibernating adults become more active in late April to mid-May when the soil surface temperatures rise above 65°F (18.3°C). The adults wander about in search of suitable grasses and crops on which to feed. After feeding for a short period of time, the female begins to insert one to three eggs in a feeding hole made in grass stems. The overwintered females may continue laying eggs to the end of June, but most eggs are laid by mid-June. Laboratory-kept females have been known to lay over 200 eggs, usually two to five per day. The eggs hatch in 6 days, depending on the temperature, and the young larvae begin to tunnel up and down the stem. If a stem is hollowed out while the larva is small, an exit hole is formed and the larva will bore into another stem. Eventually the larva becomes too large to fit inside the grass stems and it drops to the ground to begin feeding externally on the grass crowns, roots, and stem bases. This is the point at which significant damage to the turf is noticed, especially if the turf is under drought stress at this time. After 35–55 days, the larva is fully grown and pupates in a cell made of soil near the surface. The pupa gradually darkens and the reddish-brown, tineral adult emerges in 8–10 days. The new adults appear to be most common in late July into early August. These new adults feed, store fat, and generally seek suitable sites for overwintering. In prolonged summer conditions, some of the summer females lay eggs for a partial second generation. If these larvae are not able to mature, pupate, and emerge by the first hard freeze, they die. Adults have been observed trying to fly in September and October, but no great distances were covered.

Control options: See: Billbugs: Introduction.

Option 1: Cultural control—use resistant turf varieties: Many of the original improved Kentucky bluegrass cultivars, "Touchdown," "Merion," "Nugget," "Adelphi," "Baron," "Cheri," and "Newport" are quite susceptible to billbug attack. It is felt that the larger stems of these aggressive cultivars allow for better survival of the billbug larvae. The cultivars "Park," "Arista," "NuDwarf," "Delta," "Kenblue," and "South Dakota Certified" appear to be finer textured and are resistant or more tolerant to attack. Most perennial ryegrasses with endophytes are highly resistant to billbugs as are the endophyte-containing fescues. Ryegrass and fescue seed, when stored for long periods, lose the endophyte and such nonendophyte grasses are readily attacked. It is strongly recommended that golf course roughs, sports fields, lawns, and grounds should be renovated with endophytic grasses after billbug damage has occurred. Research has shown that 35–40% of the stems within a stand must harbor endophytes to suppress billbug activity below damage levels.

Option 2: Biological control—fungal diseases: Billbug adults and larvae seem to be susceptible to the entomophagous fungi, *Beauveria* and *Metarhizium*. However, these fungi rarely kill enough billbugs to have a significant reduction in turf damage. While commercial preparations of both fungi are available, replicated field trials have not provided consistent control.

Option 3: Biological control—parasitic nematodes: The entomopathogenic nematodes, *Steinernema carpocapsae*, *Steinernema glaseri*, and several *Heterorhabditis* species, have been used to infect billbug larvae in the laboratory. In field trials, *S. carpocapsae* has provided satisfactory control of adults and larvae. For best results, obtain freshly produced nematodes, apply them in early morning or evening to avoid direct sunlight, and irrigate in immediately. Keep the area regularly irrigated for a week to 10 days.

Option 4: Chemical control—spring adults: This is the most commonly used strategy. Contact and/or stomach insecticides are applied at the time when the adults come out of hibernation and are migrating in search of oviposition sites. This may be from late April to mid-May. Research indicates that the adults become active when the soil surface temperature approaches 67–69°F (18.3–20°C).

Sampling

There are several methods for determining spring adult activity. The simplest is to use pitfall traps. Small plastic cups placed inside holes made by using a standard golf course, 4.25-inch (11.4 cm) cup cutter is an easy method of making pitfall traps. These can be placed along the turf near or in flower beds so that they are out of the way. Linear pitfall traps are more efficient for monitoring billbug adult activity, but these require more time and space to install. They consist of a 3-ft long section of 3-inch diameter PVC pipe in which a half to 1-inch wide groove has been cut. The tube is buried in the ground to the level of the groove and a plastic water jug is attached to one end. Billbugs fall into the pipe and travel back and forth until they drop into the jug. Adult billbugs can be easily counted by inspecting these traps 2–3 times a week. The simplest method of sampling billbug adult activity is to watch driveways and sidewalks for the migrating adults. This works well on hot sunny days but may miss the first activity period by a couple of weeks.

Degree-Day Timing

A degree-day model using the average method of calculation, a March 1 starting date, and a threshold temperature of 50°F (10°C) predicts that the first adult activity should occur between 280 and 352 cumulative $DD_{base50F}$ (155–195 $DD_{base10C}$) and the 30% first activity (the time when the last surface insecticide would be effective) should occur between 560 and 624 $DD_{base50F}$ (311–347 $DD_{base10C}$).

Insecticides and Application

Pyrethroid insecticides are currently used for control of spring adult billbugs, but there is good evidence that applications of neonicotinoids, targeted for the larvae, also have a significant impact on the adult billbugs. Indoxicarb and chlorantraniliprole also seem To have an impact on billbug adults.

Option 5: Chemical control—larval control: Since the billbug larvae spend considerable time burrowing up and down grass stems, systemic insecticides have the potential to control these larvae. Each insecticide is taken up by the grass plants at slow to moderately rapid rates. This determines how early the pesticide should be applied. Each insecticide also has a residual action period that can range from 2 to 6 weeks after the application.

Sampling

Early detection of summer larvae is difficult. However, by mid-June, damaged stems of turf can be pulled out to reveal the sawdust-like frass characteristic of this pest. By late June and into July, the larvae are usually large enough to see in the soil and thatch by cutting into the soil and thatch layer of the turf. A golf course cup changer makes a convenient billbug larval sampling tool. Pull a plug where a frass-filled stem is found. Turn the plug open and split the soil toward the surface. Billbug larvae are easily exposed near the soil–thatch interface.

Degree-Day Timing

The larvae begin to emerge from the stems and are thus exposed to insecticide residues within the thatch and upper soil zones between 925 and 1035 $DD_{base50F}$ (513–575 $DD_{base10C}$). They can be controlled from this time until significant visual damage occurs between 1330 and 1485 $DD_{base50C}$ (739–825 $DD_{base10C}$).

Option 6: Cultural control—masking of damage: With light to moderate billbug infestations, much of the damage can be masked with adequate irrigation and fertilization. In fact, during years when soil moisture is normal to above normal in June, visible billbug damage is rare, as is the case in irrigated turf. The critical period for extensive damage appears to be when Kentucky bluegrass has set new tillers, but the tillers have not yet established independent root systems. When billbug larvae kill the crowns and drought ensues, the parent plant and all the tillers die. When moist soil is present, the tillers establish and cover up the death of the parent plant.

Hunting Billbug

Species: *Sphenophorus venatus vestitus* Chittenden (Phylum Arthropoda, Class Insecta, Order Coleoptera, Family Curculionidae) (Figure 3.69).

Distribution: Several subspecies of *S. venatus* are found all across North America. *S. venatus vestitus* is most commonly found from Maryland across to Kansas and south. In states west of the Continental Divide, the hunting billbug can be found in Arizona to Utah, and in the southern range, it is often found with the Phoenician billbug that appears to have a similar life cycle.

Hosts: Attacks zoysiagrass where it may be called the "zoysia billbug" and often damages bermudagrass. This pest has also been found in bahiagrass, centipedegrass, and St. Augustinegrass as well as some cool-season turf, such as Kentucky bluegrass.

Damage symptoms: Patches of turf turn yellow and then brown. In many zones, larvae are able to overwinter and these larvae will continue to feed on the roots and stolons of dormant zoysiagrass and bermudagrass. Winter-damaged turf may appear as irregular dead patches in the spring green-up period. Hunting billbug larvae appear to feed on ground stolons, which results in numerous small segments that can be brought to the surface by rubbing your hand back and forth in the dead areas of turf. Inspection of these pieces will reveal that they have been scooped out on one end.

Description of stages: All stages are typical for weevils.

Eggs: The elongate, bean-shaped white eggs are inserted into stems of the preferred grasses.

Larvae: First instars are about 1/16 inch (1.5 mm) long, and have the typical weevil larval form. Mature larvae have tan to brown head capsules with black mandibles. Fully grown larvae are 1/4–3/8 inch (7–10 mm) long.

Pupae: The pupae are first cream colored, changing to a reddish brown just before the adult emerges. The pupae have the distinct snout of the adult weevil.

Adults: The adults are generally larger and more robust than the bluegrass billbug. Adults range from 1/4 to 7/16 inch (6–11 mm), and often have a coating of soil that adheres to the body surface. Clean specimens have numerous visible punctures on the pronotum with a distinct Y-shaped, smooth, raised area behind the head. This area is enclosed by a shiny parenthesis-like mark on each side.

Life cycle and habits: Recent studies of the life history of this pest have occurred in Arkansas, Florida, and North Carolina, but little is known about the biology of this pest elsewhere. In the northern part of its range, this pest overwinters as dormant adults in the soil. In southern states, the adults may be found walking and feeding all year, whenever temperatures are high enough for activity. It is stated that most of the eggs are inserted into grass stems or leaf sheaths in the spring when zoysia and bermudagrasses are well out of winter dormancy; however, there is some evidence that the adults may actually lay some eggs in the soil or thatch adjacent to stems. The eggs take 3–10 days to hatch and the young larvae mine down the inner surface of the leaf sheaths or bore into stems. As the larvae increase in size, they drop into the ground and feed on roots and stolons. Mature larvae pupate after 3–5 weeks of feeding. The pupae take 3–7 days to mature. Because of the extended period of oviposition, larvae may be found from May into October. In the warm-season turf zones, new adults often occur in July and August and these can lay eggs for a second generation as well as overwinter. Larvae produced in this second generation are able to survive and continue feeding as long as the soil does not freeze. Pupation occurs in the early spring and the new adults join the other adults that have successfully overwintered. In the northern zoysia-growing regions, most damage occurs in August and one generation appears to be the norm. In Florida, up to five generations per season can occur with considerable overlap of stages. Damage in Gulf states may occur in late summer, after the turf has slowed its growth or in the early spring.

Control options: See: Billbugs: Introduction, and Bluegrass Billbug.

Option 1: Cultural control—reduce transport of infested turf: The hunting billbug has been a severe pest of zoysiagrass and bermudagrass sod farms. In these situations, the adults or larvae may be transported to new sites when lawns are sodded. Be sure to obtain certified pest-free sod.

Option 2: Chemical control—time insecticides for first generation: Overwintered adults and newly emerging adults are usually present, feeding and beginning to lay eggs in May into early June. Application of a neonicotinoid insecticide at this time can stop adult oviposition, and the systemic and thatch-adsorbed residues can kill any larvae that may already be present.

Annual Bluegrass Weevil (= Hyperodes Weevil)

Species: *Listronotus* (= *Hyperodes*) sp. near *anthracinus* (Dietz). Some confusion exists concerning the taxonomic definition of this pest. It has been placed in the genera *Hyperodes* and *Listronotus*. During the time when it was placed in *Hyperodes*, turfgrass managers adopted this name for the common name, though most taxonomists agree that this species should now be in the genus *Listronotus*! (Phylum Arthropoda, Class Insecta, Order Coleoptera, Family Curculionidae) (Figure 3.70).

Distribution: *Listronotus anthracinus* has been collected in most northern states east of the Mississippi river, but the *Listronotus* causing turf damage has been regionally found damaging turf from the New England states, across to northeastern Ohio, down to West Virginia, and back across Delaware. It is also known in southeastern Canada.

Hosts: The primary host of this pest is annual bluegrass, *Poa annua* L., but the adults and larvae will feed on creeping bentgrass.

Damage symptoms: Highly maintained bentgrass with a predominance of *P. annua* in golf greens, aprons, tees, and short-cut fairways, as well as grass tennis and croquet courts, is attacked. Light infestations cause a slight yellowing and browning of the turf. Moderate infestations cause small irregular patches of dead turf, and heavy attack kills turf in large areas. Damage begins to become obvious in late May or through June, and this damage can be initially

FIGURE 3.70
Annual bluegrass weevil adult feeding on grass blades.

mistaken for disease, black turfgrass ataenius attack, or annual bluegrass senescence. Larvae and pupae in the soil–thatch zone should be easy to find when turf is moderately to heavily damaged.

Description of stages: This pest has a complete life cycle (egg, larvae, pupae, and adults), with the larvae being typical for weevils, a C-shaped legless grub.

Eggs: The eggs are oblong, about 1/32 inch (1 mm) long, and change from yellow to smokey black when about to hatch.

Larvae: The larvae are C-shaped, legless, and have a creamy-white body. The head capsule is light brown in young larvae but becomes darker in older larvae. Newly hatched larvae are about 1/32 inch (1 mm) long and grow to about 3/16 inch (4.5 mm) when mature.

Pupae: Pupae are about 1/8 inch (3.5 mm) long and have the beak, wing pads, and legs evident. They are first creamy white and turn red-brown before the adult emerges.

Adults: Adults are 1/8–5/32 inch (3.5–4.0 mm) long, and have the combined length of the snout and prothorax distinctly shorter than the length of the wing covers. This is different than billbugs that have combined snout and prothoracic lengths equal to the wing cover length. This *Listronotus* is black-brown with the wing covers coated with fine yellowish hairs, yellowish scales, and scattered spots of grayish-white scales. Newly emerged adults (callow adults) are orange-brown in color and require several days before becoming fully pigmented.

Life cycle and habits: This pest overwinters in the adult stage in protected areas near places where short-cut *P. annua* is cultivated. They prefer to hide in tufts of bunch grasses or in thatchy turf in golf course roughs and among leaf litter around bushes and trees. These adults become active in early spring, usually in late March into early April, and seek out actively growing *P. annua*. It is evident that overwintered weevils feed on *Poa* leaf blades as they migrate in search of short-cut *P. annua*. After maturing eggs, the females extend their ovipositors down the lower leaf sheaths and deposit from one to nine eggs; usually two to three eggs are laid per stem. The eggs take 4–5 days to hatch and the young grubs chew their way into the stems. Once inside, the larva feeds up and down the stem leaving tightly packed sawdust-like frass behind. The stems react to this feeding by having the leaves turning yellow and gradually wither. The larvae may move from stem to stem, hollowing out the centers. When the larvae are too large to fit inside stems, they feed externally, especially at the crown. This is the time when extensive turf death occurs. During the first generation, this damage occurs when *P. annua* often displays summer dieback or decline. The five larval instars are passed in 3–4 weeks and the fully developed larva digs about 3/16–1/2 inch (5–10 mm) into the soil to form an earthen pupal cell. This cell is formed by the circular wiggling of the larva that packs the soil. The cell formation takes 3–4 days and the pupa remains in the cell for 5–7 days before molting into the adult form. The freshly emerged adult is light colored, soft, and remains in the pupal cell for 5–6 days to harden before emerging. These light-colored adults are called callow or teneral adults and their presence indicates the completion of a generation. The entire cycle takes 25–45 days, depending upon food and temperature. The mature adults hide during the daytime but come to the tips of the grass stems to feed at night. This feeding continues through July and early August. Many northeastern states experience two to four generations of this weevil per season, though there is some debate as to whether these are distinct generations or are the result of the first generation of adults having a prolonged oviposition period during the summer. When fall conditions arrive, the adults seem to disappear, apparently going to their overwintering sites. The fall adults are known to both walk and fly to overwintering sites.

Control options: Few natural controls appear to exist and this pest appears to be a major pest only where *P. annua* dominates the turf being cultivated.

Option 1: Cultural control—destroy overwintering sites: When it was found that this pest generally overwinters in woodlots and high-cut turf, it was recommended to remove leaf litter and cut turf in roughs lower in the fall may reduce populations. This has not proved to be an effective control.

Option 2: Chemical control—early spring application of insecticides targeting adults: Applications of contact and stomach insecticides when adult weevils first begin to migrate has been shown to significantly reduce subsequent damage. The key to this strategy is timing of the applications and location of the applications! Early recommendations were to make these applications between the time when *Forsythia* is in full bloom until flowering dogwood is in full bract. Linear pitfall trapping data indicates that this may be too late in many locations as the first adult movement can occur as early as the time when *Crocus* and daffodils bloom. Chlorpyrifos and bifenthrin have been the two insecticides most commonly used against these migrating adults. Unfortunately, up to five applications of bifenthrin per year have been documented and populations of the annual bluegrass weevil have been identified that are now resistant to all pyrethroid insecticides! Because of this, it is highly recommended to alternate chemistry if additional adulticide treatments are needed during the season. Some of the neonicotinoids and indoxicarb are potential alternate insecticides. Recent recommendations are to treat roughs and border areas of tees, fairways, and greens first for migrating adults. If monitoring indicates that adults are still migrating after 2–3 weeks, a second application may be necessary and this treatment would now include the tee, fairway, and green surfaces.

Sampling

Adult annual bluegrass weevils can be easily sampled for by using linear pitfall traps (Figure 3.71) and soap flushing. The linear pitfall traps appear to be more precise in detecting the moving beetles while soap flushing can provide data on the relative abundance of the weevils per unit area of the turf. Several traps should be used in areas known to be regularly attacked by this pest. Locate one trap along the border of harborage areas (like a woodlot or tall grass field) and the rough. A second trap should be placed inside the rough but next to the fairway or tee or green approach. This arrangement will allow the turf manager to detect the first movement of the weevils out of overwintering sites and detect the extent of movement into vulnerable areas (tees, fairways, and greens). Soap flushes are difficult to perform in high-cut turf (higher than 2 inches [5 cm]) and detection of adults in the short-cut turf is likely to be too late to initiate adult control treatments.

Option 3: Chemical control—applications targeting larvae: Several systemic insecticides have been shown to be effective against the larvae burrowing into annual bluegrass stems. Several neonicontinoids (most notably clothianidin) have demonstrated efficacy against the weevil larvae. Applications made at the time when the adults are at peak egg-lay apparently reduce adult activity and are

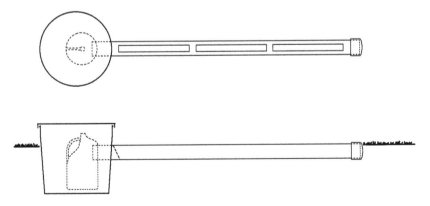

FIGURE 3.71
Diagram of a linear pitfall trap constructed using PVC pipe and some plastic containers.

effective against the young larvae feeding within and around the stems. Where annual bluegrass weevils have extended migration periods (as evidenced by pitfall trapping records), split applications of neonicotinoids plus a pyrethroid, with first application made when the first weevil adults are detected and the second application made 3 weeks later, have been highly effective. Chlorantraniliprole is also recommended for use as a larval systemic insecticide.

Option 4: Chemical control—combined program, adults and larvae: Effective management of the annual bluegrass weevil on golf courses generally requires a multiapplication approach. Early migrating adults are first targeted using the tactics described in Option 2. About 2–3 weeks after this first adulticide application or applications, Option 3 is used to control any larvae that may have resulted from adult weevils that survived the adulticide treatments. After this treatment or treatments, the turf is sampled for larval populations as they begin to finish their development. If larvae are detected, additional larvicide treatments are applied to avoid additional turf damage. This is followed by additional sampling of adult populations (detergent flushes) and treating with adulticides using alternate insecticide modes of action.

Sampling

There are several methods recommended for sampling of the annual bluegrass weevil larvae and pupae. Whether you use a soil sampler or a golf course cup changer, sufficient samples need to be taken that yield several larvae and/or pupae to assess the development stage of the weevil. Visual inspection of these samples has been shown to be largely inadequate, especially when young larvae are present. Using a salt water solution resolves this issue. Mix four cups of common table salt in one gallon of water. Take three to four soil sampler samples (remove the excess soil) or one-fourth a cup changer sample (remove the excess soil and cut the turf-thatch in quarters) and place in a quart canning jar, then fill 2/3 with the salt solution and shake vigorously. Allow to sit for 3–5 min and inspect the surface of the water for the eggs, larvae, pupae, and adults of the weevil that tend to float in the strong salt solution. Some superintendents put their samples into buckets or Even heavy plastic bags with the salt water.

Option 5: Chemical control—rescue treatments: Once the larvae have nearly matured and damage is evident, superintendents always want to know what to do to "rescue" their turf! At best, this will always be a difficult situation as there are few products that will provide sufficient control to halt further damage. If the larvae are still feeding *outside* the stems, trichlorfon, chlorpyrifos, indoxicarb, or clothianidin applications have been shown to significantly reduce larval populations. Keys to successful applications are preirrigation to ensure that the thatch and upper layers of soil are moist and immediate irrigation after the application to move the insecticide to the target pests. However, the turf manager must also pay attention to the damaged areas that may require renovations (removal and replacement of sod) or reseeding.

Option 6: Biological control—insect parasitic nematodes: Several of the entomopathogenic nematodes have been shown to be potential control agents, but consistent control is difficult to achieve. Keys to success are to obtain fresh nematodes, apply them in the late afternoon (to avoid sunlight), immediate postapplication irrigation, and keeping the treated area moist for several days after the application.

Cranberry Girdler

Species: *Chrysoteuchia topiaria* (Zeller) (Phylum Arthropoda, Class Insecta, Order Lepidoptera, Family Pyralidae) (Though technically a sod webworm, the cranberry girdler is more subterranean in habit and rarely feeds on turf leaves. It is also called the subterranean sod webworm) (Figure 3.72).

Distribution: Found across North America but it is more common in the cool-season and transition turfgrass zones.

Hosts: Prefers cool-season grasses such as Kentucky bluegrass, bentgrass, and fine fescues.

FIGURE 3.72
Cranberry girdler larva in thatch.

Damage symptoms: The larvae feed in the soil–thatch zone, at the crown of grasses. This causes small circular areas of dead turf. Heavy populations can cause general turf death similar to grub damage.

Description of stages: The stages are typical of moth complete life cycles, though the larvae do not have the dark spots diagnostic of most sod webworms.

Eggs: They are somewhat oval with broadly rounded ends and are approximately 0.017 × 0.013 inch (0.43 × 0.33 mm). Only slightly raised longitudinal ridges and no cross ridges are present, as in other sod webworm eggs.

Larvae: This webworm's larvae are not typical for sod webworms because they lack the distinct dark spots over the body. Cranberry girdler larvae are dirty-white in color, with a tan head capsule. The mature larvae are 5/8–3/4 inch (16–20 mm) long.

Pupae: The pupae are located in a tough silken case attached to the base of plants, or in the soil. The pupae are typically light chestnut-brown in color and are 5/16 × 3/32 inch (7–9 × 2.5 mm).

Adults: The adult moths have a wing span of 9/16–13/16 inch (15–20 mm), and are quite colorful for sod webworms. The forewings have a series of longitudinal brown and cream stripes following the wing veins, a silver chevron across the wing tip followed with ocher, a series of three black spots, and a tip fringe of silver scales.

Life cycle and habits: Adult cranberry girdlers are active fliers from late June to mid-August. The adults emerge in the evening or at night to expand and harden their wings. The females begin to release a sex attractant that calls in receptive males for mating. Mated females generally do not begin laying eggs until the following day. On the second evening, fertilized females locate suitable habitat to deposit eggs. Unlike other sod webworms that lay eggs while flying, these females usually drop their eggs into the turf while resting on a plant. The females lay about 40% of their eggs on the second night and are usually finished within a week. Each female may lay 450–500 eggs. At 72°F (22.2°C), eggs take 9–11 days to hatch. The young larvae prefer to set up residence in the thatch or upper soil surface next to grass crowns. As the larvae grow, they may actually bore into grass crowns but they prefer to feed on roots, stems, and shoots. Webbed tunnels lined with fecal material are usually evident by August and September. The larvae mature rapidly, molting 5–7 times before cool temperatures cause the larvae to begin winter diapause. Mature or nearly mature larvae spin a tough silk case, called a hibernaculum, in the soil or thatch in October for overwintering protection. By November, over 90% of the larvae will be prepupae enclosed inside the hibernaculae. In the following May, the larvae that did not mature in the fall resume feeding and may molt another time. Fully mature larvae do not feed, but pupate along with the rest of the population in May and early June. Depending on the temperature, pupae take 2–4 weeks to mature.

Control options: See: Sod Webworms: Introduction. Since this species often lives in silk-lined burrows extending through thatch layers, it is usually sheltered from surface-applied contact and stomach insecticides. Also, spring sod webworm treatments may not kill mature larvae that do not have to feed before pupating.

Option 1: Chemical control—use pheromone traps to time controls: The sex pheromone, a combination of 1.0 mg (Z)-11-hexadecenal and 0.05 mg (Z)-9-hexadecenal, is commercially available for use in sticky traps. These traps can be placed out in June and July to monitor adult activity. Controls can be applied 7–10 days after peak emergence.

Burrowing Sod Webworms

Species: Several species of *Acrolophus* are found across North America in association with turfgrasses. *Acrolophus plumifrontellus* (Clemens), *Acrolophus popaenellus* (Clemens), and *Acrolophus arcanellus* (Clemens) are the more common species encountered (Phylum Arthropoda, Class Insecta, Order Lepidoptera, Family Acrolophidae) (Figures 3.73 and 3.74).

Distribution: The above-named species are generally present east of the Rocky Mountains. Most activity is noticed in the transition and southern turfgrass zones.

Hosts: Little host specificity is known. Infestations have been recorded from Kentucky bluegrass, tall fescue, and bermudagrass turf.

FIGURE 3.73
A burrowing sod webworm adult male, *Acrolophus popeanellus*.

FIGURE 3.74
A burrowing sod webworm larva in burrow.

Damage symptoms: Few turfgrass areas have been significantly damaged. Damage often looks like sod webworm or cutworm activity. Occasionally, the larvae will build silken tubes over the surface of the turf. These are unsightly and are often pulled out during mowing. These tubes may litter a lawn after the caterpillars pupate where they look like empty cigarette papers.

Description of stages: Almost nothing is known about the immatures of burrowing webworms. Adults are often collected at lights and are relatively easy to identify.

Eggs: The eggs of *A. popaenellus* are roughly spherical and approximately 0.02 inch (0.5 mm) in diameter. They are cream colored when laid, but turn dark gray-brown after a few hours. Each egg has 18–20 longitudinal raised ridges.

Larvae: The larvae have a velvety grayish or brownish-white body with a distinctly chestnut-brown head capsule and prothoracic collar. Under a 10× hand lens, the head has several raised ridges. The hooks (crochets) on the prolegs are arranged in a pear-shaped row of large hooks surrounded with many smaller hooks. Mature larvae are 1.0 × 1/8 inch (25–30 × 3 mm).

Pupae: The chestnut-brown pupae are 9/16–13/16 inch (14–20 mm) long, with the large palps visible on the surface next to the smaller antennae.

Adults: The adults are mottled brown or reddish-brown moths with a wing span of 1.0–1⅜ inch (25–35 mm). The palps of the males of most species are very large, hairy, and curved over the head, just touching the thorax. In some species, the palps extend across the thorax. These very active moths often have the wing scales rubbed off if they have been out for some time.

Life cycle and habits: Very little is known about this group of moths. They are mainly tropical and subtropical in distribution, with species living on bromeliads and orchids. Many species in North America apparently feed on the roots and stems of grasses. Whatever little is known is from observations of attacks on corn roots on no-till or minimum-till corn fields. The adult moths are active from mid-June through July, though an occasional individual may be found through August. These adults fly at dusk until an hour or two after dark. They move very swiftly and upon landing, they quickly crawl to the ground surface. The eggs are not attached to plant material but are dropped over the ground. The eggs take about 7–14 days to hatch and the young larvae establish themselves next to a plant stem. They soon begin to burrow vertically into the soil, where they construct a silk-lined tunnel. These burrows may extend 4–24 inches (10–60 cm) into the ground, depending on the species and soil conditions. The silk tubes may be extended above the ground level and along grass stems. In very close-cut turf, these tubes are occasionally built across the turf surface. The larvae spend the winter in their burrows, but resume activity in the spring. At maturity, the tunnels may be enlarged from 5/32 to 1/4 inch (4–6 mm) in diameter. The pupa is located in the tunnel near the ground surface.

Control options: Burrowing sod webworms are rarely common enough to warrant controls in home lawns, grounds, and sports fields. However, the silk lining of their burrows may cause concern when they become entrapped in lawn mower blades after the adults have emerged. These can look like discarded cigarette wrappers. Occasionally, people will notice the silk tunnels that have been constructed to the turf surface. Burrowing webworms can be controlled like most true sod webworms. See: Sod Webworms: Introduction.

European Crane Fly, Common (or Marsh) Crane Fly, or Leatherjackets

Species: European crane fly—*Tipula paludosa* (Meigen); common or marsh crane fly—*Tipula oleracea* Linnaeus (Phylum Arthropoda, Class Insecta, Order Diptera, Family Tipulidae) (Figures 3.75 and 3.76). (Other species of *Tipula* may be found in managed turf across North America. Native species rarely attack living plant material, but when they do, they can be treated like the European crane fly.)

Distribution: European crane fly and the common crane fly are natives of Europe that have become established in the Pacific Northwest (including British Columbia, Nova Scotia, Washington, and Oregon) and northeastern North America (including Canadian

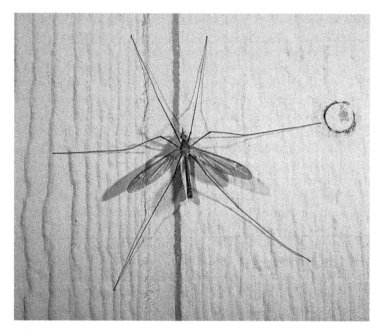

FIGURE 3.75
A crane fly adult, *Tipula* spp.

FIGURE 3.76
A crane fly larva, *Tipula* spp., showing diagnostic black head capsule and anal lobes.

provinces of Ontario and Quebec, New York, Pennsylvania, Ohio, and Michigan). The common crane fly appears to be moving into new territory faster than the European crane fly. There are *Tipula* species all across North America and local species can commonly infest turf where they feed primarily on moist, decaying thatch and organic matter. Large numbers of these crane fly larvae are often present after the turf was killed from disease, white grub attack, drought, or other causes.

Hosts: All cool-season turfgrasses.

Damage symptoms: The plant-feeding species cause major thinning of the turf, bare areas, and lodging of seed stalks and tall grasses. Large numbers of larvae cause concern when people have to walk on them to cross the turf, and the larvae may migrate from damaged turf, in mass, in search of more food. Adults look like giant mosquitoes and can cause fear in people who are unaware that they do not have biting or sucking mouthparts.

Description of stages: These flies have a complete life cycle.

Eggs: Black, 3/64 inch (1 mm) long, elongate-oval eggs with one side flattened and the other pointed; often described as tiny black jellybeans.

Larvae: The larvae are maggot-like but possess a distinctive black head capsule and finger-like projections that arise from the tip of the abdomen. The larvae are initially light brown, but as they grow, they turn gray to grayish-brown, developing a tough, but flexible exoskeleton. This is where the name "leather jacket" comes from. The larvae have four instars, and at maturity they may exceed 1.0 inch (25 mm) in length. The head capsule may be exposed during feeding or moving but is often withdrawn if disturbed. The larvae have two typical spiracular plates (breathing holes) at the end of the anal segment. These plates are surrounded by six fleshy anal lobes. The shape and form of the spiracular plates and fleshy lobes are used in the identification of species.

Pupae: The translucent, brownish pupae have the legs, wind pads, and antennae glued down, and they are formed just below the soil surface. These 1inch (25 mm) long pupae wiggle to the surface at emergence time.

Adults: These large crane flies have a 13/16 inch (20 mm) long, slender body with very long legs, almost 4 inches (10 cm) from the front to back leg. Adults are brownish-tan with smoky-brown wings. Adults can be identified by wing, head, and genitalic features, but you should consult an expert for species determinations.

Life cycle and habits: The primary difference between the European and common crane flies is when the adults fly—European crane flies take flight only in late summer and the common crane fly has two generations with spring and late summer flights. Both species of crane flies require mild, winter temperatures, cool summers, and average annual rainfall of at least 24 inches (60 cm). Adult flies look like giant mosquitoes, as they emerge from lawns, pastures, and roadsides. The European crane flies are generally active from mid-August to late September and the common crane flies can fly from late April into early May, then again in mid-August into September. The adults may gather in large numbers on the sides of homes and other structures. These adults cannot bite or sting. The adults mate and females begin to lay eggs. European crane fly females lay all their eggs within 24 h after emerging while common crane flies can take 3–4 days. The eggs swell by absorbing soil moisture and European crane flies hatch in 1.5–2 weeks while common crane fly larvae hatch in a week. The brownish maggots begin feeding by rasping and nibbling on turf foliage, stems, and roots. By the time winter cold hits, the European crane fly larvae have reached the third instar while the common crane fly larvae generally over-winter in the fourth, and last, instar. These larvae may feed slowly during the winter, even under the cover of snow. In the spring, the common crane fly larvae pupate and the first adults emerge in late April. European crane fly larvae continue to feed, molt one more time and then enter a period of summer resting, called estivation, before transforming into pupae and emerging as new adults in mid-August. The common crane fly adults lay eggs in April and May and these larvae continue feeding until they pupate in early August. When an adult is ready to emerge from the pupal case, the tail of the pupa is rotated, which pushes the head end out of the turf canopy. These pupal cases are easily observed in short-cut turf.

Control options: Because these pests were imported from Europe, probably in the 1960s, few natural biological control agents are to be found. Even where natural controls are present in Europe, the larvae periodically damage turf. The obvious difference in the two species' life cycles suggests that two pesticide applications may be needed, but a single fall application can catch the young larvae of both species, thereby eliminating the need for a spring treatment.

Option 1: Cultural control—habitat modifications: Crane fly eggs and first instar larvae are very susceptible to desiccation. Refrain from watering turf in September. Ryegrasses with endophytes seem to be less preferred as food plants.

Option 2: Biological control—parasitic nematodes: The commercially available, entomophagous nematode *Steinernema carpocapsae* has been effective when used at the rate of 22.2 billion infective larvae per acre; lower rates of 1–2 billion reduce crane fly larvae by about 50%. Other strains or species are being developed, so check with your supplier to determine the best nematodes to use.

Option 3: Chemical control—fall insecticide applications: Several insecticides have provided good control of crane fly larvae but the timing of the applications is critical for success. Applications need to be made about 2–3 weeks after the peak adult flight and before cold winter temperatures begin.

Option 4: Chemical control—spring insecticide applications: Applications of insecticide targeting overwintered larvae have not been consistent. However, spring applications made in late May into early June specifically for control of the common crane fly can be successful.

Frit Fly

Species: *Oscinella frit* (Linnaeus), though other stem-boring fly larvae are known to attack grasses (Phylum Arthropoda, Class Insecta, Order Diptera, Family Chloropidae) (Figure 3.77).

Distribution: This pest is a native of Europe, where it commonly attacks field crops. It has been found in virtually every state across North America.

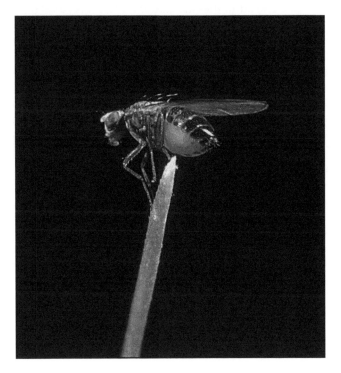

FIGURE 3.77
Frit fly, *Oscinella frit*, resting on the tip of a grass blade.

Hosts: This fly has a wide host range in the grass family. It is occasionally a pest in bentgrass and rarely a pest in Kentucky bluegrass and ryegrasses in the United States.

Damage symptoms: Individual grass stems are killed by the burrowing activity of the larvae. Usually the seed head stems are destroyed, but when large populations are present, entire areas can be killed. This pest is most common in bentgrass areas where seed stems are allowed to form, such as along the collars of golf greens and margins of fairways.

Description of stages: Like all flies, this pest has a complete life cycle with egg, larval, pupal, and adult stages.

Eggs: The translucent shiny cream-colored eggs are slightly bean-shaped and about 1/64 inch (0.4 mm) long.

Larvae: The maggot-like larvae have no legs and no head capsule. However, the anterior end is pointed and has a pair of tiny black hooks used for rasping food. The mature larva is about 1/8 inch (3 mm) long.

Pupae: The pupae are light reddish-brown, oval, and about 5/64 inch (2 mm) long, slightly smaller than a sesame seed. Under magnification, only ring-like segments can be seen.

Adults: The adult flies are rather undistinguished black flies with yellowish or white markings. They are about the size of the small flies that gather around decaying fruit, approximately 3/32 inch (2–2.5 mm) long.

Life cycle and habits: The larvae generally overwinter in a small tunnel eaten out of a grass stem. In the spring, when the grass resumes growth, the maggot continues to feed by rasping inside the stem. As the maggot tunnels downward, it may pass nodes, killing the stem from that point outward. The maggot matures in several weeks and forms a pupa inside the stem or in the duff surrounding the grass plant. The adult flies emerge in a week or two and fly in search of new grass plants. The flies can often be seen in considerable numbers resting on the tips of grass blades in the morning or evening sun. They also tend to alight on lighter-colored surfaces such as golf balls or equipment placed on the turf. The new adults insert eggs in the space between a leaf and stem and these eggs take about a week to hatch in warmer weather. Several generations can occur over a season and activity seems to be greatest when there is cool moist weather. When seedhead stalks emerge, the larvae of this fly causes the tips to die and turn white. These withered seed heads are called "frits."

Control options: Reasonable control of this pest has been obtained in Europe by using pesticides in the spring when the adults are active or by using systemics when the larvae are tunneling in the stem. It is suggested that some of the neonicotinoid insecticides should be able to control the larvae, but data are lacking.

Soil-Inhabiting (e.g., Thatch- and Root-Infesting) Insects

The insects that inhabit the thatch and soil zones can be some of the most devastating pests because they destroy turf roots, crowns, and underground rhizomes and stolons. These critical parts of the turf plant can tolerate light to moderate damage but when heavy damage occurs, the plant dies.

This group of pests is also the most difficult to control because of their isolation from control materials. Pesticides and biological controls have to travel through the turf canopy, then thatch that may be compacted and impervious to water, and finally the high organic matter in the upper inches of soil. Most pesticides tend to adsorb onto the organic matter and surfaces of living plant tissues. Tiny biological controls such as entomopathogenic nematodes and bacteria are fractions of a millimeter long and quite susceptible to ultraviolet light radiation as well as immediate death from drying.

Management of thatch thickness and texture is important to managing pests that live in the thatch/soil zone. Turf managers should carefully review techniques of core aeration, top dressing, and verticutting, as well as turf cultivar selection and usage of fertilizers and irrigation.

White grubs and mole crickets are the most common and serious pests in this group. White grubs can literally graze off all the roots, so that the turf lifts up like a loose carpet. Mole crickets appear to be more selective in their feeding with some species eating plant materials and others being primarily insect predators. Both tunnel extensively in the soil–thatch zone and this can break the connections of root hairs to soil. This results in root desiccation and dieback. There are many other insects in this zone, most of which are beneficial or inconsequential. However, some scales (ground pearls and others), root aphids, mealybugs, and wireworms can cause noticeable damage from time to time. Unfortunately, little is known about the life histories of these occasional pests.

Sampling and monitoring are extremely important when dealing with soil-inhabiting pests. Early detection is imperative. By the time you see visible turf symptoms, considerable and possibly irreparable damage may have occurred. Loss of turf and the need for renovation indicate that the turf manager did not perform good sampling and monitoring procedures. Be alert and aware!

White Grubs: Introduction

Species: White grubs (grubworms or simply, grubs) are the C-shaped larvae of a large group of beetles called scarabs. Many species of scarabs are found in North America, and several of these commonly attack turfgrasses. The most important species are Japanese beetle, *Popillia japonica* Newman; May or June beetles, *Phyllophaga* spp.; northern masked chafer, *Cyclocephala borealis* Arrow; southern masked chafer, *Cyclocephala lurida* Burmeister; southwestern masked chafer, *Cyclocephala pasadenae* Casey; western masked chafer, *Cyclocephala hirta* LeConte; black turfgrass ataenius, *Ataenius spretulus* (Haldeman); and the green June beetle, *Cotinis nitida* (Linnaeus). Other, more localized, white grub pests are European chafer, *Rhizotrogus majalis* (Razoumowsky); the Asiatic garden beetle, *Maladera castanea* (Arrow); and Oriental beetle, *Exomala orientlis* (Waterhouse). Over the last decade, white grubs of scarab subfamily Dynastinae (rhinoceros beetles) have been found damaging turf in southern states. The two species most commonly cited are the sugarcane beetle, *Euetheola humilis* (Burmeister), and the sugarcane grub, *Tomarus subtropicus* (Blatchley) (Phylum Arthropoda, Class Insecta, Order Coleoptera, Family Scarabaeidae) (Figures 3.78 and 3.79).

Distribution: White grubs are perennial pests of all the cool-season and transition-zone turfgrasses. Occasionally, white grubs will attack southern grasses when the turf is highly maintained and cut short. May–June beetles, masked chafers, green June beetles, and the black turfgrass ataenius can be found across most of North America. The Japanese beetle is present in eastern states, though

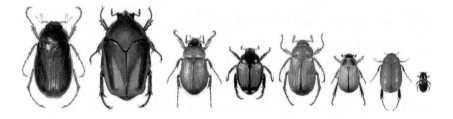

FIGURE 3.78
Adults of common white grub genera and/or species that attack turfgrasses in North America: (left to right) May–June beetle, *Phyllophaga* spp.; green June beetle, *Cotinis nitida*; masked chafer, *Cyclocephala* spp.; Japanese beetle, *Popillia japonica*; European chafer, *Rhizotrogus majalis*; Oriental beetle, *Exomala orientalis*; Asiatic garden beetle, *Maladera castanea*; and black turfgrass ataenius, *Ataenius spretulus*.

FIGURE 3.79
Larvae of common white grub genera and/or species that attack turfgrasses in North America: (left to right) May–June beetle, *Phyllophaga* spp.; green June beetle; masked chafer; Japanese beetle; European chafer; Oriental beetle; Asiatic garden beetle; and black turfgrass ataenius.

it has recently moved to states west of the Mississippi river. The European chafer, Asiatic garden beetle, and Oriental beetles are fairly restricted to the northeastern states and eastern Canada.

Hosts: All species of turfgrass may be infested.

Damage symptoms: White grubs eat organic matter, including the roots, rhizomes, and crowns of plants that are located in the soil–thatch zone. Therefore, damage first appears to be drought stress. Heavily infested turf first appears off-color gray-green, and wilts rapidly in the hot sun. Continued feeding will cause the turf to die in large irregular patches. The tunneling of the larvae causes the turf to feel spongy underfoot, and the turf can be rolled back like a loose carpet. Grub populations may not cause observable turf injury, but predatory mammals such as skunks, raccoons, opossums, armadillos, feral pigs, and moles or birds may dig in search of a meal.

Life cycles and habits: Scarabs have a complete life cycle with egg, larval, pupal, and adult stages. Japanese beetles, masked chafers, European chafers, Asiatic garden beetles, Oriental beetles, and green June beetles have annual life cycles over most of their ranges. It appears that some masked chafer species, and the sugarcane beetle may have two or more generations per year in Florida. The May–June beetles usually take 2–3 years to develop in the northern states but some southern species have annual cycles. The black turfgrass ataenius has two to three generations per summer. Most turf scarabs overwinter as larvae but the black turfgrass ataenius and May–June beetle adults overwinter.

Identification of species: The adults are easily identified to genus but the grubs are the stage usually found in the turf. The grubs are identified by the form, shape, and arrangement of bristles (the raster) on the last abdominal segments. A 10× to 15× hand lens is usually adequate for identification and the common white grub groups can be identified using a raster pictorial key (Figure 3.80).

Control options: White grubs seem to be periodic pests, attacking turf areas irregularly from year to year. In other situations, white grub attack is often repeated. Studies have shown that where Japanese beetle or masked chafers have damaged turf, there is an approximate 80% chance of repeating the damage in the following year. The major factors influencing development of damaging numbers of grubs are organic matter availability (especially thatch) and soil moisture present at the time of egg lay. Recently established turf rarely has grub populations for the first 4–6 years. However, once organic matter levels in the upper inch of the soil increase and a thatch layer is present, grubs can be perennial pests for several years. In general, in years with normal or above normal rainfall, grub populations tend to increase. Irrigation can counter the effects of

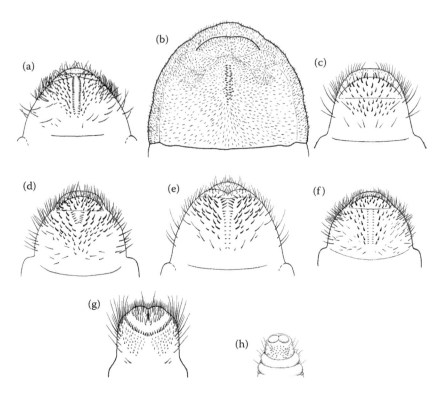

FIGURE 3.80
Diagrams of rasters of common white grub genera and/or species that attack turfgrasses in North America: (a) May–June beetle, *Phyllophaga* spp.; (b) green June beetle; (c) masked chafer; (d) Japanese beetle; (e) European chafer; (f) Oriental beetle; (g) Asiatic garden beetle; and (h) black turfgrass ataenius.

below-normal rainfall. Well-maintained turf next to ornamental plants favored by the adult beetles (especially Japanese beetle and May–June beetles) seems to be more commonly attacked. However, masked and European chafer adults do not feed, and these pests build up in well-watered and maintained turf. Black turfgrass ataenius and green June beetle adults seem to be highly attracted to turf with decaying thatch layers.

Option 1: Cultural control—host plant modifications: Certain species of scarab adults prefer specific host plants. Where Japanese beetles are common, do not plant roses, grapes, and lindens along high-maintenance turf areas. May–June beetle adults prefer oaks, and the green June beetles feed on ripening fruit such as peaches. Tall fescues appear to be less severely damaged when compared to Kentucky bluegrass, bentgrass, and perennial ryegrass. This is thought to be due to the thicker and tougher root systems that tall fescues possess.

Option 2: Cultural control—water management: Practically all white grub species require moist soil for their eggs to develop. The young larvae are also very susceptible to desiccation. In areas where turf can stand some moisture stress, do not water in July and early August when white grub eggs and young larvae are present. On the other hand, moderate grub infestations can be outgrown if adequate water and fertilizer is applied in August through September and again in May when the grubs are feeding. This latter strategy is not preferred because mammals may dig up the turf, or irrigation bans may occur.

Option 3: Natural control—parasites: Several parasitic wasps, *Tiphia* spp. and scoliids, attack white grubs and may effectively reduce populations in certain areas. Masked chafers and green June beetles are the species most commonly attacked. However, these parasitic wasps may take 2–3 years to build up effective populations, during which time turf damage may occur. Several parasites have been imported for the control of the Japanese beetle, but most are poorly established and restricted to only a few states.

Option 4: Biological control—milky diseases: Several strains of the bacterium *Paenibacillus popilliae* have been found that infect and kill white grubs. However, the commercial preparation of this bacterium is extracted from Japanese beetle grubs and is most active for this species. This bacterium is picked up by feeding grubs, and causes the body fluids to turn a milky white before grub death. Fresh bacterial preparations should be used and 3–5 years are needed to provide lasting controls. Replicated field trials in Kentucky and Ohio indicate that the currently available products do not perform sufficiently to suppress grub populations below damaging levels. Other surveys of this pathogen have revealed that most of the grub species are naturally infected by strains of this bacterium and all except for the black turfgrass ataenius appear to be weak pathogens (resulting in less than 25% mortality). Adding commercial preparations does not change this mortality level.

Option 5: Biological control—parasitic nematodes: Insect parasitic nematodes in the genera *Steinernema* and *Heterorhabtitis* have been shown to be effective against white grubs. Field trials of *S. carpocapsae* strains have generally resulted in less than 50% control, though some *Heterorhabdtitis species* have achieved 80% control or better. It has also been shown that available species do not appear to be effective from one season to the next.

Option 6: Chemical control—preventive pesticide applications: Where white grub damage appears to be repeated, applying pesticides for control of anticipated grub populations is a useful strategy. In fact, neonicontinoid insecticides and anthranilic diamides appear to work best when applied before egg lay through the egg hatch period. Many of these same pesticides appear to have residues in the soil–thatch zone that remain effective for 50–100 days after application. Where multiple pests occur in the same turf (like billbugs, chinch bugs, and/or caterpillars), application of one of these pesticides, targeting billbugs or chinch bugs, can also control the new generation of white grubs that arrive 1–2 months after the application.

Option 7: Chemical control—curative pesticide applications: When using the IPM approach, sampling of grub populations are recommended before deciding if controls are needed. Unfortunately, most white grub species are difficult to sample until the grubs have reached the second instar stage. At this time, an insecticide that has curative ability will be needed. When organophosphate and carbamate chemistry dominated the grub control products, curative control was relatively easy to achieve. The neonicontinoids registered for white grub control also appear to be able to control second instar grubs at satisfactory levels, but this efficacy drops off when the white grubs become third instars. Curative applications are often made in mid- to late August when conditions can be hot and dry. To achieve maximum efficacy, irrigation of the area to be treated should be done the day before the application. This will moisten the thatch and upper inch of soil. After the application, sufficient irrigation should be applied as soon as it is possible to move the residues into the soil–thatch zone.

Option 8: Chemical control—late-season, rescue pesticide applications: Occasionally, damaging populations of white grubs may go undetected until September or October. By this time, the annual white grubs are third instars and may be 30–50 times the body weight of a newly hatched grub. These mature grubs may stop feeding in preparation for digging down into the soil for the winter. Chemical control of these large grubs is difficult, at best. In many cases, animal digging is the real problem as the grub populations may be below turf-damaging levels. At present, only one insecticide, trichlorfon (= Dylox®) has been a consistent control pesticide. This still has to be irrigated well after the application to keep the grubs near the soil–thatch interface and to wash in the pesticide. Of the neonicotinoids, clothianidin (= Arena®) and dinotefuran (= Zylam®) have also provided good control of these late-season grubs. Where animal digging is an issue, application of the human waste-based fertilizer, Milorganite, has been effective in discouraging animals for 10–14 days after an application.

Option 9: Chemical control—spring pesticide applications: As with the late fall rescue applications, spring treatments are often ineffective.

Though the grubs feed in the spring, the grubs are large and may not be feeding on a consistent basis. The same pesticides recommended for rescue treatments are recommended if spring treatments are deemed necessary.

Maximizing Control of White Grubs with Insecticides

Adult Sampling

Adult activity of May–June beetles, masked chafers, European chafers, Oriental beetles, and Asiatic garden beetles can be monitored using light traps. Useful predictive data can be obtained by monitoring beetle captures one to two times a week. Simply plot the number of beetles collected over the date sampled on graph paper. If the number of beetles collected drops for 7–10 days in a row, you can assume that the peak emergence and oviposition time has passed. Most species have eggs that hatch within 14–21 days. Therefore, to have the maximum levels of grub insecticides in the soil–thatch interface when the first instars emerge and surface to begin feeding, applications 3–4 weeks after the peak adult activity was noted.

Grub Sampling

White grub populations should be assessed when the grubs are large enough to be easily seen (early to mid-August for most of the annual grubs). The simplest monitoring tool to use is a golf course cup changer. This tool pulls a 4.25-inch-diameter plug of turf and underlying soil. Each sample is approximately 1/10 ft^2, so the average grubs per sample multiplied by 10 will yield a good approximation of the grubs per square foot. For golf courses, three people with cup changers and a recorder are recommended. The three grub samplers distribute themselves equally across the fairway and pull plugs every 10–20 paces. Each sample is turned upside down and the soil is split to the thatch zone. If grubs are present, they will be found in chambers at the soil–thatch interface. Turn the plug 90° and split it again. Record the number of grubs found, including zeros on a map of the fairway. Such a sampling crew can survey a typical 18-hole course in 1–2 days. The same method can be used on lawns and sports fields, but one or two samplers can be used. Pull plugs on a transect or grid pattern. Populations between 10 and 15 grubs per ft^2 can cause significant turf damage later in the fall, September and October. Of course, populations occasionally reach 40–60 grubs per ft^2 and these levels can cause damage by late August.

Time spent doing grub sampling can usually be reduced by sampling only in the most likely turfgrass habitats. Sample areas that experienced grub damage in recent years first.

Insecticides and Application

Most of the insecticides registered for grub control can produce 90–95% control when applied at the right time and properly irrigated after the application. Where irrigation is available, liquid sprays are usually the least expensive and effective treatments. Where irrigation is not available, or where home owners cannot be trusted to irrigate in an application, granular formulations are better choices. Granular formulations can retain the insecticide until rainfall occurs.

Table 3.1 shows published data on the performance of currently registered insecticides applied during different intervals. Notice that some insecticides lose efficacy when applied in May (the residues do not persist until July and August when the new grubs hatch) while others retain their efficacy. On the other side of the season, some grub insecticides drop in efficacy when applied from mid-August into early September (when second instar grubs are present), while other products perform very well. Unfortunately, there are few data on the efficacy of grub insecticides when applied in late September and early October (rescue treatments) but trichlorfon, clothianidin, and dinotefuran are insecticides normally recommended for this time.

Studies have established that 95–99% of any pesticide used for grub control ends up in the thatch! Pre- or postirrigation does not seem to change this binding. If the thatch layer is 1 inch thick or more, the grubs will probably not contact challenging doses of the insecticides. Where thick thatch layers are encountered, it is recommended that dethatching be performed. It has also been shown that core aerification or spiking does not improve the movement of surface-applied insecticide into the soil–thatch zone.

When organophosphate and carbamate insecticides dominated the grub control products, several of the insecticides began to work poorly after two or more seasons of use. Grub resistance was suspected, but the actual problem was that of microbial degradation (= enhanced degradation). It was found that these insecticides were susceptible to additional degradation by bacteria and/or fungi that actually were able to use the compounds as food sources! These microbes tend to build up if a pesticide is used continuously. To reduce the chances of creating enhanced microbial degradation problems, use a pesticide only once, when needed, and alternate pesticide chemistries.

In general, irrigating after an insecticide application is made will improve the performance for soil insect control. It is also generally recommended that grass clippings be returned to the turf for one to two mowings after a grub insecticide application. Do not wait more than 30 days to recheck the grub infestation, especially if the original population was high. If the grub population has not been reduced below six grubs per ft^2, consider reapplication of another pesticide. Remember, the smaller the grubs, the easier they are to kill with insecticides.

TABLE 3.1

Comparison of Grub Insecticide Efficacy by Time of Application (Using Japanese Beetle and Masked Chafer Data)

Insecticide	Rate (lb. ai./a.*)	Average % Control (# Tests)				
		May	June	July	To August 16	To September 10
Carbaryl (= Sevin)	4.0	–	–	34.3 (2)	–	81.4 (1)
	8.0	–	–	56.1 (2)	–	77.8 (2)
Chlorantraniliprole (= Acelepryn)	0.1	95.0 (12)	94.0 (12)	92.5 (7)	–	–
	0.15	98.7 (8)	94.9 (7)	94.3 (6)	–	–
	0.2	94.8 (4)	99.8 (4)	98.2 (5)	–	98.9 (1)
	0.25	99.0 (4)	99.2 (5)	96.2 (7)	–	94.2 (1)
Clothianidin (= Arena)	0.2	62.5 (2)	79.2 (1)	97.7 (5)	100 (1)	100.0 (1)
	0.25	99.9 (5)	90.0 (1)	98.8 (5)	–	84.0 (3)
	0.3	93.0 (1)	99.6 (3)	99.0 (2)	83.3 (2)	97.0 (1)
	0.4	100 (1)	–	98.2 (2)	84.7 (1)	80.5 (2)
Halofenozide (= MACH2)	1.5	88.5 (8)	94.4 (23)	88.8 (21)	89.6 (19)	77.7 (27)
	2.0	80.5 (4)	63.7 (9)	94.3 (13)	75.9 (6)	–
Imidachloprid (= Merit)	0.25	97.0 (3)	90.9 (6)	97.2 (14)	89.5 (1)	–
	0.3	79.1 (26)	89.7 (45)	95.4 (55)	92.5 (32)	92.7 (41)
	0.4	81.0 (2)	94.0 (2)	96.3 (9)	78.6 (3)	99.5 (2)
Lambda-cyhalothrin (= Triazicide)	0.04	–	–	17.0 (2)	15.8 (1)	–
Permethrin	0.26	–	46.0 (2)	20.7 (3)	39.6 (2)	–
Thiamethoxam (= Meridian)	0.2	59.9 (8)	96.8 (14)	95.8 (26)	92.9 (15)	85.2 (12)
	0.26	83.5 (6)	94.9 (4)	99.8 (6)	94.6 (10)	89.7 (6)
Trichlorfon (= Dylox/ Proxol)	8.0	–	–	19.3 (2)	66.4 (9)	78.7 (29)

Source: From studies published in *Arthropod Management Tests* (1976–2011), using Japanese beetle and masked chafer efficacy data where checks had 4+ grubs per square foot and significant results. (Studies from Shetlar, 1999–2010, were used that were not published in AMT.) Compiled by D. J. Shetlar, September 2011.

*lb.ai./a. = pounds of active ingredient per acre.

Black Turfgrass Ataenius

Species: *Ataenius spretulus* (Haldeman) (Phylum Arthropoda, Class Insecta, Order Coleoptera, Family Scarabaeidae) (Figure 3.81).

Distribution: The species is found in all states east of the Rocky Mountains as well as California. Similar species in this genus as well as *Aphodius* are known to attack turfgrasses worldwide. *Aphodius granarius* (L.) is the species most commonly found. The easiest method of distinguishing *Ataenius* from *Aphodius* is the examination of the hind leg. The hind leg of *Aphodius* has two bumps or spurs on the tibia (the long segment just before the

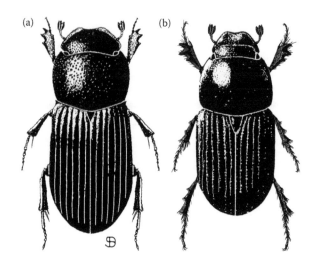

FIGURE 3.81
Diagrams of adults of (a) *Ataenius spretulus* and (b) *Aphodius granarius* showing spurs on hind tibia.

small series of small tarsal segments). *Ataenius* adults have a smooth tibia.

Hosts: Commonly attacks bentgrass, annual bluegrass, and Kentucky bluegrass, especially on golf courses.

Damage symptoms: This grub usually damages cool-season turfgrasses on golf courses. Turf first wilts and does not respond long after irrigation. The turf is easily peeled back because of the lack of roots. Death of turf in irregular patches occurs in June or August. *Aphodius* appears to have only one generation per year and the larval damage often occurs a bit later than normal *Ataenius* damage.

Description of stages: The larvae look like typical white grubs but the adults are more elongate than other turf-infesting scarabs. Species of *Ataenius* and *Aphodius* require an expert for accurate species identification.

   Eggs: The pearly-white eggs are approximately 0.02 × 0.03 inch (0.52 × 0.72 mm), and are laid in clusters of 11–12 eggs. The eggs expand slightly after absorbing water from the soil.

   Larvae: The C-shaped white grubs are very small but third instars can be separated from other grubs using a 10× hand lens. The tip of the abdomen has two distinct anal pads and the few raster bristles are scattered at random. Smaller grubs have these characters but a microscope may be needed to see them. Full-grown larvae are approximately 5/16 inch (8 mm) long.

Pupae: The small, 5/16 inch (6–8 mm) long pupae are first white but become red-brown on maturity.

Adults: The small, shiny black beetles are 1/8–7/32 × 1/16–3/32 inch (3.6–5.5 × 1.7–2.4 mm). The prothorax has small pits scattered over the surface, and the wing covers have distinct longitudinal grooves. Newly emerged adults are reddish to dark chestnut-brown, but these become black in a few days.

Life cycle and habits: There are many species of *Ataenius* in North America and most are dung feeders. However, a couple of species, especially *A. spretulus* and *A. strigatus*, feed on decaying humus as well as living plant roots. In western states, *Aphodius* species may be found with *Ataenius*. *A. spretulus* was first recorded as damaging golf greens in the 1930s, but major outbreaks of this pest began to occur in the 1960s and 1970s on golf course greens, tees, and aprons. Damage may extend into the fairways. Optimum habitat for this grub seems to be short-cut turf with a moist compacted thatch layer. Adult black turfgrass ataenius beetles overwinter 1–2 inches (2.5–5.0 cm) in the soil under leaf litter and plant material along the edges of fairways and in wooded roughs. The adults emerge in early spring, usually when spring crocus and red buds are in bloom. These adults warm in the spring sun and fly to turf areas where they dig into the thatch. Upon finding a suitable egg-laying site, usually in April and early May, the females lay clusters of 11–12 eggs in the thatch just above the soil. Most of the eggs are present from mid-May to early June, during which time they take about a week to hatch. The tiny white grubs feed on the organic material in the thatch, including grass roots. Grub populations of 200–300 per ft² (2000–3000 per m²) are common and grub populations of 50 per ft² (540 per m²) can severely damage the turf. The first-generation larvae take about 4 weeks to mature. These mature grubs dig into the soil and make a compact pupal chamber. The pupa takes a little more than a week to mature and becomes reddish brown before the adult emerges. Young adults are also reddish brown and these "callow" adults may emerge and crawl or fly about before becoming completely black in a few days. The first-generation adults are active in July, laying a new batch of eggs. The second generation of grubs can severely damage the turf in August when rainfall is scarce. Because the summer adults may lay eggs over an extended period of time, a few larvae can be found in September and October. If the larvae do not mature by winter, they die. The second-generation adults emerge in late August through September and these seek overwintering sites, usually near the edges of wooded lots.

Control options: See White Grubs: Introduction. This grub has two generations per year and a treatment may be needed for either generation, unless a long residual product is used.

Option 1: Chemical control—timing using indicator plant phenology: Control of the first generation of grubs is best achieved if pesticides are applied when black locust (*Robinia pseudoacacia*) trees or Vanhoutte spirea (*Spirea × vanhouttei*) shrubs are in full bloom. The second generation of larvae can be controlled when the Rose-of-Sharon (*Hibiscus syriacus*) begins to bloom.

Option 2: Biological control—milky diseases: *Ataenius* and *Aphodius* grubs have their own strains of these diseases. None are commercially available. Infected grubs will not feed and these often remain alive, though not feeding, after an insecticide application. Do not assume that the insecticide did not work if milky *Ataenius* are present.

Asiatic Garden Beetle

Species: *Maladera castanea* (Arrow) (Phylum Arthropoda, Class Insecta, Order Coleoptera, Family Scarabaeidae) (Figure 3.82).

Distribution: Introduced from Japan in the 1920s. Most common in north-eastern United States from New England states across to Ohio and down into South Carolina. Also occurs in Ontario province, Canada.

Hosts: Larvae occasionally attack turf but seem to prefer more varied habitats that contain a variety of roots from weeds, flowers, and vegetables. The adults feed on over 100 species of plants, preferring asters, dahlias, mums, roses, and a variety of trees and vegetables.

FIGURE 3.82
Pair of mating Asiatic garden beetles hanging from a leaf at night.

Damage symptoms: The grubs cause typical damage to turf—wilting and irregular patches of dead turf. However, since this species likes roots of other plants, the grubs may be clustered around weedy areas. Grubs may also congregate next to flower beds where plants preferred by the adults are growing. The adults strip foliage from plants, leaving a ragged appearance. They do not skeletonize like Japanese beetles. Flowers often have the petals eaten off.

Description of stages: The stages are typical of a scarab having a single generation per year. Since this beetle is rather small, most of the stages are likewise diminutive.

Eggs: The eggs are laid in clusters of 3–15 and are loosely held together by a gelatinous material. The individual eggs are oval and about 3/64 inch (1 mm) long. After absorbing water, the eggs, become spherical.

Larvae: Newly hatched larvae are about 1/16 inch (1.4 mm) long and have light brown head capsules. Full-grown larvae are 5/8 inch (15–18) long when stretched out. These grubs can be identified by the longitudinal anal slit and transverse, curved row of brown spines making up the raster pattern.

Pupae: The pupae remain in the last larval skin and are about 3/8 inch (8–10 mm) long. At first they are white and gradually turn tan.

Adults: The adults are 3/8 inch (7–10 mm) long and broadly wedge shaped. They are chestnut-brown and often have a slight iridescent sheen to the elytra when alive. The abdomen protrudes slightly from under the wing covers, and the undersurface of the thorax has an irregular covering of short yellow hairs. The hind legs are distinctly larger and broader than the others.

Life cycle and habits: The adult beetles may be active from late June to the end of October, but most of the adults are found from mid-July to mid-August. The adults emerge at night and fly actively when temperatures are above 70°F (21.1°C). When the temperature drops below this, they tend to walk up the plants or grasses to feed rather than fly. The adults are strongly attracted to lights. During the day, the beetles hide in the soil around favored food plants. After feeding several nights, the females begin laying eggs in small clusters. The females tend to search out turf and pastures for egg laying and generally deposit the eggs 1–2 inches (2.5–5 cm) deep in the soil. The females lay eggs over several weeks and the average number of eggs is 60. The eggs normally hatch in 10 days during summer temperatures. The young larvae dig to the soil surface, where they feed on roots and decomposing organic material. Most first instar larvae are found in August and early September. The second instars are found in September, and many do not reach the third instar until the following spring. About half the population overwinters as second instars and the remainder are partially developed third instars. As cool October temperatures arrive, the larvae burrow down 8–17 inches (20–43 cm) to pass the winter. The larvae return to the soil surface in the spring and all seem to mature by mid-June, at which time they pupate 1.5–4 inches (4–10 cm) in the soil in compacted earthen cells. The pupal stage is relatively short, lasting 8–15 days. The adult remains in the old pupal skin, changing from white to the mature chestnut-brown for a few days before digging to the surface.

Control options: See: White Grubs: Introduction. Generally, because of its small size, grub populations below 20/ft² (215/m²) are not severely damaging to turf if water and fertilizer are available. Damaging populations tend to build when there are several years of rainy summer weather.

European Chafer

Species: *Rhizotrogus* (= *Amphimallon*) *majalis* (Razoumowsky) (Phylum Arthropoda, Class Insecta, Order Coleoptera, Family Scarabaeidae).

Distribution: This European pest was first detected in Newark, New York, in 1940. Since then, the pest has spread into Massachusetts, New Hampshire and upper New York, west to Michigan, and south from West Virginia across Maryland. It also occurs in Ontario and Quebec provinces.

Hosts: The grubs feed on a wide variety of plant roots and organic matter in the soil. They are known to feed on the roots and thatch of all cool-season turf grasses.

Damage symptoms: Typical grub damage of thin turf, wilting, and death in irregular patches can be found in the fall and early spring.

Description of stages: This pest has stages typical of an annual white grub.

Eggs: The freshly laid eggs are oval, approximately 0.03 × 0.02 inch (0.73 × 0.49 mm), and a shiny, milky white color. After absorbing water, the eggs become dull gray and swell to 3/32 × 7/64 inch (2.0 × 2.7 mm).

Larvae: The first instar larvae are about 5/32 inch (4 mm) long, and these grow to approximately 11/16 inch (17 mm) when fully grown third instars. All three instars are typical C-shaped white grubs that can be identified by the raster that has two parallel rows of bristles that diverge laterally at the anus. These grubs are slightly smaller than the May/June beetles. May/June beetle grubs have two parallel rows of bristles on the raster that do not diverge at the anus.

Pupae: The pupa looks like most scarab pupae but is slightly larger than that of the Japanese beetle and smaller than the May/June beetle pupae.

Adults: The adults look much like some of the light-colored June beetles. However, the European chafer is 7/16 inch (13–14 mm) long, shorter than most May/June beetles, and the wing covers have distinct longitudinal grooves (striatae). The most diagnostic characteristic is the absence of a tooth on the tarsal claw of the middle leg. The May/June beetles have a distinct tooth on each claw.

Life cycle and habits: The adults emerge from the pupal cells in mid-June and continue mating and oviposition until late July. Most activity occurs from the last week of June through the second week of July. The adults emerge at sundown and fly to nearby trees and shrubs silhouetted against the sky. Here, large numbers fly, with a considerable buzzing noise, for 20–35 min. When the sky is truly dark, the adults settle on the foliage and begin copulation. Copulation continues and pairs of mating adults fall from the trees. By daybreak, the adults will have returned to the soil. Cool or rainy nights greatly reduce flight and mating activities. Apparently, adults may return several times for mating, but eventually females dig into the soil to lay eggs. Each female lays 15–20 eggs in 2–5 days. The eggs are usually laid singly in compacted cells of soil, 2–6 inches (5–15 cm) deep. The eggs swell as they absorb soil moisture and hatch in about 2 weeks. The first instar may remain deep in the soil if surface soil moistures are low. Eventually, the young larvae move to the surface and feed on plant roots. If food is sufficient, the first instar matures in about 3 weeks and the second instar takes about 4 more weeks to mature. The third instars feed for a period in the fall before moving down for the winter. This pest moves up, down, and sideways depending upon soil moisture and food availability. This grub may feed longer than many other species in the fall before moving down. This species is also one of the first to return to the surface in the spring, often in March. Pupation occurs in mid-May, 2–6 inches (5–15 cm) in the soil.

Control options: See: White Grubs: Introduction. This pest is controlled like most white grub species; however, the grubs quickly react to dry soil conditions and heat by moving deeper into the soil. This species is also the last grub to move down in the fall and has been seen at the soil surface when snow melts in mid-February.

Green June Beetle

Species: *Cotinus nitida* (Linnaeus) (Phylum Arthropoda, Class Insecta, Order Coleoptera, Family Scarabaeidae) (Figure 3.83).

Distribution: This native of North America is commonly found from southern Pennsylvania across to Oklahoma and south. It is most

FIGURE 3.83
Green June beetle adult feeding on blackberries.

commonly a turf pest in the transition zones of Tennessee and
Kentucky to the Carolinas.

Hosts: This pest has grubs that feed on the roots of many species of
turfgrasses and field crops. It seems to prefer areas with high
organic matter, especially heavily manured areas such as livestock pastures.

Damage symptoms: The larvae occasionally attack turf sufficiently to kill it
in irregular areas. Since the larvae feed on grass blades at night,
a general thinning of the turf is common. The larvae and adults
make burrows in the turf, throwing up mounds of soil. This activity resembles the soil mounds made by ants or ground-nesting
wasps. The larvae also have the alarming habit of crawling about
on their backs after rains. These migrating grubs creep across
sidewalks, driveways, and often end up in swimming pools and
garages.

Description of stages: This grub has a life cycle typical of annual grubs.

Eggs: The eggs are planted in loose masses of 10–30. They are
oval when laid, 1/16 × 5/64 inch (1.5 × 2.1 mm), but upon absorbing
water become spherical, 7/64 inch (2.5 × 2.8 mm) in diameter.

Larvae: The grubs are somewhat atypical for scarabs. They have
the normal C-shape, but usually move by stretching out on their

backs to creep along with an undulating motion. Because of this mode of movement, the legs are considerably smaller than other white grubs. This type of movement is diagnostic. First instars are 1/4 inch (6–6.5 mm) long, second instars are 5/8 inch (15–17 mm) long, and mature third instars are 1¾ inch (45–48 mm) long.

Pupae: The pupa is formed in a distinct earthen cell and measures approximately $1.0 \times 1/2$ inch (25×13 mm). They are first light yellowish brown but gain some of the metallic green sheen of the adult before emerging.

Adults: The adults are about 13/16–1.0 inch (20–25 mm) long, and generally are a striking velvety green color with orange-yellow margins. The lower surface is a shiny metallic green. The head has a distinctive flat horn.

Life cycle and habits: The green adults begin to emerge in late June but are most common in July and August. This species flies during the daytime and makes a considerable buzzing noise, which alarms some people. The adults may congregate around seeping wounds of trees, and are very common on ripe grapes, figs, peaches, plums, melons, some vegetables, and ears of corn. Once a feeding site is established, several adults may be actively flying about or attempting to mate. Newly emerged females call males to the ground for mating by producing an attractive odor. Females ready to lay eggs dig a burrow into the ground, throwing up a small mound of soil. The females excavate a small cavity in the soil of 2–5 inches (5–13 cm) in which they lay 10–30 eggs. These eggs are packed into a ball of soil about the size of a walnut. Females may emerge and dig a new burrow elsewhere but more commonly they continue in the first burrow to dig one to two additional egg chambers. The eggs swell as they absorb moisture and the grubs hatch out in 2–3 weeks. The young grubs are very active and can move a considerable distance in a short period of time. Since the eggs are clustered, the larvae first appear to work in groups or colonies. The grubs feed on organic material, preferring manure and rotting plant remains. By the time the grubs are second instars, they have constructed individual burrows that average 6–12 inches (15–30 cm) deep. At night, the grubs deposit soil around the burrow entrance, making small mounds 2–3 inches (5–8 cm) in diameter, which resemble ant mounds. The grubs may also emerge and crawl about on their backs during warm, wet evenings. This is especially common after a warm afternoon thunderstorm. As winter temperatures arrive, the grubs dig deeper and remain inactive at the bottom of the burrow. These grubs may become active at any time when the temperature rises. Larval digging may occur anytime during the late summer, through winter, and into spring. By late May and early June, the grubs have matured and they construct an earthen cell glued together with a secretion. In about 3 weeks, the adults break out of the pupal cell and dig to the surface.

Control options: See: White Grubs in Turfgrass.

Option 1: Cultural control—reduce organic fertilizers: This pest prefers soils with very high organic content and is encouraged by the addition of animal manures. Do not apply manures or organic fertilizers with uncomposted manure to turf in the spring or early summer as these may be attractive to egg-laying females.

Option 2: Chemical control—avoid irrigation when possible: When pesticides are used, the green June beetle grubs surface at night to feed and this is where they will come into contact or consume the pesticide. This has an unfortunate result as the grubs usually remain on the surface! Having hundreds of third instar grubs on the surface after an application can result in considerable odor problems. When using pesticides to control this pest, try to target the first or second instars.

Japanese Beetle

Species: *Popillia japonica* Newman (Phylum Arthropoda, Class Insecta, Order Coleoptera, Family Scarabaeidae) (Figure 3.84).

Distribution: This import is generally found east of a line roughly running from Michigan, southern Wisconsin, eastern Nebraska and Kansas into Arkansas, and across to Georgia. It is also a pest in Ontario province in Canada. Occasional introductions are made into

FIGURE 3.84
Japanese beetle adults skeletonizing leaves.

western states such as California when the adult beetles or larvae are shipped in commerce. The original population was detected in New Jersey in 1916, having been introduced from Japan.

Hosts: The adult beetles are general herbivores and are known to feed on over 400 species of broadleaf plants, though only about 50 species are preferred. The grubs will also feed on a wide variety of plant roots, including ornamental trees and shrubs, garden and truck crops, and turfgrasses. They seem to especially relish Kentucky bluegrass, perennial ryegrass, tall fescues, and bentgrass.

Damage symptoms: The adults are skeletonizers, that is, they eat the leaf tissue between the leaf veins but leave the veins behind. This leaves the appearance of lace and the remaining tissue withers and dies. Often, the adults will attack flower buds and fruit. The grubs can kill small seedling plants but most commonly damage turf. The turf first appears off-color, as if under water stress. Irrigation of this turf causes a short, but not long-lasting response or no response at all. The turf feels spongy underfoot and can be easily pulled back like an old carpet to reveal the grubs. Large populations of grubs kill the turf in irregular patches.

Description of stages: The life stages of the Japanese beetle are typical of white grubs.

Eggs: The white oval eggs are usually about $1/16 \times 3/64$ inch (1.5×1.0 mm). They are placed in the soil where they absorb moisture and become more roundish.

Larvae: The larvae are typical white grubs that can be separated from other soil-dwelling white grubs by the presence of a V-shaped series of bristles on the raster. First instar larvae are about $1/16$ inch (1.5 mm) long, while the mature third instars are about $11/4$ inch (32 mm) long.

Pupae: The pupae are cream colored first and become light reddish-brown with age. The average pupa is about $9/16 \times 9/32$ inch (14×7 mm).

Adults: The adults are a brilliant metallic green color, generally oval in outline, $5/16$–$7/16 \times 3/16$–$9/32$ inch (8–11×5–7 mm) wide. The wing covers are a coppery brown color and the abdomen has a row of five tufts of white hairs on each side. These white tufts are diagnostic. The males have a sharp tip on the foreleg tibia, while the female has a long rounded tip.

Life cycle and habits: Larvae that have matured by June pupate and the adult beetles emerge during the last week of June through mid-July. On warm sunny days, the new male beetles crawl onto low growing plants and rest for a while before taking flight. When virgin females emerge, they produce a sex pheromone that attracts dozens of males to the spot of emergence. This mounding of beetles is often noticed in the early afternoon. After mating, the females seek suitable food plants and begin to feed as soon as possible. The feeding damage caused by these early arrivals releases plant odors that attract more adults to the area. This causes hundreds of adults to gather in masses on the unfortunate plants first selected. In cool weather, the adults may feign death by dropping from the plants but normally they will take flight when disturbed. Subsequent mating attempts occur on the food plants, and several matings by both males and females is common. After feeding for a day or two, the females leave the feeding sites in the afternoon and burrow into the soil to lay eggs at a depth of 2–4 inches (5–10 cm). Females may lay one to five eggs scattered in an area before leaving the soil the following morning or a day or two later. These females return to feed and mate. This cycle of feeding, mating, and egg laying continues until the female has laid 40–60 eggs. Most of the eggs, 75% of a population, are generally laid by mid-August, though adults may be found until the first frost of fall. If the soil is sufficiently moist, the eggs will swell in a few days and egg development takes only 8–9 days at 80–90°F (26.7–32.2°C) or as long as 30 days at 65°F (18.3°C). The first instar larvae dig to the soil surface, where they feed on roots and organic material. If sufficient food and moisture are available, the first instars can complete development in 17 days at 78°F (25.6°C), or as long as 30 days at 68°F (20°C). The second instars take 18 days to mature at 78°F (25.6°C) and 56 days at 68°F (20°C). While this development is going on, the grubs tunnel considerably laterally in search of fresh roots. This leaves the turf in a very spongy condition. Generally, most of the grubs are in the third instar by early fall and are ready to dig into the soil to hibernate. The grubs move down 4–8 inches (10–20 cm) into the soil as cool temperatures arrive. At this depth, the soil rarely gets below 25°F (–3.9°C) and the grubs survive with no difficulty. The grubs return to the surface in the spring as the soil temperature warms. Generally, the grubs can be expected to be active at the surface when the surface soil temperatures are about 60°F (15.6°C), usually in mid-April. The grubs continue their development in the spring and the few second instars seem to mature in time to pupate along with the third instars. The mature grubs form a prepupa in mid-June. This form empties its gut and has a translucent appearance. The pupa is formed in the split skin of the prepupa in an earthen cell 1–3 inches (2.5–7.5 cm) in the soil.

Control options: See: White Grubs in Turfgrass. Since this pest is important in agriculture, considerable effort has been placed on developing control strategies.

Option 1: Cultural control—habitat modification: Since the eggs and young grubs are very susceptible to dry soils, do not irrigate during the time when the eggs and first instar larvae are developing. However, if natural rainfall occurs, this tactic will not work. Do not plant highly desirable adult Japanese beetle food plants near the turf. This is especially possible on golf course fairways.

Option 2: Biological control—natural parasites: Several parasitic wasps, *Tiphia* spp., and a parasitic fly, *Hyperecteina aldrichi*, are known to attack the Japanese beetle. These parasites do not seem to reliably reduce the populations, though the fly has become more plentiful in the New England states over the last decade. The *Tiphia* appear to be more efficient in southern states.

Option 3: Mechanical control—trapping: Several traps have been developed to capture the adults. These traps generally use a mixture of attractive plant odors and the sex pheromone. Test data indicate that these traps do not reduce grub populations in moderate to heavy infestation zones. In some cases, the traps can actually cause more damage to ornamental plants near the traps. Trapping is useful in monitoring and detecting populations only.

Northern Masked Chafer

Species: *Cyclocephala borealis* Arrow (Phylum Arthropoda, Class Insecta, Order Coleoptera, Family Scarabaeidae). The western masked chafer, *C. hirta* LeConte, is common from western Kansas into central California and is similar to the northern masked chafer (Figure 3.85).

FIGURE 3.85
Northern masked chafer adults mating at night in turf.

Distribution: A native of the Americas from Canada to South America. Commonly a pest in the cool-season turf areas from New England across to Wisconsin.

Hosts: Commonly attacks Kentucky bluegrass and perennial ryegrass. Adults do not feed.

Damage symptoms: Typical grub damage to turf may occur in the fall or spring. Heavy infestations cause the turf to appear off-color due to water stress, but this turf dies in irregular patches. Infested turf feels spongy underfoot and is easily lifted because of the absence of roots.

Description of stages: Typical annual scarab life cycle with a complete metamorphosis.

Eggs: The eggs are barrel-shaped, pearly white, and approximately $1/16 \times 3/64$ inch (1.68×1.3 mm) when freshly laid. After absorbing soil moisture, the eggs increase slightly in diameter and are more rounded.

Larvae: Newly hatched first instar larvae are about 3/16 inch (5 mm) long and grow to about 7/8 inch (23 mm) when mature third instars. This larva looks very much like *C. lurida* and can be distinguished by characteristics of the mouthparts. The grub can be field identified as being a *Cyclocephala* by the irregular pattern of large bristles comprising the raster.

Pupae: The pupae are first creamy white in color and gradually turn reddish brown before the emergence of the adult.

Adults: The adults are yellowish ocher-brown and have darker markings on the heads and eyes from which the name "masked" chafer is derived. The body is covered with fine hairs, which is a good field character for distinguishing this species from the southern masked chafer. Males are slightly larger, $11/32 \times 1/4$ inch (11.8×6.76 mm), and darker than females, $3/8 \times 1/4$ inch (11.0×6.66 mm). Males have large front tarsal claws and a longer antennal club. This club is longer than the combined length of the other antennal segments and separates it from *C. lurida* males in which the club is shorter than the combined length of the other antennal segments.

Life cycle and habits: Adult beetles usually begin emergence in late June and are active into mid-July. Males come to the soil surface after dark before females emerge. This species apparently is active later at night than the closely related, *C. lurida*. Maximum activity occurs around midnight. Unmated females come to the soil surface releasing a sex pheromone that attracts the males. Many males often cluster around calling females, and the successful male clasps the female with his modified front legs. Mated females and males fly at night and are strongly attracted to lights. The males tend to fly within 2 ft (60 cm) of the ground, while females seem to fly at higher altitudes. Neither males nor females feed on plant material. Mated females dig down 4–6 inches (10–15 cm) and lay an average of 11–12 eggs. If soil moistures are sufficient, the eggs swell within a few days and hatch in 20–22 days. The young larvae burrow to the soil surface in search of plant roots. The larvae also eat general organic material. The larvae grow rapidly when adequate moisture and roots are present and third instars are common by September. It is during this time of the season that most of the damage occurs. As the soil temperatures begin to drop in the fall, the larvae begin to dig downward to hibernate. Larvae may dig down 18 inches (45 cm) but most are within 12 inches (30 cm) of the surface. Winter mortality may be heavy, with over 50% dying. Grubs surviving the winter return to the soil surface in late April and May to feed. The larvae again move down slightly in late May and early June to pupate. The pupa is formed within the old larval exoskeleton, which splits down the center line.

Control options: See: White Grubs: Introduction.

Option 1: Cultural control—withhold irrigation: Since the eggs require moisture for development, restricting irrigation in July and early August may significantly reduce egg survival.

Option 2: Chemical control—use degree-day timing: The first adults of the northern masked chafer first emerge between 898 and 905 cumulative degree-days (base 50°F) and 90% of the adult flight occurs between 1377 and 1579 cumulative degree-days. Applications of short residual insecticides would work best about 20–30 days after the 90% flight period.

Southern Masked Chafer

Species: *Cyclocephala lurida* Bland (= *C. immaculata*) (Phylum Arthropoda, Class Insecta, Order Coleoptera, Family Scarabaeidae). The southwestern masked chafer, *C. pasadenae* Casey, occurs in western Texas to southern California and is very similar to the southern masked chafer in biology.

Distribution: A native of North America, but has also been collected from Central and South America. Commonly a pest from southern Pennsylvania across to Nebraska and south.

Hosts: Commonly attacks turfgrasses in the transition zones (Kentucky bluegrass and tall fescues) and in southern bermudagrass areas.

Damage symptoms: Turf begins to show drought stress in late fall or spring and does not rapidly recover after rain or irrigation. Heavy infestations result in turf dying in irregular patches. Birds, skunks, raccoons, and opossums commonly dig up turf along the edges of the dead patches.

Description of stages: Typical annual scarab life cycle with a complete metamorphosis.

Eggs: The eggs are oval when laid, and are about 1/16 inch (1.7 mm). These pearly-white eggs absorb moisture from the surrounding soil and increase to 3/64 inch (2.1 mm) in diameter and have a nearly spherical shape.

Larvae: First instars are about 31/16 inch (4.5 mm) long at hatching and reach 7/8–1.0 inch (22–25 mm) when maturing as third instars. The mouthparts must be dissected to distinguish this species from *C. borealis*. The larvae have the irregular pattern of bristles on the raster, which is typical of all *Cyclocephala* larvae.

Pupae: The 11/16 × 5/16 inch (17 × 8 mm) pupae are first creamy white and gradually change to reddish-brown just before the adult emerges.

Adults: The adults are a dull dark yellow-ocher and have darker brownish-black markings on the heads and eyes. The thorax and wing covers do not have conspicuous hair, which distinguishes this species from *C. borealis*. The adults are 7/16–9/16 × 7/32–9/32 inch (11–14 × 6–7 mm). Males have an enlarged fifth tarsal segment on the forelegs, which is used to grasp the female. However, the antennal club of males is shorter or equal to the combined length of the other antennal segments; *C. borealis* males have a longer club.

Life cycle and habits: Adult beetles usually begin emergence in mid-June and are active into mid-July. Males come to the soil surface after dark before females emerge. This species apparently is active earlier in the evening than its sibling species *C. borealis*. Males begin to emerge just before sunset and skim the ground surface in search of unmated females. Unmated females come to the soil surface, climb upon a grass blade and release a sex pheromone. Many males often cluster around calling females, and the successful male clasps the female with his modified legs. Mated females and males fly at night and are strongly attracted to lights. The males tend to fly within 2 ft (60 cm) of the ground, while females seem to fly at higher altitudes; most of the activity is finished by midnight. Neither males nor females feed on plant material but merely mate and disperse at night. Mated females dig down 4–6 inches (10–15 cm) and lay 11–14 eggs. If soil moistures are sufficient, the eggs swell within 8 days and hatch in 14–18 days at 70–75°F (21.1–23.9°C). The young larvae burrow to the soil surface in search of plant roots. The larvae also eat general organic material. The larvae grow rapidly when adequate moisture and roots are present. The second instars are reached in 20–24 days at 80°F (26.7°C), and third instars are common by September. It is during this time of the season that most of the damage occurs. As the soil temperatures begin to drop in the fall, the larvae begin to dig downward to hibernate. Larvae may dig down 12 inches (30 cm), but most are within 3–6 inches (7.5–15 cm), at least in southern states. Grubs surviving the winter return to the soil surface in late April and May to feed. The larvae again move down slightly in late May and early June to pupate. The pupa takes about 17 days to mature.

Control options: See White Grubs: Introduction and Northern Masked Chafer.

Option 1: Chemical control—use degree-day timing: The first adults of the northern masked chafer first emerge between 1000 and 1109 cumulative degree-days (base 50°F) and 90% of adult flight occurs between 1526 and 1679 cumulative degree-days. Applications of short residual insecticides would work best about 20–30 days after the 90% flight period.

Oriental Beetle

Species: *Exomala* (= *Anomala*) *orientalis* (Waterhouse) (Phylum Arthropoda, Class Insecta, Order Coleoptera, Family Scarabaeidae) (Figure 3.86).

Distribution: This native of Japan was first detected in Connecticut in 1920. It has since moved, mainly in soil of nursery stock, to New York, Pennsylvania, Ohio, New Jersey, most of the New England states, and down to the Carolinas.

FIGURE 3.86
Oriental beetle adults range from dark brown forms to light tan forms and the adults are fond of feeding on flower petals at night.

Hosts: Readily attacks the roots of cool-season turfgrasses. The adults feed little and are occasionally found on flowers, especially daisies.

Damage symptoms: Typical grub damage consisting of wilting turf that turn into irregular dead patches. This pest prefers to feed on turf in sunny areas. Adult feeding damage is rarely detectable.

Description of stages: This pest has a cycle normal for an annual white grub.

Eggs: The white eggs are first oval in shape, being 3/64 × 1/16 inch (1.2 × 1.5 mm), and swell to 1/16 × 5/64 inch (1.5 × 1.9 mm) after a few days in moist soil.

Larvae: The C-shaped white grubs are approximately the same size as Japanese beetle grubs. Mature grubs are about 1 inch (25 mm) long. The anus is transverse and the raster has two parallel rows of about 14 short, stout spines. These might be confused with young May/June beetle larvae. However, May/June beetle larvae have a Y-shaped anal opening.

Pupae: The pupa is about 3/8 × 3/16 inch (10 × 5 mm). They are first cream colored and turn to light brown. The tip of the abdomen has a thick fringe of hair-like setae.

Adults: The adults are 3/8 inch (9–10 mm) long and vary considerably in markings on the thorax and wing covers. Individuals may be entirely brownish-black to entirely straw colored except for a brown head and mark on the pronotum. Usually the head is solid dark brown, the pronotum is dark in the center outlined in straw color, and the wing covers have longitudinal grooves and are mottled with patterns of dark brown on straw.

Life cycle and habits: The adults begin to emerge in late June and some individuals may be found into September. Most of the adults are active in July. The beetles may fly short distances in the morning and evening and are commonly found on flowers, where they are chewing on the petals. Some of the adults are attracted to lights but never in large numbers. A few days after mating, the females burrow into the soil to lay eggs. The eggs are laid in small groups between 3 and 9 inches (7.5–23 cm) in the soil. Females lay an average of 26 eggs. The eggs must be in moist soil so that water can be absorbed and development continued. At normal soil temperatures in July and August, the eggs take 18–24 days to hatch. The young grubs move to the soil surface to feed on roots and organic material. The second instar larvae are found in 3–4 weeks and these usually molt into third instars in another 3–4 weeks. The majority of the larvae overwinter as third instars but some overwinter in the second instar. Larvae burrow down late October and November and return to feed the following April and May. The pupae are present from mid- to late June and usually take about 2 weeks to mature.

Control options: See: White Grubs: Introduction.

Sugarcane Beetle and Sugarcane Grub

Species: The rhinoceros beetle subfamily of scarabs, Dynastinae, includes the masked chafers (in the tribe Cyclocephalini) and another tribe of shiny black or chestnut-brown beetles that can be called sugarcane beetles and grubs (tribe Pentodontini). A black beetle called the sugarcane beetle, *E. humilis* (Burmeister), has been associated with corn, sweet potato, and sugarcane crops, though the grubs prefer grasses for development. Another representative of this group has been called the sugarcane grub, *T. subtropicus* (Blatchley), in Florida where it also damages sugarcane and the larvae can be found in turfgrass (Phylum Arthropoda, Class Insecta, Order Coleoptera, Family Scarabaeidae) (Figure 3.87).

Distribution: *E. humilis* has been found south of a line running from North Carolina, across Tennessee into southern Texas. *T. subtropicus* is most commonly encountered in Florida but records indicate its presence along the Gulf Coast.

FIGURE 3.87
Sugarcane beetles are shiny black scarabs that are often attracted to lights at night.

Hosts: The adults of these beetles often dig into soil and feed on tubers of sweet potatoes, the developing stalks of corn, rice, and cotton, as well as the rhizomes of sugarcane. The larvae of both species develop in soils covered by grasses, both pastures and managed turfgrass.

Damage symptoms: The larvae feed like most white grubs by ingesting organic matter, including the roots of grasses. Unfortunately, the larval raster pattern is very similar to masked chafer grubs and the larvae are probably misdiagnosed as chafer grubs. Damaged turf wilts and can be pulled back like a loose carpet and animal digging is common.

Description of stages: These insects have typical white grub stages, though complete descriptions of all the stages are lacking in the literature.

Eggs: Not well described in the literature, but appear to be oval when laid and rounded after absorbing moisture from the soil; appear to be laid in groupings.

Larvae: Typical scarab grubs with brown to red-brown head capsules. Newly hatched *E. humilis* are about 1/4 inch (4.8 mm) long and fully developed third instars can be 1⅜ inch (32 mm) long. Mature larvae of *T. subtropicus* are much larger, being nearly 2 inch (50 mm) long. The raster area of both species has 50 or more stout spines, which can look like the pattern found on the *Cyclocephala* grubs. An expert is needed to examine the mouthparts for species determination, but the large size is a good field indication that the grubs are not masked chafers.

Pupae: Pupae are formed in soil chambers; *E. humilis* pupae are 3/4 inch (19 mm) long and *T. subtropicus* pupae are nearly an inch (22 mm) long.

Adults: Adults of *E. humilis* are 5/8 × 1/2 inch (14 × 10 mm), and adults of *T. subtropicus* are 7/8–1 × 1/2–5/8 inch (21–25 × 12–14 mm). Both are shiny black beetles, strongly domed in shape with noticeable lines and pits on the wing covers. There are several beetles in this tribe that look alike, so an expert should be consulted to determine the species.

Life cycle and habits: Both these beetles have annual life cycles over their ranges. Like the May/June beetles, adults overwinter in soils and they may fly from April through June. Mated females burrow into fertile soil, usually covered by grasses or turf. The females lay small clusters of eggs, which absorb moisture from the soil and become roughly spherical before the embryo can develop. After 3 weeks, the eggs hatch and the young grubs begin to feed on organic matter in the soil, including the roots of grasses. Development is fairly rapid and the grubs complete their development in 2–3 months. Most of *E. humilis* larvae finish their development by early August and after a 2-week pupation period, the new adults emerge in late

August through September. These adults are highly attracted to lights at night and can become a nuisance in parking lots, around buildings and schools. The adults seek root crops, especially sweet potatoes, and they can burrow into the bases of corn and sorghum stalks, sugarcane, and rice. Apparently, some of the *T. subtropicus* larvae do not complete development and may overwinter before pupating early the following spring.

Control options: See: White Grubs in Turfgrass. There have been few field studies dedicated to the control of either of these grubs, but most states that have them recommend using grub insecticides applied in late June through July to eliminate the young grubs as they feed.

May and June Beetles, *Phyllophaga*

Species: Over 150 species of *Phyllophaga* are known in North America, but of these, only about 25 have been found attacking turfgrasses. An expert is needed to make species identification of the adults and larvae (Phylum Arthropoda, Class Insecta, Order Coleoptera, Family Scarabaeidae).

Distribution: Some species, such as *Phyllophaga hirticola* (Knoch), *Phyllophaga crenulata* (Froelich), *Phyllophaga tristis* (Fabricius), and *Phyllophaga ephilida* (Say), are generally distributed from the Great Plains east. *Phyllophaga crinita* Burmeister is an important pest in Texas and Oklahoma, and can also be found in Louisiana to Georgia. *Phyllophaga latifrons* (LeConte) commonly attacks St. Augustinegrass in Florida. Only a few species like *Phyllophaga anxia* (LeConte) and *Phyllophaga fervida* (Fabricius) are found all across North America. Many species of this group appear to prefer lower-maintenance turf and are relatively intolerant to the pesticides used in high-maintenance turf. Also, with shifts from organophosphate and carbamate insecticides, May/June beetle outbreaks are appearing in Wisconsin, Illinois, and Michigan. In southern and western states, *Phyllophaga* are often present in large populations where they may occur with masked chafer species.

Hosts: In general, *Phyllophaga* grubs will feed on the roots of many types of plants, including turfgrasses. The adults feed on the foliage of many trees and shrubs.

Damage symptoms: Damaged turf wilts as if under drought stress, and eventually this turf dies in irregular patches. Some southern grasses may not show these symptoms readily because of moist sandy soil and deep roots. However, digging by mammals and birds can be more of a problem.

Description of stages: The life stages of *Phyllophaga* are typical of white grubs.

Eggs: The white oval eggs are usually about 5/64 inch (2 mm) long and are placed single in earthen cells. The eggs absorb moisture and become more roundish.

Larvae: *Phyllophaga* larvae are very difficult to separate to species, and an expert is needed to make a proper identification. However, almost all *Phyllophaga* larvae have a broadly Y-shaped anus and two parallel rows of bristles pointing toward each other on the raster. Full-grown grubs are about 13/16–1³⁄₁₆ inch (20–30 mm) long.

Pupae: The pupae are white at first and become brownish with age. Most pupae are 5/8–1.0 inch (15–25 mm) long.

Adults: The different species all have the same general shape but may differ considerably in size and color. An expert is needed for species identification as male and female genitalic structures are used for accurate determination. Adults may be a light ochre brown, as in *P. crinita*, to almost black, as in *P. anxia*. Adults are usually 19/32–1.0 inch (15–25 mm) long, depending on the species.

Life cycle and habits: Northern species take 3–5 years to complete development, usually 3 years in New York across to Nebraska. Species in the middle states and south may take 1–2 years to develop. *P. crinita* and *P. crenulata* typically have annual cycles in Oklahoma through Texas. *P. latifrons* in Florida also has an annual cycle. The adults begin to emerge in April and May in southern states, and usually May and June for the rest of the United States. This is why they are called "Maybugs" and "Junebugs." *P. crinita* often emerges in July and early August in Texas and Oklahoma. The adults emerge from the soil at dusk and fly to trees and shrubs for a meal of foliage. If the evening temperatures are high enough, the adults continue feeding until dawn, when they return to the soil for hiding. Large populations can severely defoliate preferred plants such as oaks and maples. The adults are also highly attracted to night lights and often alarm people with their buzzing flight. After feeding several nights and mating, females burrow into moist soil to a depth of 2–6 inches (5–15 cm) where they lay 20–30 eggs individually packed into balls of soil. These eggs must absorb moisture from the surrounding soil to develop. After 20–40 days, the eggs hatch and the young larvae burrow upward in search of plant roots. Each species seems to be able to feed on a variety of plant roots, though certain plants are preferred. Annual cycle *Phyllophaga* develop rapidly during the summer and fall, and continue to feed as long as the soil temperatures do not get too cold. Those species usually pupate in February and March. The *Phyllophaga* with 2-year cycles generally reach the second instar by fall, and this stage overwinters. These grubs return to the soil surface the following spring and feed until late summer, when they pupate. The pupae usually transform into adults by late fall but these adults overwinter until the next spring.

The 3-year *Phyllophaga* remains as larvae for 2 years and do not pupate until the following summer. Again, the pupae may transform into the adult stage by early fall but these adults delay emergence until the following spring. In the far northern states, some *Phyllophaga* may need three to four summers to complete larval development before they pupate.

Control options: See: White Grubs in Turfgrass. The northern *Phyllophaga* rarely reach damaging populations in turfgrass situations. The southern *Phyllophaga* that damage turfgrasses can be treated like annual white grubs except that spring treatments are not beneficial, since most of the pests are in the pupal or adult stages.

Mole Crickets: Introduction

Species: Four species of mole crickets may be found in North American turfgrass. The tawny mole cricket (previously misidentified as the Changa or Puerto Rican mole cricket), *Scapteriscus vicinus* Scudder, and the southern mole cricket, *Scapteriscus borellii* Giglio-Tos (= *S. acletus* Rehn and Hebard), are the most damaging species. The short-winged mole cricket, *Scapteriscus abbreviatus* Scudder, occurs occasionally in pest levels and the native mole cricket, *Neocurtilla hexadactyla* (Perty), can be present in damaging populations in southern states and is an occasional pest in northern states (Phylum Arthropoda, Class Insecta, Order Orthoptera, Family Gryllotalpidae).

Distribution: Turf-damaging mole crickets are found south of a line running from mid-North Carolina through mid-Louisiana and into southeastern Texas. The native mole cricket is found all through the eastern half of the United States and is more of an occasional curiosity than a turf pest.

Hosts: Most damage occurs to southern grasses such as centipedegrass, bahiagrass, bermudagrass, and St. Augustinegrass.

Damage symptoms: These pests tunnel through soil, like their mammal counterpart. This tunneling breaks up the soil around turf roots and the turfgrass often dies due to desiccation. The trails themselves are considered unsightly and interfere with ball roll on greens. During mating and overwintering, the adults often throw up mounds of soil around their permanent burrows. At this time, the adults are not feeding extensively enough to kill large patches of turf. Real damage occurs in the summer months when the nymphs are actively feeding on the turfgrass roots. Heavy infestations during this period may result in large dead patches and exposed soil. St. Augustinegrass does not show as severe a response to mole crickets, possibly because of its different growth habits.

Life cycles and habits: These pests have gradual life cycles with eggs, nymphal, and adult stages. The nymphs look like the adults but

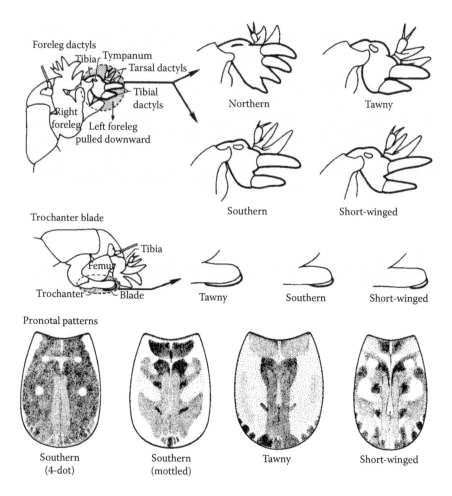

FIGURE 3.88
Leg characteristics and pronotal patterns used to identify mole cricket adults. (Florida Agr. Exp. Sta.)

are smaller and do not have wings. Most species mature by fall but do not lay eggs until the following spring.

Identification of species: The four turf-infesting mole crickets can be identified using the following key (Figure 3.88):

   1a. Four dactyls or "claws" on the foretibia; hind femur longer than pronotum .... species of *Grylotalpa* and *Neocurtilla*
   b. Two dactyls on the foretibia; hind femur shorter than pronotum, (species of *Scapteriscus*) 2
   2a. Tibial dactyls widely separated, by more than 0.30 mm, gap appearing U-shaped 3

     b. Tibial dactyls narrowly separated, by less than 0.30 mm, gap appearing V-shaped .... *S. vicinus* Scudder, tawny mole cricket

   3a. Forewing short, extending no more than one-third the abdomen length; hind wings hidden .... *S. abbreviatus* Scudder, short-winged mole cricket

     b. Forewing reaching end of abdomen; hind wings extending beyond tip of abdomen ....*S. borellii* Giglio-Tos, southern mole cricket

Control options: These turfgrass pests are some of the most difficult to control. Their ability to tunnel, fly, and eat a wide range of foods contributes to the problems. They can tunnel downward and escape most pesticide applications, especially during dry spells. Spring and fall flights can occur over several months and adults may travel up to 6 miles from pastures and road rights-of-way. New fly-ins can regularly reinfest treated areas, which is especially troublesome on golf course greens and tees.

Option 1: Biological control—encouragement of natural agents: Mole crickets are cannibalistic and many young nymphs perish in this manner. Predacious insects, birds, and mammals prey on mole crickets but may dig up turf in the effort. The parasitic wasp, *Larra bicolor* Fab. (Hymenoptera: Sphecidae), when initially introduced was only locally active, but newer introductions from South America have resulted in wider establishment and this parasite now takes a substantial toll on mole crickets from northern Florida into Alabama. A parasitic tachinid fly, *Ormia depleta* (Wiedemann), has been introduced into Florida from Brazil and it is currently established in the southern half of Florida. This is often called the red-eyed fly. It has become apparent that these parasites often require plants that flower at a proper time to provide nectar for adult parasite survival.

Option 2: Biological control—parasitic nematodes: Initial studies with insect parasitic nematodes, species of *Steinernema* and *Heterorhabditis*, resulted in poor control and nematode survival in the field. However, with the discovery of *S. scapterisci*, a species of nematode that could survive from year to year in Florida and have significant impacts on imported mole crickets was found. This nematode has been extensively released in pasture areas where it provides satisfactory suppression of mole crickets. However, on high-maintenance turf, additional mole cricket controls are often needed. *S. riobravis* is another commercially available species of nematode that has significant mole cricket activity.

Option 3: Chemical control—use of food-attractant baits: Baits are most effective if applied to kill mole crickets that are actively feeding at the soil surface. This usually occurs when the soil is warm and following a rainy period or a period of regular irrigation. When nymphs and adults come to the surface to feed, they will find the bait. Adult mole crickets are often less liable to pick up the bait, especially during spring mating periods. Most baits should be applied while the soil is moist, but not when the grass foliage is wet. The bait granules can stick to the foliage of wet plants and do not drop to the soil surface where they are needed. Also, avoid irrigating for several days after applying a bait, as this may leach out the pesticide or degrade the bait attractiveness.

Option 4: Chemical control—targeted insecticides for fall and spring, migrating adults: Control of mole crickets at the time when they are migrating is very difficult because the adults are not often actively feeding. In the spring, the females are migrating from one area to another in search of mates and suitable egg-laying areas. Once in a suitable spot, females may burrow deep into the soil, well out of the reach of most insecticides. Insecticides containing acephate seem to discourage mole cricket surface activity, but probably do not actually kill the mole crickets. Insecticides containing fipronil seem to be more effective in providing actual control.

Option 5: Chemical control—assess activity and target new nymphs: Best control of mole crickets is a result of sampling where adult activity is occurring in the spring egg-laying period and applying insecticides that effectively control new nymphs as they surface to begin feeding. Applications of various neonicotinoids (or neonicotinoids plus a pyrethroid combination products), fipronil, or encapsulated acephate have been the most effective products when used at this time.

Monitoring Spring Adults

Monitoring adult activity in the spring attempts to locate where the most eggs will be laid. Research has shown that males often congregate in areas of moist soil and habitat most suitable for females to lay eggs. The males sing from these locations, which attract more males and females. Monitoring can be as simple as noting where extensive male singing chambers are visible to using grid systems to measure the actual amount of mole cricket tunneling in a given area. The grid system uses a plastic or wooden frame (say 1 yard square) that is strung up in 1 foot squares. This is tossed in the area to be sampled and each square that has a mole cricket tunnel running under it is scored. Obviously, a score of 9 means every square had a tunnel while a score of 1 means light activity.

Where heavy activity occurs, this will be a likely place for the females to lay the majority of eggs. Female egg-laying status can also be assessed. A detergent disclosing solution can be used to force mole crickets to the surface.

Use about one tablespoon of household detergent in a gallon of water and use two gallons to flush a 1 yd^2 area of infested turf. In dryer weather, two flushes may be needed to force adult mole crickets up. Female mole crickets will have distended abdomens (and they have no dark mark at the bases of their front wings). Using a pocket knife, make a slit in the side of the female cricket's abdomen and squeeze. This will expose the eggs within the ovaries. If the eggs are clear and white in color, they are not ready for laying. If the eggs have a distinctive amber look, they will be laid within the week. Apply insecticides for nymph control as soon as most of the females have amber eggs.

Option 6: Chemical control—targeted insecticides for summer nymphs: The majority of mole cricket nymphs are present from June into September, though nymphs are found in some areas all through the year (like in south Florida). During this summer period, it is easier to use insecticides to reduce the general population and reduce the potential for damage. The young nymphs are usually at the surface or within a few inches of the surface, and are constantly feeding. Thus, they are more likely to come into contact with any insecticides present.

Sampling for Summer Nymphs

Summer nymph populations are often missed when the nymphs are small. These tiny insects make minute trails on the surface of bare soil, but these burrows can be completely hidden by turf cover. Because of this, a detergent disclosing solution (see above) must be used to detect their presence. By August, the nymphs have matured considerably and their activity is easy to visually detect. This late detection should be avoided because larger nymphs are more difficult to kill and considerable damage may have occurred.

Option 7: Chemical control—late-season treatments for large nymphs and new adults: Occasionally, mole cricket populations will be missed and damage may not occur until September and October. By this time, the nymphs are large and/or adults are present. These larger crickets tend to burrow deeper into the soil, and if it is dry, they will only surface after irrigation or a rain event. At this time, only fipronil products appear to be effective as well as some of the baits. To maximize insecticide efficacy, irrigate the turf to coax the mole crickets to the surface or watch for rain events that may bring them up.

Tawny Mole Cricket

Species: *Scapteriscus vicinus* Scudder (originally misidentified as *Scapteriscus didactylus*, the Changa or West Indian mole cricket)

(Phylum Arthropoda, Class Insecta, Order Orthoptera, Family Gryllotalpidae) (Figure 3.89).

Distribution: Introduced into Georgia around 1899; now found throughout Florida and south of a line running from North Carolina, across Georgia, and into southern Texas; a native of coastal South America from northern Argentina through Brazil and in Columbia, Panama, and Costa Rica.

Hosts: This is a pest of vegetable crops and can inhabit all the warm-season and transition-zone grasses.

Damage symptoms: Produces typical mole cricket damage to turf. This species is primarily a plant feeder and it prefers to feed on roots, but it also damages turf by tunneling and throwing up mounds of soil around its permanent burrows. Loosened and uprooted turf usually withers and dies, leaving trails of dead turf. Heavy populations can completely kill turf, leaving bare patches of soil.

Description of stages: This pest has a typical gradual life cycle.

Eggs: Oblong, translucent amber eggs are placed in clusters in chambers 3–10 inches (7.5–25 cm) below the soil surface.

FIGURE 3.89
Top view of the tawny mole cricket adult male and female.

Nymphs: These stages look like miniature adults but they lack wings. Early instars do not have visible wing pads and later instars begin developing visible wing pads. The nymphs molt 6–8 times; the exact number is not known. Even at this early age, this species can be identified by the V-shaped space between the tibial dactyls.

Adults: Fully grown adults are about 1¼ × 3/8 inch (30–34 × 8–10 mm). They are a light tawny brown and have the obviously modified front legs and enlarged thorax typical of mole crickets. The forewings are shorter than the abdomen, and males have a dark rasp at the forewing base. The V-shaped space between the tibial dactyls is a characteristic of the species.

Life cycle and habits: Females are attracted to the calls of males in the spring when temperature and moisture allow. Egg laying may begin in March in south Florida, but 75% of the eggs are laid between May 1 and June 15. During the mating and oviposition period, the adults are extremely active, emerging from the soil at dusk to crawl about, tunnel, and fly. Flight in mass commonly occurs after rainfall, and adults are attracted to lights. Mated females dig down a few inches into suitable soil and construct an egg chamber in which they lay an average of 35 eggs. Females may construct three to five chambers and lay 100–150 eggs in total. The eggs hatch in about 20 days and the young nymphs dig upward to seek out food. These nymphs prefer to feed on plant roots but will eat small insects, and can be cannibalistic. Larger nymphs will often attack smaller nymphs and eggs. Only 10–15% of the adults and nymphs have been found with insect parts in the gut. The nymphs continue to feed, molt, and grow through the summer months. By the end of October, 85% of the nymphs have reached adulthood. The remaining nymphs continue development very slowly and most mature by the following spring. Adult and nymphal behaviors are highly regulated by temperature and soil moisture. Most feeding occurs at night, especially after rain showers or irrigation during warm weather. These insects may tunnel 20 ft per night in moist soil. During the day, individuals tend to return to a permanent burrow, which is deeper in the ground. The crickets may remain in these permanent burrows for considerable periods during cool winter temperatures or dry spells. Male mole crickets locate preferred habitats in the spring and call with their toad-like trill (3.3 kHz with 60 cycles/s at 77°F [25°C]) from the entrance of their burrows. This occurs for about an hour shortly after sunset. Mole crickets can fly more than 6 miles in a night and may fly more than once. This accounts for the great difficulty in eliminating populations in any given area.

Control options: See: Mole Crickets: Introduction.

Southern Mole Cricket

Species: *Scapteriscus borellii* Giglio-Tos (= *S. acletus* Rehn and Hebard) (Phylum Arthropoda, Class Insecta, Order Orthoptera, Family Gryllotalpidae) (Figure 3.90).

Distribution: This pest was likely introduced into Georgia or Florida around 1900; now found south of a line running from North Carolina through mid-Louisiana and into eastern Texas; populations are also found in Arizona and southern California; a native of southern South America.

Hosts: Apparently, this species prefers to prey on other insects and even each other. However, considerable feeding on plants occurs; 41% were found to have plant material in the gut. Often found in bahiagrass, bermudagrass, and St. Augustinegrass and can also damage centipedegrass and zoysiagrass.

Damage symptoms: Produces typical mole cricket damage, though most of the damage is due to tunneling. Nymphs and adults tunnel and throw up mounds of soil. Uprooted turf soon wilts and dies from desiccation.

Description of stages: This pest has a typical gradual life cycle.

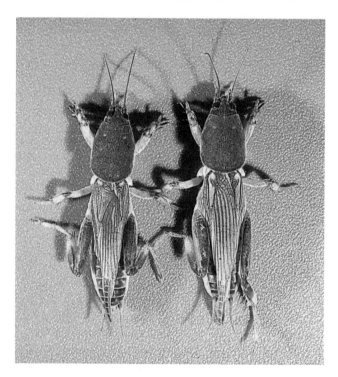

FIGURE 3.90
Top view of the southern mole cricket adult male and female.

Eggs: Elongate oval, translucent white eggs are placed in clusters of up to 35 in chambers underground.

Nymphs: The nymphs look like small adults but do not have functional wings. The wings develop as pads, which enlarge with each molt. This species can be identified by the U-shaped space between the tibial dactyls. Some populations have mottled patterns on the pronotum, while other populations have a dark background with four small pale spots.

Adults: Adults are about $1\frac{1}{4} \times 3/8$ inch (32×9 mm). They are more of a gray-brown color when compared to tawny mole crickets and have either two or four pale spots on the prothorax. The forewings are shorter than the abdomen, and the hindwings extend just beyond the tip of the abdomen. The U-shaped area between the tibial dactyls is diagnostic of the species. Males also have the darker rasp and file on the forewing base.

Life cycle and habits: This species has a cycle much like the tawny mole cricket. Egg laying occurs from mid-March (in Florida) into June, but some adults continue laying eggs into September in the northern range. The nymphs mature rapidly through the summer but only about 25% reach adulthood by winter. Some males call during the fall and mating at this time may occur. The nymphs complete development slowly during the winter, with more adults appearing from February through April with a peak in May. Nymphs and adults of this species have been found during much of the year. Most activity is found after rain or irrigation, and a spring flight period occurs during mating and oviposition. Males produce a bell-like trill (2.8 kHz with 135 cycles/s) for an hour or longer just after sunset. The southern mole cricket is more of an omnivore than the tawny mole cricket. Only 5% have been found with plant material alone in the gut, while 59% had only animal material in the gut and the rest had a combination. Thus, it is suspected that turfgrass damage caused by the southern mole cricket is due to its tunneling and soil mounding rather than actual root feeding.

Control options: See: Mole Crickets: Introduction.

Short-Winged Mole Cricket

Species: *Scapteriscus abbreviatus* Scudder (Phylum Arthropods, Class Insecta, Order Orthoptera, Family Gryllotalpidae) (Figure 3.91).

Distribution: South and central Florida in the United States; native to Argentina, Paraguay, and Brazil; also introduced into Puerto Rico, Virgin Islands, Cuba, Nassau, and Haiti.

Hosts: Seems to prefer plant material for food.

Damage symptoms: Produces typical mole cricket damage but the damage is often attributed to the nymphs of one of the other species of

FIGURE 3.91
Top view of the short-winged mole cricket adult.

Scapteriscus. This pest does not fly and thus tunnels to disperse, but it can be transported in soil that is moved.

Description of stages: This pest has a typical gradual life cycle.

Eggs: Similar to other mole crickets.

Nymphs: Similar to other mole crickets. Difficult to distinguish from the southern mole cricket, which also has a V-shaped space between the tibial dactyls. However, the ocelli are smaller in the short-winged mole cricket.

Adults: Similar to other mole crickets except that the hindwings extend no more than one-third the length of the abdomen. The hindwings are very small and the ocelli are smaller than in *S. borellii*.

Life cycle and habits: This species is probably the least studied of the turf-infesting mole crickets. It was apparently more of a pest before the tawny mole cricket was imported. However, in recent years more short-winged mole crickets have been found in turf, and its spread seems to be increasing because of transporting soil during road and building construction. This species cannot fly and must naturally spread by crawling or tunneling. The males do not make a mating call chirp but apparently produce low chirping sounds if a prospective mate is found. This pest seems to have a cycle similar to the southern mole cricket; that is, most reach adulthood by the fall.

Control options: See: Mole Crickets: Introduction.

Ground Pearls

Species: Two species of *Margarodes* are involved; usually *Margarodes meridionalis* Morrill is the species found attacking turfgrasses

FIGURE 3.92
Several ground pearl cysts recovered from the soil of infested bermudagrass.

FIGURE 3.93
Athletic field with irregular circles of turf damaged by ground pearls.

(Phylum Arthropoda, Class Insecta, Order Hemiptera, Family Margarodidae) (Figures 3.92 and 3.93).

Distribution: Generally found attacking warm-season turfgrasses from southern California across to North Carolina.

Hosts: Attacks bermudagrass, centipedegrass, zoysiagrass, and St. Augustinegrass. Most commonly damages bermudagrass and centipedegrass.

Damage symptoms: Irregular patches of the turf appear unthrifty, and over a year or two thin out or die. Damage often occurs in irregular circles that can be mistaken for fairy ring or other disease attack.

Description of stages: Very little is known about this group of pests and publications are scant. Apparently, undergoes a modified gradual life cycle such as is found in some mealybugs and scales.

Eggs: Light pinkish eggs are deposited in a mass of white waxy threads.

Nymphs: The first nymphal instar, called the crawler, is about 0.008 inch (0.2 mm) long. This stage attaches to a grass root and begins to cover itself with a hard coating of yellowish to light purple wax. This is the ground pearl. The encysted nymphs range from 1/32–5/64 inch (0.5–2.0 mm) in diameter.

Adults: The adult females are pinkish sac-like forms, about 3/32 inch (2.3 mm) long, and have well-developed front legs and shorter second and third legs. The males are rarely seen, but are tiny white to pinkish gnat-like insects with two pairs of wings and white waxy filaments arising from the abdomen tip.

Life cycle and habits: Because these insects are essentially subterranean for most of their lives, little is known about their exact behavior. Generally, it is thought that there is only one generation per year. Mature females occur in late spring, during which time they emerge from their hard waxy cysts and crawl to the soil surface. Here they may mate with the tiny winged males, but mating may not be required. Once mated, the females dig 2–3 inches (5–7 cm) deep into the soil and form a cotton-like mass of wax (ovisacs) in which they lay about 100 eggs. The eggs hatch into crawlers that disperse in the soil in search of grass roots. When suitable grass or plant roots are found, the crawlers attach themselves and secrete a hard waxy coating that becomes the pearl stage. The nymphs continue to develop inside this cyst and overwinter attached to the root.

Control options: Since these pests are protected during most of their life cycle inside the waxy cyst, no insecticides have been found to be effective.

Option 1: Cultural control—maintain healthy turf: Watering during drought periods and proper fertilization programs have been the most effective methods used to counter damage from ground pearls.

Nuisance Invertebrate, Insect, and Mite Pests

When considered as an ecosystem, the turfgrass habitat is a rich source of plant and general organic matter that can be used by general herbivores. We usually consider these herbivores as "pests" when they cause noticeable damage to the look of the turf. These herbivores are merely one part of the complete ecosystem. Turf herbivores, their remains, and the organic matter found in the turf canopy, thatch, and upper soil levels provide abundant food resources for a variety of other animals. Still other animals use the insulating quality of the turf canopy and thatch and the soil beneath to build nests or resting places. When these animals associated with the turf ecosystem cause concern or problems with the human turf managers and users, the animals become "nuisance" pests. In short, nuisance pests rarely damage the turf.

Each nuisance pest may be viewed in dramatically different ways, depending upon the use of the turf and personal views or needs of the owner, manager, and users of the turf. Earthworms are generally considered to be highly beneficial because of their ability to assist in the decomposition of thatch, conditioning, and aeration of the soil. However, earthworms collect grass leaves and smaller stems in the fall and mix them with castings around their base burrows. During the winter months, these mounds become hardened and result in "lumpy" lawns that are not appreciated by most home owners. Earthworms on putting greens leave mounds of castings, which can dull mowers and interfere with ball roll. Therefore, though earthworms are generally beneficial, they can be nuisance pests under certain conditions. Millipedes, centipedes, and many bees and wasps live in the turf habitat. Though most of these animals will not hurt people, most people fear them because of a lack of knowledge and appreciation. These are nuisance pests that many users of turf demand to be controlled.

In most instances, turf managers need to try to educate the general users of turf about the benefits of the numerous organisms that are considered nuisance pests. On the other hand, fire ants, ticks, fleas, and chiggers may use the turf ecosystem while actively attacking humans and their pets. Because these pests do not actively feed on and damage turf, they are here considered nuisance pests.

Earthworms

Species: Several species are common across North America. The common earthworm or night crawler, *Lumbricus terrestris* Linnaeus, and the red earthworm, *Lumbricus rubellus* Hoffmeister, are common large species; both are nonnatives from Europe. Smaller species belong to the genera *Allolobophora* and *Eisenia* (Phylum Annelida, Class Oligochaeta, Family Lumbricidae) (Figures 3.4 and 3.94).

FIGURE 3.94
Earthworm castings on the surface of a golf course green in Florida.

Distribution: Certain species are more common in local areas, but earthworms are present in all of North America. Most of the species that leave castings on the soil surface are nonnatives that now dominate urban habitats.

Hosts: Earthworms feed on decaying organic material such as decaying thatch and leaf litter.

Damage symptoms: Earthworms are considered beneficial because of their ability to decompose plant remains and incorporate these into the soil. However, on close-cut, well-maintained turf, such as greens and tees or in grass tennis courts, worm castings may be a problem. Several species of earthworms deposit soil and fecal material at the ground surface. These piles of soil/feces, called castings, can harden in clay soils, thereby ruining the desired smooth surface. This activity in lawns produces what is called "lumpy" ground under the turf canopy. On golf courses and sports fields, earthworm castings can interfere with ball roll, muddy equipment and clothing, as well as dull mowers.

Description of stages: Earthworms are elongate, slender invertebrates with many ring-like segments. The young and adults look alike and growth is accomplished by adding segments to the body. Depending on the species, earthworms can be 3/4–8 inches (2–20 cm) long. Earthworms can be distinguished by the presence of setae (spike-like hairs). These can be seen under a microscope if the worm is small, or felt by running the finger along the lower side of a mature worm. Earthworms move by contracting and expanding the body, while nematodes wiggle.

Eggs: The eggs are laid in a gelatinous cocoon that is 3/32–9/32 inch (1–7 mm) in diameter. These cocoons are translucent and blood vessels and/or the young worms can be seen developing inside. A cocoon may contain one or several eggs, but usually only one worm hatches.

Immatures and adults: Both these stages look alike. The sexually mature worms have a swollen band on the body, the clitellum that is used to secrete the cocoon.

Life cycle and habits: Earthworms are active anytime during the year when temperatures are above freezing. During the warmer months, the earthworms come to the soil surface to pick up parts of dead plant material. They may pull grass clippings and dead leaves down their holes for food. The worms may also feed on organic material in the soil as they tunnel. While tunneling, the worms ingest great quantities of soil and this is often deposited, excreted, at the soil surface as castings. This tunneling and soil deposition is one of nature's ways of aerating and mixing soils. During rainy nights, the worms often leave the soil to crawl about in search of mates and food. If two sexually mature earthworms come together, they often mate. Earthworms are hermaphroditic (having male and female sex organs). During copulation, earthworms join by attaching to each other through mucus bands. Sperm from each worm travels down ducts to enter the female pores of their current mate. The worms do not fertilize themselves. After mating, the fertilized eggs are deposited into a band-like mucus cocoon that slips past the egg ducts and off the tip of the worm. These cocoons are deposited in the soil and if moisture is sufficient, the eggs will mature in 2–3 weeks. The young worms emerge and immediately begin to feed in the soil. The young worms take 2–3 months to become sexually mature under ideal conditions. Some worms become inactive during dry weather and form a mucus-lined chamber deep in the soil.

Control options: Earthworms are beneficial animals and no controls are recommended. Though no pesticides are recommended or labeled for control of earthworms, certain insecticides (carbaryl, chlorpyrifos, fenvalerate, guthion, methomyl, nicotine, and propoxur) and fungicides (azinphos-methyl) are highly toxic to them. Repeated use of these products may decrease earthworm populations.

Option 1: Cultural control—reduce effects of soil mounding: Soil mounding in lawns commonly occurs during the cool, fall and spring periods. These "lumpy" lawns can be reduced by rolling with a heavy lawn roller, especially after a rain or irrigation softens the soil.

Option 2: Cultural control—topdress with sharp sands: In areas where old foundry sands can be obtained, this can be used as a topdressing to discourage earthworm activity. These sands are fractured in the heat of foundry operations, which produces sharp edges that can damage earthworm cuticle and gut linings.

Slugs and Snail

Species: Several species of slugs and snails may be found in turf but most feed on other plants. Common slugs are gray garden slug, *Agriolimax reticulatus* (Muller); spotted garden slug, *Limax maximus* Linnaeus; and tawny garden slug, *Limax flavus* Linnaeus. The brown garden snail, *Helix aspersa* Muller, is occasionally found crossing turf in search of more palatable food (Phylum Mollusca, Class Gastropoda) (Figures 3.2 and 3.3).

Distribution: Slugs and snails can be found worldwide from tropical to temperate zones.

Hosts: Each slug and snail has preferred food plants, though some are even carnivores. Slugs and snails in turfgrasses are usually feeding broadleaf weeds such as clovers and ground ivy. Some of these pests will rarely feed on southern grasses such as St. Augustinegrass.

Damage symptoms: Because slugs and snails do not normally attack turfgrasses, they are considered nuisance pests. They leave slime trails on sidewalks and over grass blades. They are also no fun to step on with bare feet.

Description of stages: Slugs and snails go through a gradual growth and the immatures look like the adults.

Eggs: The translucent white or amber-colored eggs are 3/16–7/32 inch (5–5.5 mm) in diameter and are laid in clusters of several dozen.

Immatures and adults: Slugs are merely snails without a shell. Young slugs are 3/16–9/16 inch (5–15 mm) long and most adult slugs are 1³⁄₁₆–1⁹⁄₁₆ inch (30–45 mm) long, but some large species can get to be 2¾–3½ inch (70–90 mm) long. They are usually gray or brown with spots or mottling. Snails may have flattened or elongate shells and tropical forms may be brightly colored. Both snails and slugs move on a flat foot that secretes a mucus trail.

Life cycle and habits: The gray garden slug will be used as an example. This species overwinters in the adult stage and sometimes in the egg stage, as long as freezing temperatures are avoided. In the spring, the adults become active and lay clusters of eggs in crevices in the soil or under boards, rocks, or leaf litter. The eggs take about 20 days to hatch at 80°F (26.7°C) and twice as long at 50°F (10°C). The young slugs feed on decaying plant material as well as leafy green plants. The slugs can become sexually mature in 10–15 weeks. Slugs and snails are generally creatures of the night that hide under litter, rocks, or boards during the day. Slugs and snails also prefer moist environment. Slugs may become dormant in dry summer periods by hiding in crack and crevices and losing up to 20–40% of their body moisture. Snails often crawl upon vertical surfaces and form a mucus plug over the shell opening, which reduces water loss during dry periods.

Control options: These pests generally do not have to be controlled in turf unless flower beds are nearby or weeds are present.

Option 1: Cultural control—modify the environment: Removing plant residues from flower beds and controlling the weeds in lawns will help reduce slug populations. The practice of using mulches around trees and shrubs also contributes to slug survival. Try to create landscaped areas that allow the surface to dry rapidly after watering or rains.

Option 2: Chemical control—use molluscicides: Several pesticides are registered for slug and snail control. These are usually applied in a bait form.

Spiders and Tarantulas

Species: Several hundred species of spiders may be found in the turf. Most common groups are the wolf spiders, family Lycosidae, the grass and funnel-web spiders, family Agelenidae, and the sheet-web spiders, family Linyphiidae. The tarantulas found in the southwestern states are merely large, hairy spiders in the family Theraphosidae. Dangerous spiders such as the recluse spiders, *Loxoceles* spp., and the widows, *Latrodectus* spp., are not associated with turf (Phylum Arthropoda, Class Arachnida, Order Araneae) (Figure 3.5).

Distribution: Different species of spiders are located across the world.

Hosts: Spiders are predators on other arthropods, and species can be found in all turfgrass habitats.

Damage symptoms: Spiders do not damage turf and are generally considered beneficial in their activities of reducing other arthropod populations.

Description of stages: Spiders undergo little change during their development. Young spiders look like the adults but may have slightly different color patterns. All spiders have four pairs of legs, no antennae, and the cephalothorax narrowly joined to the abdomen.

    Eggs: The eggs are usually round and deposited in clusters in a silken sac.

    Immatures and adults: It is beyond the scope of this work to describe all the forms of spiders found in the turf. Usually, wolf spiders are medium to large spiders, 9/16–1¾ inch (15–45 mm) in diameter, with gray or brown and often with darker stripes on the thorax. Wolf spiders do not usually spin webs but may dig a burrow in the ground. Funnel-web spiders cover the turf with a sheet of webbing that ends in a funnel-like tunnel that leads into a burrow. The sheet-web spiders include the bowl and doily spiders. These tiny spiders, 5/32–3/8 inch (4–10 mm), construct a bowl-shaped web between grass blades and irregular webbing over the bowl. These are often visible in the morning dew. Tarantulas merely construct a burrow in the ground.

Life cycle and habits: Each spider has a unique life cycle but generally a single generation is completed per year. The tarantulas often take 2 years to mature. The wolf spiders are active predators and do not spin a web. The females often carry the egg case with them and the young ride on her back for a while before foraging on their own. Tarantulas are also active predators but usually stay near their base burrow. The other spiders build different types of web nests that are used to trap their insect prey.

Control options: No controls are recommended because spiders are beneficial. Occasionally, wolf spiders will enter a house in search of food. In the southern states, the tarantulas, because of their large size, cause concern but their bite is no worse than a bee sting. In fact, tarantulas tame quickly in captivity and make interesting pets. If spiders need to be controlled, spot treatments or mechanical crushing is sufficient.

Chiggers

Species: *Trombicula alfreddugesi* (Oudemans), *Trombicula batatas* (Linnaeus), and *Acariscus masoni* (Ewing) are common species. Others in these genera may occasionally cause problems. (Phylum Arthropoda, Class Arachnida, Order Acarina, Family Trombiculidae) (Figure 3.95).

Distribution: *T. alfreddugesi* is found from Canada to South America, while *T. batatas* and *A. masoni* are more common in the southern states.

Hosts: These mites are parasitic on reptiles, amphibians, mammals, and birds in their larval stage. The nymphs and adults are apparently free living and do not take a blood meal. Chiggers can be found in any turf, but seem to prefer taller grasses and weeds where mammals and birds frequently nest.

Damage symptoms: No damage is done to turf. The larvae cause itchy, red welts on the skin of humans when they attempt to feed.

Description of stages: Chiggers have typical mite stages of eggs, larvae, nymphs, and adults.

    Eggs: The eggs are microscopic, roundish, and reddish when about to hatch.

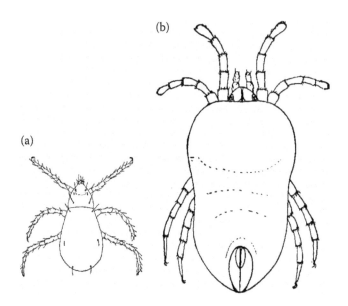

FIGURE 3.95
Chigger larva (a) and adult female (b). (USDA.)

Larvae: The larval chigger is orange-yellow or light red and has only three pairs of legs. They are about 0.006 inch (0.15 mm) long and barely visible with the unaided eye.

Nymphs: The nymphs are 3/64 inch (1–1.2 mm) long, bright red, and covered with complex flattened hairs. These now have four pairs of legs. Under a microscope or hand lens, these mites have puffy or lobed cephalothoracic areas.

Adults: The adults look like the nymphs but are slightly larger.

Life cycle and habits: Apparently, the adults and nymphs overwinter. These are general scavengers or predators. Eggs are laid in the spring and hatch in a week. The young larvae climb onto animals or on the clothing of people. Here, they seek out a suitable soft spot and often bury their head into the base of a hair follicle. They inject anticoagulant saliva and begin to suck blood. The saliva is allergenic and causes severe itching and a red swelling. Often, the mites are trapped in the hair follicle and cause even more inflammation. Successfully fed larvae drop off the host and enter into a transforming state. They then molt into the nymph that has four pairs of legs and this, in turn, molts into the adult, usually without feeding. The complete life cycle may take 50 days, but usually only one cycle occurs per year.

Control options: Chiggers are very difficult to control because of their ability to feed on so many hosts. Most people can adequately protect themselves by properly applying repellents that state that they are effective against chiggers. Most well-maintained lawns do not get chigger populations. Chiggers prefer tall grasses and weedy areas where rodents or ground-nesting birds frequent.

Option 1: Cultural control—modify habitats: Since chiggers do best in areas where grasses are allowed to grow tall and other shrubby weeds grow, keep these areas mowed and remove weedy growth where possible.

Option 2: Chemical control—area sprays: Several insecticides are registered for chigger control. Apply these materials evenly over the lawn and surrounding plant material. Reapplications may be necessary if small rodents are present. Be sure to check labels to ensure that the formulation is registered for use on turf so as to avoid photoxicity issues.

Ticks

Species: The most common tick encountered in lawns and grounds is the American dog tick, *Dermacentor variabilis* (Say). Other ticks occasionally found around homes and grounds are the brown dog tick, *Ripicephalus sanguineus* (Latreille), and the lone star tick, *Ambylomma americanium* (Linnaeus). The blacklegged tick, *Ixodes scapularis* Say (commonly called the deer tick), that is associated with Lyme disease is not usually associated with well-maintained turf (Phylum Arthropoda, Class Arachnida, Order Acari, Family Ixodidae) (Figures 3.7 and 3.96).

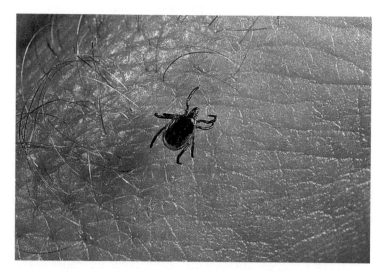

FIGURE 3.96
The blacklegged tick is a common vector of Lyme diseases.

Distribution: The American dog tick is most common in the states east of the Rocky Mountains as well as California, Idaho, and Washington. The brown dog tick is cosmopolitan, but prefers the warmer states, and tropical and semitropical countries. The lone star tick is found from Texas and Oklahoma across to the Atlantic Coast and south into Central and South America. Blacklegged ticks are found in northeastern North America.

Hosts: Ticks do not attack turfgrasses but are blood-sucking parasites of animals. Tall grasses and weeds provide shelter for ticks and they may develop large populations in these areas. The brown dog tick prefers to feed on dog blood in all its stages. It rarely attacks other animals or humans. The American dog, lone star, and blacklegged ticks prefer small rodents in the larval and nymphal stages but adults will feed on humans, dogs, and livestock.

Damage symptoms: No damage is done to turfgrasses. Ticks are known to vector several diseases, including Rocky Mountain spotted fever that is present across North America, viral encephalitis, and tularemia. The blacklegged tick is the major vector of Lyme disease, though the American dog tick can also harbor this disease. The American dog tick and lone star tick can also cause tick paralysis, a neurotoxic reaction to the saliva of ticks. This paralysis is found in certain people, especially when these ticks attach to the skin on the back of the neck or head.

Description of stages: Ticks lay eggs that hatch into six-legged larvae, called seed ticks. Seed ticks molt into normal eight-legged nymphs that molt into adults.

Eggs: The eggs are shiny and oval, about 1/64 inch (0.4 × 0.5 mm), yellowish to brown and laid in clusters of 500 to thousands.

Larvae: The larvae or seed ticks are 0.02–0.04 inch (0.5–1.0 mm) long and have only three pairs of legs.

Nymphs: The nymphs have four pairs of legs and are smaller than the adults, being 1/16–3/32 inch (1.5–2.5 mm) long.

Adults: The adults are easily identified by looking at the shape of the mouthparts, anal openings, and markings on the top cephalothoraxic shield and abdomen. Male ticks have a hard shield, the scutum, covering the abdomen, while females have the scutum extending only halfway down the abdomen. The lone star tick is the easiest to identify because it usually has a white spot at the base of the scutum. The brown dog tick is uniformly reddish-brown, and the American dog tick is reddish-brown with white mottling. Unfed adult ticks are 1/8–1/4 inch (3–6 mm) long, but engorged females may be 3/8–5/8 inch (10–15 mm) long.

Life cycle and habits: These three ticks are similar in that they are three-host ticks. That is, they have to feed three separate times on the blood of a host before the cycle can be completed. Some other ticks are two-host ticks or four-host ticks. The adults of these pest ticks grab hold of the hair of a passing animal and climb to a suitable place to insert the mouthparts and begin feeding on blood. Males hunt for females attached to an animal and mate on the host. After feeding for several days, the fully engorged females drop off and seek a protected spot in leaf litter, a crack, or crevice in the ground, or under furniture and the like in a house. The engorged female may deposit 1000–5000 eggs over a 3-week period. The eggs are held together by a gelatinous material that reduces desiccation. The eggs hatch in a month and the seed ticks (larvae) seek a host. Ticks look for hosts by climbing to the tips of tall grasses or the leaves of bushes and shrubs. Here they hold on with a couple of legs and hold the other legs outward. The legs have hook-like claws that grab hold of the fur of a passing animal or the clothing of humans. After feeding for several days, the seed ticks drop off the host and find a secluded place to molt into the nymph. Seed ticks rarely survive winter temperatures. The nymphs must again find a suitable host and feed before dropping to molt into the adult stage. Most ticks, outside, overwinter in the adult stage. These adults become active when the spring temperatures get to be 70°F or higher during the day. Usually, only one generation per year is found, but the brown dog tick may breed continuously in tropical areas or in dwellings.

Control options: Ticks are like fleas in that a well-planned set of controls must be followed. Preventing ticks on domestic pets should be done under the advice of a veterinarian. Pet bedding and living spaces should be carefully treated to reduce larval and nymphal populations of ticks. In landscapes, reduce the use of thick mulch layers, overgrown and unmanaged areas that encourage small rodent populations—mice, voles, rats, chipmunks, squirrels, and so on. Small rodents are keys to the survival of tick larvae and nymphs. Generally, lawn and garden treatments are most effective during the spring periods when the newly active adults are looking for a host. The brown dog tick may be established inside a home and a pest control professional will be needed for proper treatments. Often, additional treatments will be necessary if pets are allowed to roam in weedy or wooded areas and if tick eggs hatch at a later date.

Sowbugs and Pillbugs (Isopods)

Species: Several species may be present. Usually the pillbug is *Armadillium vulgare* (Latrelle) and sowbugs may be species of *Oniscus* and *Procellio* (Phylum Arthropoda, Class Crustacea, Order Isopoda) (Figure 3.10).

Distribution: Sowbugs and pillbugs are found worldwide.

Hosts: Sowbugs and pillbugs rarely feed on turfgrasses but they may feed on the grass clippings and germinating seed of newly planted turf. Usually, they are scavengers, feeding on decaying organic material or young tender plant parts—roots, and stems of seedling plants.

Damage symptoms: Broadleaf plants may have cotyledons eaten off or young leaves will have ragged edges. These are considered nuisance pests when they invade homes in search of cool moist places to hide.

Description of stages: These crustaceans go through a gradual metamorphosis similar to some insects, that is, the immatures look like the adults. Pillbugs are able to roll into a compact ball when disturbed. Sowbugs are more active and cannot roll into a ball.

Eggs: The small white eggs are retained in a pouch carried by the female underneath her body.

Immatures: The young isopods have less than seven pairs of legs and are carried in the brood pouch. After emerging from this pouch, they look like the adults but they are smaller.

Adults: The adults are elongate oval and 3/16–3/8 inch (5–10 mm) long. They may be solid or mottled gray or brown, and have a pair of conspicuous antennae and seven pairs of legs.

Life cycle and habits: Depending upon the species and temperatures, one to three broods per year may be found. Generally, isopods are continuous breeders and all stages can be found any time of the year. When mature, the adults mate and a female may produce 25–200 eggs held in a brood pouch. The young stay in the pouch for 1–2 months, apparently obtaining food from their mother. After emerging, the young may take several months to a year to mature. Sowbugs and pillbugs may live for 2–3 years. Most species molt in two stages, the front half of the exoskeleton is shed and then the last half. This is why some individuals will look like they are lighter colored on one half of the body.

Control options: Since these arthropods require moist environments and organic material for food, modifying the habitat is one of the best strategies. In general, these animals are considered beneficial because they help decompose waste plant material.

Option 1: Cultural control—modify the environment: Sowbugs and pillbugs are generally a problem where mulches are used around ornamentals and flower beds. They can also feed on grass clippings if they are piled in an area or are allowed to pile up in the turf. Reduce mulching or encourage drying of the soil surface to help reduce populations.

Option 2: Chemical control—use pesticides: Several insecticides will also kill this crustacean. Check for labels that include sowbugs or pillbugs.

Centipedes

Species: Several species of centipedes in different families may inhabit turf or the underlying soil. Adult centipedes with 15 pairs of short legs are in the order Lithobiomorpha and usually in the genus *Lithobius*. Adults with 21–23 pairs of legs are in the order Scolopendromorpha. This family includes the large tropical species that can get to be over 6 inches (150 mm) long. The burrowing centipedes belong to the order Geophilomorpha that have 29 or more pairs of legs and are very slender and elongate (Phylum Arthropoda, Class Chilopoda) (Figure 3.13).

Distribution: Centipedes are found worldwide.

Hosts: These arthropods are predators and are found in turfgrasses where they are hunting other arthropod prey.

Damage symptoms: These cause no damage to turf but are considered nuisance pests if they happen to crawl inside a house in search of food. All centipedes have poison fangs beneath the mouth opening. However, only the larger individuals, greater than 1½ inch (40 mm), have the ability to pierce human skin. The poison is painful but not generally dangerous to people unless they are hypersensitive.

Description of stages: Centipedes are easily separated from millipedes by their long antennae, rapid running ability, and the presence of only one pair of legs per trunk segment attached to the side of the body. Little is known about the life stages of centipedes.

Eggs: The spherical, white eggs are about 5/64–5/32 inch (2–4 mm) in diameter and are generally laid singly. Some species have females that dig a burrow and protect an egg cluster.

Immatures and adults: The immatures look much like the adults but are often lighter colored. Most above-ground centipedes are 13/16–1½ inch (20–40 mm) long, while the burrowing centipedes may be up to 3 inches (70 mm) long but only 1/8 inch (3 mm) wide.

Life cycle and habits: Little is known about the life cycles and habits of centipedes. Most species overwinter as adults in protected places in the soil, under rocks, or under bark of dead trees. The females lay eggs during the spring and summer and these hatch in several weeks. The young mature over a season while actively preying on insects, spiders, and other arthropods.

Control options: No controls are needed unless these arthropods are beginning to invade houses. Even then, simple crushing and removal are usually adequate. The greatest problem with these animals is the erroneous fear that centipede bites are extremely dangerous to people. See: Millipedes, Control options.

Millipedes

Species: Many species can be found. Usually two to three species are present in any turf habitat. The garden millipede, *Oxidus gracilis* Kotch, is a common species with a flattened dorsum. Other flattened species are usually in the genus *Polydesmus* while cylindrical species are usually in the genus *Callipus* (Phylum Arthropoda, Class Diplopoda) (Figure 3.14).

Distribution: *Oxidus gracilis* is worldwide in distribution and other species may have regional territories.

Hosts: Millipedes are general scavengers, usually feeding on decaying animal and plant materials.

Damage symptoms: Because these arthropods rarely attack living plant material, they are considered nuisance pests. Occasionally, they will feed on the roots and leaves of seedling plants.

Description of stages: Millipedes can be separated from centipedes by the presence of what appears to be two pairs of legs per visible body segment. Be careful not to count the legs on the first few segments because the second pair of legs may be absent or reduced for copulatory behavior. They have a simple life cycle in which the young look like the adults but have fewer pairs of legs.

Eggs: Millipede eggs are usually laid in clusters. They are white to brown and about 1/64 inch (0.4 mm) in diameter. Often, they are sticky and soil readily adheres to them.

Immatures and adults: Newly hatched millipedes have only three pairs of legs but can be separated from insects by not having a conspicuous abdomen without legs. As the young grow, segments and legs are added. Adults may be 9/16–1⅜ inch (15–35 mm) long.

Life cycle and habits: Little is known about the life cycles of millipedes. Apparently, most overwinter as immatures or adults. Adults lay eggs in the moist spring and fall months in clusters of 20–100. These clusters are usually in cavities in the soil or under rocks or pieces of wood. The eggs hatch in 1–3 weeks and the young take 21–25 weeks to mature. The adults can live for one to several years.

Control options: These nuisance pests are considered beneficial unless they invade homes in search of cool, moist places to hide. Since they feed on decaying organic material, removal of this food will reduce populations.

Option 1: Cultural control—modify habitat: Mulching around flower beds and ornamentals often allows millipede populations to increase. Allow these areas to dry periodically and remove any grass clippings or leaf litter from next to buildings.

Option 2: Chemical control—use pesticides: Several insecticides will also kill this group of arthropods. Check insecticide labels for an appropriate product. General turf sprays for this group is not recommended. Treat around homes and buildings by applying a residual material next to doors and porches where millipedes may attempt to enter.

Earwigs

Species: Several species may be present, though the European earwig, *Forficula auricularia* Linnaeus, ringlegged earwig, *Euborellia annulipes* (Lucas), or the shore earwig, *Labidura riparia* (Dallas) are species commonly seen in turf (Phylum Arthropoda, Class Insecta, Order Dermaptera) (Figures 3.97 and 3.98).

Distribution: The European and ringlegged earwigs have been transported around the world and are found in the United States along the eastern and western coasts. The shore earwig is found in tropical and subtropical areas of the world and is restricted to the Gulf states across to Arizona and California.

Hosts: Earwigs are scavengers and predators and can be found in turfgrass areas where suitable food can be found.

Damage symptoms: Earwigs rarely damage turf and are usually considered beneficial except when they become a nuisance pest. Earwigs usually prey on other insects or feed on freshly killed

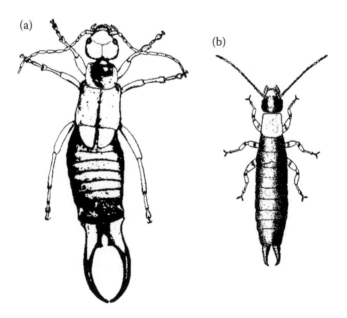

FIGURE 3.97
Diagrams of the European earwig (a) and ringlegged earwig (b). (UDSA and Ohio Agr. Exp. Sta. respectively.)

FIGURE 3.98
Mound of soil pushed out on a golf course green by the seashore earwig adult.

plant and animal materials. Their secretive behavior, hiding in cracks and crevices and inside flowers, causes some alarm when people discover them. Their forceps-like tails are used for defense and larger ones can lightly pinch the skin. These forceps are not poisonous, though many earwigs release a foul odor if handled. In southern states, large numbers of these pests may enter homes and hide in flower pots, in clothes dropped on the floor, and in the crack between door and window sills. In spite of their namesake, earwigs do not enter the ears of humans in an attempt to eat brains! However, earwigs have been known to enter ears of campers at night, apparently in search of a dark place to hide.

Description of stages: These insects have a gradual metamorphosis with egg, nymphal, and adult stages. The most characteristic feature of an earwig is its forceps-like cerci at the tip of the abdomen.

Eggs: The eggs are white, oval, and about $1/32 \times 3/64$ inch (0.8×1 mm). They are placed in clusters in a chamber in the ground.

Nymphs: The nymphs look like the adults but usually have thin narrow forceps, lighter color, and no wings.

Adults: The adults are brown to black and are often marked with yellow or orange. They may be 5/8–1.0 inch (16–25 mm) long and males generally have large curved forceps, while females usually have thinner ones. Some species have both winged and wingless adults in the same population.

Life cycle and habits: Female earwigs are protective of the eggs and young. Females of the European earwig usually lay 30–90 eggs in a brood chamber in the fall or early winter. They will clean the eggs to reduce molds and fungi, and attempt to protect eggs from predacious insects. The eggs usually hatch in 70–90 days and the young nymphs travel only short distances from their nest. By the time they are third instars, the nymphs are more venturesome and will forage considerable distances from their original nest. Often, they get lost but seem to be welcome in the nests of other earwigs. Development takes 2–3 months in cool weather. In hot dry summer months, the earwigs are less active and hide in cool, moist areas. Sometimes, European earwig females will lay another batch of eggs in the spring. The ringlegged earwigs and shore earwigs continuously breed during the warmer months. They may produce one to six clutches of eggs per year. The ringlegged earwig nymphs take about 90 days to mature, and the shore earwigs take 40–60 days. The shore earwigs are not as protective of the nymphs as other species and the females often eat the young after the brood chamber is opened.

Control options: Because earwigs are generally beneficial in their predatory activities, no controls are recommended for general turf treatments. However, when these insects invade homes, barrier and house invasion treatments may be necessary.

Option 1: Cultural control—reduce habitat suitability: Earwigs need other insects for food and require moisture. Reduce reliance on mulches and remove excess debris from around the home, that is, boards, leaf litter, rocks, and so on can greatly reduce earwig populations. Irrigate turf areas only and try not to wet the soil next to structures. Make sure rain downspouts drain well away from the house.

Option 2: Chemical control—barrier insecticide treatments: Several residual insecticides are useful in reducing earwig invasion into homes. Treatment of the turf is rarely necessary to reduce earwig invasion.

Bigeyed Bugs

Species: Several species of the genus, *Geocoris. Geocoris bullatus* (the large bigeyed bug), *Geocoris uliginosus*, and *Geocoris punctipes* are commonly found in turf (Phylum Arthropoda, Class Insecta, Order Hemiptera, Family Lygaeidae) (Figure 3.27).

Distribution: *G. bullatus* and *G. uliginosus* are widely distributed in the United States, from coast to coast and in southern Canada. *G. punctipes* is found south of a line from New Jersey west to southern Indiana and Colorado, also southwest through Arizona and California.

Hosts: These can be abundant predators of pests of turf as well as field crops. Tough primarily predators, bigeyed bugs often feed on plants but have not been shown to cause any damage.

Damage symptoms: Bigeyed bugs in turf are often associated with chinch bug populations, and because of similarities in size and shape are blamed for damage. Bigeyed bugs are considered important beneficial predators.

Description of stages: These bugs have a simple life cycle.

Eggs: Most species lay ovoid or subcylindrical eggs, slightly less than 3/64 inch (1 mm) long, and have a typical circle of 6–10 peg-like projections on the anterior end. The eggs range from white to yellowish tan, but the eggs of *G. bullatus* have been stated to be light pink.

Nymphs: Five nymphal instars are found in all species. The nymphs are elongate oval, like many plant bugs, and they can resemble chinch bug nymphs except for the obvious eye characters and color. Bigeyed bugs have laterally bulging eyes, which distinctly curve backward to overlap the front of the thorax. The head is about as wide as the body proper. *G. uliginosus* nymphs have blackish-brown head, thorax, and wing covers, while *G. bullatus* and *G. punctipes* nymphs are straw colored with irregular darker spots.

Adults: Bigeyed bug adults are about 1/8 inch (3–4 mm) long, oblong-oval bugs with heads much broader than long, and possessing conspicuous eyes that curve back around the front of the thorax. Adults run and fly rapidly if disturbed. *G. uliginosus* adults are almost entirely black with light areas along each side. Adults of *G. bullatus* and *G. punctipes* are mottled buff or straw colored with numerous small punctures.

Life cycle and habits: Most bigeyed bugs have several overlapping generations per year. In northern states, *G. bullatus* usually overwinters as eggs, and *G. uliginosus* and *G. punctipes* seem to overwinter as adults. In general, these bugs appear to require warmer temperatures to have rapid development. Using *G. uliginosus* as an example, adults become active in the spring when daytime temperatures exceed 75°F (23.9°C). Apparently, these active adults emerge from their overwintering hiding places such as cracks and crevices in the soil or in litter around plants. The adults feed on small arthropod prey such as mites, aphids, other bugs, moth larvae, and insect eggs. They also take moisture from plant tissues or surface dew. After finding suitable prey populations, females attach their eggs to plant stems and leaves. These eggs hatch in 11–23 days at 70–80°F (21.1–26.7°C). Hatching may not occur if the average temperature drops below 60°F (15.6°C). The nymphs also prey on small insects and mites and take about 30–40 days

to mature when growing at 75°F (23.9°C). However, considerably longer times are needed at lower temperatures. Adults live
40–80 days, depending on environmental variables and females
can lay 100–300 eggs during this time. Thus, two to four generations can occur per season, depending on the temperature. Since
these predators are not prone to exclusively inhabit turf, they do
not seem to control surface-feeding turfgrass pests. In fact, they
often appear to arrive in chinch bug-infested turf after damage
is apparent. It is at this time that the bigeyed bugs are confused
with chinch bugs because of their similarities in size and shape.
Some evidence indicates that the bigeyed bugs may eventually
decimate the chinch bug population but usually after considerable damage has occurred.

Control options: Bigeyed bugs are considered beneficial predators, though
they often pierce plant tissues with their sucking mouthparts to
obtain liquid. At present, no system has been developed to encourage an early buildup of bigeyed bugs to control greenbugs or
chinch bugs.

Leafhoppers

Species: Many different species can be found feeding on the different species of turfgrasses. Species may be large and green like
the sharpshooters, *Draeculacephala* spp., to small, yellow or
brown species like the gray lawn leafhopper, *Exitianus exitiosus*
(Uhler), and the painted leafhopper, *Endria inimica* (Say). Other
leafhoppers without common names are often in the genera
Agallia, *Forcipata*, *Latalus*, and *Polyamia* (Phylum Arthropoda,
Class Insecta, Order Hemiptera, Family Cicadellidae) (Figures
3.99 and 3.100).

Distribution: Leafhoppers can be found across the United States in turfgrass habitats.

Hosts: All species of turfgrasses have leafhopper species present.

Damage symptoms: Leafhoppers are not considered to be important pests
of turfgrasses, though thousands of individuals can be found in
a square yard of turf. The slight damage observed is yellowing
and spotting of leaves and stems where these insects pierce the tissues with their sucking mouthparts. Often, homeowners become
concerned with the large numbers that literally jump and fly in a
cloud in front of their feet.

Description of stages: These insects have a gradual metamorphosis with
egg, nymphal, and adult stages.

    Eggs: The elongate, white, slightly bean-shaped eggs are about
3/64 inch (1 mm) long. They are inserted into slits cut into stems or
placed into leaf sheaths.

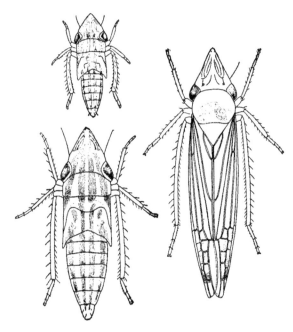

FIGURE 3.99
A sharpshooter leafhopper, *Draeculacephala mollipes*. Young nymph (top right), mature nymph (lower right), and adult. (USDA.)

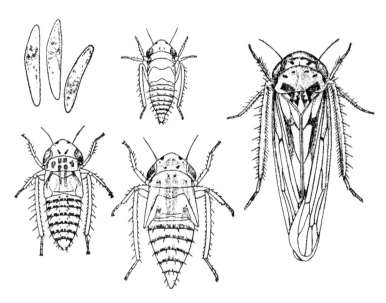

FIGURE 3.100
Gray lawn leafhopper, *Exitianus exitiosus*. Eggs (top left), three nymphal instars, and adult (right). (USDA.)

Nymphs: The wedge-shaped nymphs, with bristle-like antennae, have jumping hind legs, but often they will move sideways or run to the other side of a leaf to avoid observation. The nymphs are often lighter colored than the adults.

Adults: The adults are also wedge shaped and jump and fly rapidly if disturbed. An expert is needed to make species identification.

Life cycle and habits: Most leafhoppers overwinter in the egg or adult stages, depending on the species. Those that overwinter as adults become active when the turf comes out of winter dormancy. These adults feed and mate, and females may insert 75–200 eggs into grass stems over several weeks. The eggs hatch in a week or two and the young nymphs feed at the base of leaves on along the stems. Most nymphs go through five instars in a couple of weeks to a month. The new adults begin another generation and the total population explodes by midsummer. Depending on the leafhopper species and location, up to five generations per year may be found.

Control options: These surface insects are not usually controlled because little damage is caused by their feeding. However, large populations are considered a nuisance and some turf discoloration may occur in hot, dry periods. Little information exists on the exact biology of leafhoppers in turf and control techniques.

Option 1: Chemical control—general cover sprays: General cover sprays are needed only when the leafhoppers are too much of a nuisance. Often, repeat treatments will be needed as the adults can easily fly from one lawn to another.

Ground Beetles

Species: Many species with several shapes and forms can be found in turfgrasses. The seedcorn beetle, *Stenolophus lecontei* Chaudoir, occasionally digs into golf greens (Phylum Arthropoda, Class Insecta, Order Coleoptera, Family Carabidae) (Figures 3.26 and 3.101).

Distribution: Many species are found throughout the world.

Hosts: These are generally predaceous beetles, though some feed on plant materials.

Damage symptoms: No damage is done to turfgrasses, and these are considered beneficial. However, homeowners are often upset when these fast-running beetles enter a home. The seedcorn beetle occasionally burrows into golf greens, throwing up a tiny mound of soil. Hundreds of these can be found the morning following a flight, often in midspring or late summer. As far as can be determined, these beetles appear to be looking for germinating seeds and they will depart when they find no seeds.

FIGURE 3.101
Seedcorn beetle adult recovered from golf course green.

Description of stages: These elongate beetles have a complete life cycle with egg, larval, pupal, and adult stages. This is a very large group with over 2500 species described from North America.

    Eggs: The eggs are generally oval and cream colored. Each species has a different-sized egg.

    Larvae: The larvae may be of various sizes and color patterns but they are usually light brown with darker spots. They have an elongate fleshy abdomen, well-defined running legs, and projecting mouthparts. Most ground beetle larvae have a pair of pointed projections extending from the tip of the abdomen.

    Pupae: The pupae are formed in cells in the soil and are generally cream colored.

    Adults: The adults are usually brown or black, though some have iridescent colors or sheens. They may be 1/8–1.0 inch (3–25 mm) long and often have longitudinal ridges on the wing covers, projecting mandibles, thread-like antennae, and long running legs. An expert is needed to identify species.

Life cycle and habits: Most ground beetles have an annual life cycle with adults and/or larvae overwintering, depending on the species. The larvae are predaceous or scavengers, feeding on a wide variety of animal and plant food. However, one subfamily appears to be seed feeders and these are occasionally attracted to turf.

Control options: Ground beetles do not need to be controlled for turfgrass protection. However, they often enter houses, and spot treatments or crushing are satisfactory. In fact, ground beetles have been shown to be important predators of eggs and young of sod webworms, cutworms, and chinch bugs. Surface insecticides, like pyrethroids, have been applied to discourage the seedcorn beetles, but effective residues are short lived.

Rove Beetles

Species: Many species of various genera may be found in turfgrass habitats (Phylum Arthropoda, Class Insecta, Order Coleoptera, Family Staphylinidae) (Figure 3.102).

Distribution: Staphylinids are found worldwide.

Hosts: Rove beetles are general predators and/or scavengers.

Damage symptoms: These beetles do not damage turf and are considered beneficial because they attack other insects or help remove organic material. Occasionally, they may dig into the soil of golf course greens and tees, leaving small mounds of soil on the surface.

Description of stages: These beetles have a complete life cycle with egg, larval, pupal, and adult stages.

> *Eggs:* The eggs are generally oval, yellowish-white, and often have fine striations on the surface.

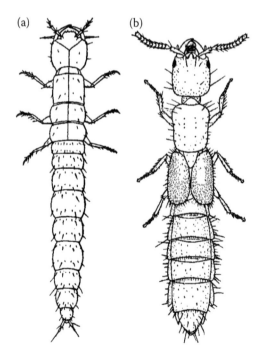

FIGURE 3.102

Diagram of a rove beetle, *Nodobius pugetanus*, larva (a) and adult (b). (Redrawn from Struble.)

Larvae: The larvae are slender with tapering light-colored bodies and darker head capsules. They may be 1/16–1⅜ inch (2–35 mm) long and often have two thin tail-like structure on the tip of the abdomen.

Pupae: The pupae are light colored and resemble the adults in form.

Adults: The adults are usually 1/8–3/4 inch (3–20 mm) long, slender and elongate, brown to black in color, and have bead-like antennae. The head projects forward and the forewings are reduced to short leathery pads that rarely reach halfway down the abdomen. They often look like earwigs, but lack the forceps on the abdomen.

Life cycle and habits: Little is known about rove beetle biology. Most are scavengers of molds, fungi, and decaying organic material. Some are predators and feed on the eggs, larvae, or pupae of sod webworms, grubs, and cutworms. Eggs are laid next to a food source and these hatch in 7–15 days. The larvae move and feed rapidly. Development may take 2–4 weeks. The larvae form a pupal cell, usually in the soil, and the pupa may take 6–14 days to mature. Rove beetles overwinter in the adult or larval stage.

Control options: No controls are needed for these beneficial insects. Adults may be attracted to lights in great numbers in the spring or summer. Some of these produce a foul odor if handled or crushed. Spot treatments or crushing is adequate for control. Other species that dig into greens and tees may need periodic sprays of surface insecticides to reduce their activity.

Fleas

Species: Usually the cat flea, *Ctenocephalides felis* (Bouche), is the one found in lawns. Occasionally, the dog flea, *Ctenocephalides canis* (Curtis), or the human flea, *Pulex irritans* Linnaeus, may be present (Phylum Arthropoda, Class Insecta, Order Siphonaptera) (Figure 3.103).

Distribution: These fleas have been distributed around the world by commerce and human activities.

Hosts: All three species may live on domestic animals and can bite humans. The larvae may feed on organic material in the turf but do feed on living plants.

Damage symptoms: No damage is done to turf but the adult fleas can bite humans and pets, causing considerable skin irritation.

Description of stages: Though highly modified for a parasitic existence, fleas have a complete life cycle with egg, larval, pupal, and adult stages.

Eggs: The eggs are shiny white, oval, and about 0.02 inch (0.5 mm) long.

FIGURE 3.103
Cat flea adult feeding on arm. Notice blood excrement droplet that serves as food for the free-living larvae.

Larvae: The larvae are slender and resemble primitive fly larvae. They are cream colored, and have long sparse hair and a distinct head capsule.

Pupae: The pupae are formed inside a dirty white, oval, silken cocoon that is about 5/32 inch (4 mm) long.

Adults: The adults are wingless, laterally compressed insects with brown to black bodies and long hind legs used for jumping. They are 3/64–1/8 inch (1–3 mm) long and are covered with short spines that hook onto animal fur, making them difficult to pull out.

Life cycle and habits: Domestic flea adults spend most of their time on mammal hosts found around the house. Mating and egg laying occur on the animal. The large eggs eventually drop off the animal. This may be inside a house, in a pet pen, or out in the lawn. The eggs cannot survive freezing temperatures. In warmer weather, the eggs hatch in 2–14 days and the rapidly moving larvae feed on organic debris, usually excrement from their host and blood pellets from adult fleas. Though the larvae prefer to be in the bedding of their host, they do well by feeding on general organic debris found in the lawns, carpeting, or upholstered furniture. The cat flea is best adapted for feeding in these nonpreferred areas. The larvae usually take 2 weeks to develop in moderate temperatures and with ample food supplies. In unfavorable conditions, the larvae may take a year to develop. Because the larvae often feed on the excrement of animals, they pick up the eggs of internal parasites, especially tapeworms. The parasites encyst inside the flea bodies and remain there until eaten by a cat or dog. After maturing, the larvae spin a tough cocoon in which they pupate. The pupa matures in 1–3 weeks during the summer months. In cool weather or when no animal activity is present, the pupae may enter dormancy for 6–12 months. The entire life cycle can be completed in as little as 2–3 weeks, but usually several months are required. This is why flea populations tend to become problems at the end of the summer after a couple of summer generations have been completed. The adult fleas must have a blood meal before eggs can be laid. Fleas do best in humidity of about 70% and temperatures between 65°F and 80°F (18.3–26.7°C).

Control options: Control of fleas takes a well-planned and executed program. In essence, the adult fleas have to be killed on their host and the larvae have to be killed in their breeding areas. This usually means that the pets must be taken to a veterinarian for thorough treatment and holding while the house (if the pet was allowed inside) and lawn are treated for adult and larval fleas. This routine may have to be repeated in a few weeks to get any fleas that hatched from eggs or emerged from dormant pupae. Once a lawn and home have been treated and declared free of fleas, a constant monitoring of fleas on the pets and quick control of any new adults is necessary to reduce the chances of a new population buildup.

Option 1: Biological control—parasitic nematodes: Field tests have shown that products containing the insect parasitic nematode, *Steinernema carpocapsae*, can control larval and pupal fleas in lawns.

Option 2: Chemical control—area insecticide applications: General contact insecticide applications can reduce flea populations, though retreatments are often necessary. Products that contain contact insects plus growth regulators for larval control are often more effective.

Option 3: Integrated control—treat pets and landscape: When flea populations are high, both pets and areas that pets frequent need to be treated. Topically applied or pill-administered systemic insecticides should be used to treat pets while contact insecticides plus growth regulators should be used in the landscape where pets frequent. Low flea infestations can usually be managed using just the systemic insecticides.

March Flies (Bibionids)

Species: Several species in the genera *Bibio* and *Dilophus* have larvae that inhabit soils around grass roots (Phylum Arthropoda, Class Insecta, Order Diptera, Family Bibionidae) (Figure 3.104).

FIGURE 3.104
March fly, *Bibio* spp., larvae, and pupae recovered from turf thatch.

Distribution: Bibionids are found around the world, but they are known to damage roots of crops and grasses in Europe, Asia, and North America. Bibionids most commonly attack roots of cool-season turf, but several species are known from southern states. In fact, the famous "love bugs" are bibionids common to the Gulf states that fly while mating.

Hosts: The larvae generally feed on decaying organic matter but are known to feed on the roots of grasses. Larvae may be found in Kentucky bluegrass, annual bluegrass, bentgrass, and fescue lawns.

Damage symptoms: Dead patches of turf, 2–10 inches (5–25 cm) in diameter, often appear in the spring after the snow cover has disappeared. Damage often resembles snow mold attack or early sod webworm damage. The dead turf easily lifts out to reveal the buff-colored, worm-like larvae.

Description of stages: This pest has a complete life cycle (egg, larval, pupal, and adult stages) with forms typical of more primitive flies.

    Eggs: The eggs are cylindrical with rounded ends, slightly curved, and 1/32 × 1/128 inch (0.5 × 0.2 mm). They are laid in clusters of several hundred and are cream colored when first deposited. The eggs often darken to a light reddish-brown before hatching.

    Larvae: The first instar is a little over 1/16 inch (1.5 mm) long and has a large, light brown head and white, tapered body covered with long bristles. As the larva grows and molts, the bristles are lost and a row of short fleshy projections develop on each segment.

The head becomes almost black and the body turns light yellow-brown when the larvae mature. Mature larvae may be 1/4–3/4 inch (6–20 mm) long, depending on the species.

Pupae: The pupa has wing pads, legs, and antennae easily observable. The abdomen is capable of twisting if the pupa is disturbed. Pupae are yellowishwhite and 5/16–19/32 inch (8–15 mm) long.

Adult: The adults are dark colored, being black or reddish with brown or yellow legs. They are rather clumsy when walking. The wings are folded flat over the abdomen and each wing has a dark spot midway down the front margin. Some species have black wings. They have short bead-like antennae, and the males have large eyes that practically cover the head. Females have much smaller eyes.

Life cycle and habits: Adult March flies usually emerge in late March to May, depending on the species, location, and weather. Considerable numbers of adults may be found in suitable habitats such as moist lands with high organic debris present. The adults are quite active on sunny days, flying short distances close to the ground. They are frequent visitors to early spring flowers and flowering fruit trees. The males with small flat abdomens readily copulate with the females that have abdomens distended with eggs. Shortly after mating, the females seek out moist, organic soil. The females then dig a burrow the size of their body by folding back their front tarsi and using the enlarged spurs on the tibia. Within a day, the females may dig 3/4–1 5/8 inch (2–4 cm) into the soil where a slightly larger chamber is made. Within this chamber, 100–400 eggs are laid. The female then backs up the tunnel or tunnels a short distance from the egg cluster to die. The eggs hatch in 20–40 days depending on soil temperatures. The newly hatched larvae have no legs but long spines and bristles aid in moving through the soil. Apparently, the larvae eat decaying plant material or plants recently damaged by other insects. After a few weeks, the young larvae molt and shed their long hairs. By this time they have moved to areas with considerable food, usually under leaf litter, piles of grass clippings, or manure. Development is rather slow and the larvae may remain dormant in the soil during periods of drought. The larvae resume active feeding and growth in the fall wet periods and they may be active under snow cover. By February and March, the larvae have molted several times and are fully grown. During the winter they may have concentrated under patches of snow mold or white grub-damaged turf. The mature larvae burrow 3/8–3/4 inch (1–2 cm) into the soil and pupate in an oblong cell. After a few weeks, the adults dig to the soil surface to begin the cycle again.

Control options: March flies are generally considered nuisance pests and rarely attack healthy turf. In fact, no pesticides are registered for control of March fly larvae.

Option 1: Cultural control—habitat modification: March flies require fairly moist turf with high organic content. By allowing the turf to dry periodically, many of the larvae can be killed. Applications of manure should be avoided when March flies are known to be active. Follow turf management practices that reduce thatch buildup.

Option 2: Cultural control—reduce turf damage: Many March flies seem to be opportunists. The larvae will often feed in areas damaged by other soil insects or diseases. Proper management of white grubs, billbugs, and diseases should reduce the chances of March fly damage.

Ants: General

Species: There are over 600 species of ants in North America and many of these may nest in turf areas. The noxious ants are harvester ants, *Pogonomyrmex* spp., that clear areas around their burrow openings in south and western states, and can bite; the turfgrass ant, *Lasius neoniger* Emery, that construct small volcano-shaped mounds in short-cut turf; and fire ants, *Solenopsis* spp., that make large soil mounds and sting aggressively (see below). Other nuisance ants are pavement ant, *Tetramorium caespitum* (Linnaeus); pyramid ant, *Conomyra insana* (Burkley); cornfield ant, *Lasius alienus* (Foerster); black field ant, *Formica subsericea*; and the crazy ant, *Paratrechia longicornis* (Latreille) (Phylum Arthropoda, Class Insecta, Order Hymenoptera, Family Formicidae) (Figures 3.105 through 3.107).

Distribution: Harvester ants are most common from Arkansas through Oklahoma and west. The turfgrass ant is generally distributed east of the Rocky Mountains, though there are other small mound-building ants in western states. The turfgrass ant is the major golf course and sports field pest in the eastern cool-season turf zone. The pavement ant is an introduced species that lives in the soil but commonly invades homes and buildings in the eastern half of North America. The black field ant makes low mounds in turf and surrounding flower beds. This species is found primarily in northeastern states. Crazy ants also make low mounds in warm-season turf zones.

Hosts: Most ants do not directly attack turf, though the harvester ants and field ants may sting and cut down plants near the burrow opening. While most of the above ants can inhabit almost any turf habitat, damage and activity is most common in short-cut turf of golf courses or skinned areas of sports turf facilities.

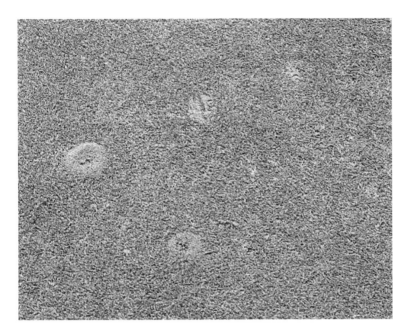

FIGURE 3.105
Mounds of the turfgrass ant, *Lasius neoniger*, on the surface of a golf course green.

FIGURE 3.106
Mound of the red imported fire ant.

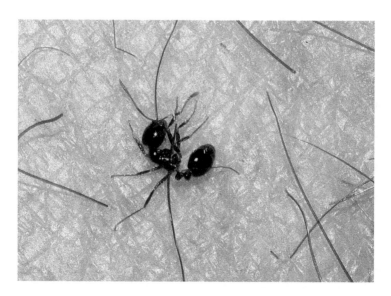

FIGURE 3.107
Red imported fire ant stinging the skin of a person.

Damage symptoms: Except as mentioned above, most ants do not attack turf directly. Mounds built on short-cut turf can kill out patches of the turf, interfere with ball roll, and dull mower blades. When ants release new reproductives (swarmers), the air can be filled with the flying ants. This can be very disruptive of sporting events and outings on lawns.

Description of stages: Ants have complete life cycles with forms associated with social insects—workers, queens, and drones. The workers are wingless and female. Queens and drones (kings) have wings, but mated queens tear off their wings upon initiating a new colony.

> *Eggs:* The translucent white eggs are usually elongate or bean shaped. They are normally kept in brood chambers where they are stacked until hatching.

> *Larvae:* The larvae are generally white or cream colored, slightly C-shaped, and lack functional legs. These are commonly sorted and kept by workers, according to size.

> *Pupae:* The pupae may spin tough, tan to yellow cocoons in which they transform into adults. Some species of ants do not form cocoons or only form cocoons for new queens and drones. The cocoons are often mistaken for eggs when an ant colony is dug up.

> *Adults:* Ants are social insects with wingless worker, wingless soldier, and winged reproductive casts. Some species may have various sizes of workers, each doing different tasks. The queens are often much larger than any other individual while the males may not be much larger than a normal worker. Most ants have a

single queen, but some species have multiple queens in the colony. Ant identification is usually done using a normal worker, and an expert should be consulted if species determination is necessary. Unfortunately, most ant colonies are most noticeable when the new reproductives are swarming. These winged ants are much more difficult to identify to species.

Life cycle and habits: All ants live in colonies of several hundred to thousands of individuals. They generally fall into three categories: plant or fungus feeders, sweet feeders, and/or fat or protein feeders. Harvester ants collect plant seeds and parts, which are chewed up and used for food. Most of the sweet feeders tend aphids, scales, and mealybugs that produce sugar-rich excrement called honeydew. The fat and meat feeders are usually general scavengers or predators that make use of other insects or animals living in the turf habitat. The turfgrass, cornfield, and yellow ants tend root aphids. The pavement and crazy ants are scavengers and predators. Most of these nuisance ants build small cone-shaped mounds of soil around the opening of their nests. Most ants produce new queens and drones in midsummer to late fall. Often, just before or after a summer rain, these new winged reproductives will emerge in mass. Hundreds of workers will spill out of the nest opening and appear to be milling about while the winged forms take flight as soon as they reach the surface. After mating, the potential queens seek out a suitable site to dig a tiny burrow. Mated turfgrass ants often throw up small mounds of soil on golf green and tees the night after a swarming. She will close herself in this burrow and remain dormant for the winter. In early spring, the female lays a few eggs and feeds the resulting larvae oral secretions derived from the breakdown of her flight muscles and stored fat. When the first workers emerge from their pupae, the brood chamber is opened to the soil surface and foraging for food begins. These first workers are often half the size of normal workers. If this female and her workers successfully establish a colony, she is then considered to be a successful queen. Initial colonies expand rapidly once a second brood of workers are reared. Most colonies take 2–3 years before they are large enough to produce new reproductives. New colonies being established in areas already occupied rarely succeed as the ants can be destroyed by the other ants.

Control options: The common nuisance ants can be very difficult to manage because they are constantly reestablishing during the swarming period or they avoid pesticide applications by living in the soil. If ants are a real nuisance, attempt to get the species identified so that you can work with the knowledge of its food and nesting habits.

Option 1: Cultural control—habitat modification: Many mound-building ants will not tolerate constant disturbing from mowing, especially lower cutting heights. On the other hand, many ants need to have access to direct sunlight to help heat the brood. In these cases, taller-cut, thick turf will discourage these ants. Harvester ants rely on seeds, so mowing and using plant growth regulators to suppress grass seed formation may discourage these ants. The aphid and mealybug tending ants must have access to these insects. Check ornamental trees and shrubs and flowers for aphid infestations tended by ants. Many of the commonly used surface insecticides eliminate the various insects that serve as ant foods.

Option 2: Chemical control—surface insecticide applications: Treatment of individual mounds of nuisance ants is usually not successful as the ants may detect the pesticide and simply close their nests until residues have dissipated. Larger-area applications are usually needed to kill the foraging workers, eliminate food sources, and reduce the vigor of the colony. Combination products using a neonicotinoids plus a pyrethroid, applied in early spring when ant colonies begin activity, have been successful in eliminating colonies.

Option 3: Chemical control—using ant baits: Baits designed for ant control generally contain a slow-acting pesticide or an IGR. These are designed to be picked up by worker ants, taken into the colony and fed to the larvae and queen. Baits are often designed to be attractive to sugar-feeding ants or to oil- or protein-feeding ants, so check the label for the ants controlled.

Fire Ants

Species: Four economically important species of fire ants are found in the United States. They are native fire ant, *Solenopsis geminata* (Fabricius); red imported fire ant, *Solenopsis invicta* Buren; black imported fire ant, *Solenopsis richteri* Forell; and the southern fire ant, *Solenopsis xyloni* McCook (Phylum Arthropoda, Class Insecta, Order Hymenoptera, Family Formicidae) (Figures 3.106 and 3.107).

Distribution: The red imported fire ant, the most important species, is a native of Brazil but was introduced into Alabama between 1933 and 1945. This pest has spread into an area roughly south of a line running from the Virginia–North Carolina border across to southern Oklahoma and down to Corpus Christi, Texas. The black imported fire ant is a native of Argentina and Uruguay that was imported into the Mobile, Alabama area as early as 1918. This ant can be found over most of the same range as the red imported fire ant, but colonies are much less common. The native fire ant was originally located across the southern states but has been displaced by the imported species. Colonies are still scattered from South Carolina across Texas. The southern fire ant is a native to Arkansas, Oklahoma, and Texas.

Hosts: Mounds may be built in any turf situation as well as other open areas.

Damage symptoms: Fire ants do not attack turf but cause problems when they build earthen mounds for warming their eggs, larvae, and pupae. They also have a sting that causes a burning and itching sensation at its minimum effect and serious welts or blisters, even allergic shock at its worst. Unsuspecting people and pets playing on lawns have been severely injured when they accidentally stand or sit on the ant mounds. These ants have a nasty habit of pouring out of the nests, climbing onto the invader without stinging, then suddenly, all the ants bite and sting at the same time.

Description of stages: Fire ants are social insects with a complete metamorphosis. Though they have egg, larval, pupal, and adult stages, the adults belong to several castes—wingless workers of several sizes, winged queens, and winged males (drones or kings). The following descriptions are based on the red imported fire ant:

Eggs: The translucent white eggs are laid in clusters that are picked up and stored by workers in the upper part of the mounds. They are elongate bean shaped and about 1/32 inch (0.75 mm) long.

Larvae: The larvae are grub-like with no legs and a small head capsule. The larvae are usually white in color. Worker larvae get to be up to 1/4 inch (6 mm) long, while the reproductives may be twice as long.

Pupae: Fire ant pupae do not form a silk cocoon like some other ant species. The pupa looks like the adult but has the antennae, legs, and wing pads (if it is a reproductive) held close to the body. The pupa is first white but becomes darker before the adult emerges.

Adults: The workers are 1/16–1/4 inch (1.6–6 mm) long, and have the typical elbowed antennae of ants and a thin waist. The constriction at the waist, the petiole, has two nodes; the antennae are 10-segmented, with the last two segments forming a distinct club. The reproductives are winged and the queens may be twice the size of the largest workers. Depending on the species and locality, fire ants may be yellow to reddish-brown to black in color. An expert should be consulted if species determination is necessary.

Life cycle and habits: Established colonies produce new queens and winged males during the warmer spring and summer months. These winged reproductives swarm periodically, usually five to nine times, often after a rain. Mated potential queens attempt to establish a new colony by digging a small hole in the soil and closing up the entrance. Inside this chamber, the queen begins to lay 15–20 eggs in 2–3 days. Eggs are added to the pile over the next week, by which time the first eggs have hatched. The first larvae feed on several of the surrounding eggs. At this time the queen picks up the young larvae and sorts them into another pile. The larvae are then fed a liquid regurgitated by the queen. Apparently, the queen uses body stores and degenerated wing muscles to provide larval food. A queen may lose over half of her body weight rearing the first brood. After 20–25 days, the larvae pupate and the resulting tiny workers emerge 4–7 days later. These diminutive first workers are about 1/5 the size of the smallest workers found in an older colony. These workers, called minimis, break open the nesting chamber and begin foraging for insect food and start to enlarge the nest. The queen is now fed and she begins to lay more eggs that are properly cared for by the workers. If adequate food and water is in the areas, the colony grows steadily over the next few months. If a colony is established in June, it may contain 6000–7000 individuals by the following December. As the soil temperatures drop, the colony growth slows. By the following June, a 1-year old colony may have 10,000–15,000 workers and is producing new winged forms. Colonies 2–3 years old may have 20,000–200,000 workers. Established mounds will have a central pile of granular soil with openings and often smaller mounds around the parameter. The mound is established as a solar collector to store and incubate the larvae and pupae. Fully mature colonies may have mounds 40–60 cm in diameter and 25–30 cm high. In clay soils, the mounds may be very hard and have been known to damage plows or mowing equipment. The mounds are generally not as tall in sandy soils. Considerable evidence exists that colonies may move their mounds and nests in search of food or when pesticides are applied. This is why mounds seem to appear overnight. The foraging workers rarely harm people, but when the mound is disturbed, the ants emerge in mass to swarm the intruder. New-born livestock have been killed by attacks of disturbed ants.

Control options: Since fire ants are important agricultural as well as urban pests, considerable effort has been expended in developing control techniques. Unfortunately, these pests have extremely great powers of reproduction and the ability to move considerable distances.

Option 1: Biological control—introduction of parasites and diseases: Various governmental agencies have evaluated and released several parasites and diseases for control of fire ants, but none has been effective over wide areas. Some of the most interesting parasites are phorid flies. The adult flies insert eggs into worker ants and the fly maggots eat the ant's internal organs before pupating in the head capsule. Pupation causes the head to drop off! Itch mites and insect parasitic nematodes have also been used, but without lasting effects. At present, fire ant scientists recommend conserving native ants that can often compete with fire ant colonies, thereby limiting their colony size.

Option 2: Cultural control—boiling water treatments: Pouring three gallons of near-boiling water into a fire ant mound successfully eliminates the colonies about 60% of the time. Obviously, this is a labor-intensive task, but works when few colonies are present. The hot water can also kill the turf, so this technique should be used in bare areas.

Option 3: Chemical control—spot treatments of mounds: Individual mounds can be treated with contact and stomach pesticides, either by drenching or injection. Workers pick up the pesticide which is passed through the colony. This is an effective method for a small number of mounds, especially in turfgrasses. However, often, the ants detect the poison before the queen is killed and they may simply move the colony to another location.

Option 4: Chemical control—use poison baits: Several ant baits have been developed that are attractive to the foraging ants. By picking up the food bait, the workers distribute the poison throughout the colony. Take care to use the bait in such a manner that birds and other animals do not feed on it. Where there are many fire ant mounds in turf, area-wide broadcast of the bait is appropriate. When few mounds are involved, just treat the turf around each mound.

Option 5: Chemical control—area treatment with insecticides: A few pesticides have been identified that are useful in controlling widespread fire ant infestations. Sprays of pyrethroids or neonicotinoids insecticides can suppress fire ant activity, but repeated applications are needed to eliminate colonies. However, products containing fipronil have been able to eliminate colonies in a few weeks after a single application.

Option 6: Integrated approach—monitoring and treating: In larger turfgrass areas, like golf course, athletic fields, parks and entire neighborhoods, an approach of area-wide treatments followed by spot treatments can be followed. The key to success of such programs is monitoring and assessing fire ant mound populations. When fire ant mounds are numerous, like 10 or more in or around a football field, an area-wide application of a bait, spray, or granule is made. After 4–6 weeks, the area is reassessed. If a few mounds persist, these should be spot treated (baits, injections, drenches, etc.). The ant populations are then monitored every few months and spot treatments are generally used. If a new outbreak occurs, the area-wide treatment is again undertaken.

Cicada Killer

Species: *Sphecius speciosus* (Drury) (Phylum Arthropoda, Class Insecta, Order Hymenoptera, Family Sphecidae) (Figure 3.108).

FIGURE 3.108
Cicada killer female returning to burrow with paralyzed cicada.

Distribution: Found in all states east of the Rocky Mountains. Most common in southern states where the annual cicadas are prevalent.

Hosts: Does not attack turf. Selects light-textured or sandy soils that are well drained. Prefers to dig burrow in sparsely vegetated areas such as golf course sand traps.

Damage symptoms: Mounds of soil piled up around a hole, approximately 1/2 inch (12 mm) in diameter. Males are very protective of territory and often buzz or dive bomb people who enter the territory being protected. Females can sting but will not do so unless severely provoked.

Description of stages: This is a wasp with a typical annual complete life cycle.

　　Eggs: The greenish-white eggs are 1/8 × 3/64 inch (3–6 × 1–1.5 mm) and are attached to a cicada buried underground.

　　Larvae: The larva is grub-like, but has no legs or well-defined head capsule. Mature larvae are 1 1/8–1 3/8 inch (28–32 mm) long.

　　Pupae: Mature larvae spin a tough spindle-shaped case in which the pupae reside. The pupae are at first white, but take on the adult colors before emergence.

　　Adults: The large adults are impressive with rusty red heads and thoracic areas. The wings are infused with orange and the abdomen is banded with black and yellow. The adults may be

1³⁄₁₆–1¾ inch (30–45 mm) long, and have a wing span of 2⅜–4
inches (60–100 mm).

Life cycle and habits: This insect overwinters as a prepupa in a cocoon
located 7–20 inches (17–50 cm) under the surface. When spring
temperatures warm the soil, the pupal stage is formed and the
adults dig to the surface in June and July. Males tend to emerge
before the females and they establish flight territories, usually
where females are to emerge and build new nests. The males
fight off any other infringing males and are known to attack any
other moving object in the area. Fortunately these males have no
sting, but unsuspecting intruders are often shocked by the loud
buzzing and strafing activities. Males have been known to dive
bomb into people's heads and backs. The females are quite docile
and after mating set out to construct a new tunnel. The tunnels
are dug straight into the soil or are angled slightly. At the end of
each tunnel are secondary tunnels that end in a chamber. The
females search tree trunks and limbs for cicadas. When a cicada
is located, the female swoops down, grabs the screaming prey,
and drops to the ground where a paralyzing sting is applied.
The paralyzed cicada is then grasped under the wasp's body and
flown back to the burrow. Each chamber is provided with one to
three cicadas. An egg is laid on a cicada in the chamber and the
chamber is sealed. Additional chambers are provided for, until
the cicada population no longer supplies victims. The wasp egg
hatches in 2–3 days and the voracious larva quickly devours the
food. Only the cicada shells remain. By the fall, the mature larvae
spin a cocoon, shrink in size, and prepare to overwinter. A single
generation occurs per year.

Control options: Though this is considered a beneficial insect, people often
get upset with the activities of the territorial males, and the mounds
thrown up by the females can be unsightly. Therefore, no controls are recommended unless this pest greatly interferes with
human activities.

Option 1: Cultural control—habitat modification: Since this insect prefers
open sandy soil to build its burrow, good turf culture that promotes the growth of thick turf will often discourage burrow
building.

Option 2: Mechanical control—male destruction: Since the males are the offend-
ing aggressors, capture them with an insect net or swat them down
and crush. This usually ends complaints by golfers and allows the
females to capture the cicadas.

Option 3: Chemical control—insecticide application: Dusting of the burrow openings with insecticide dusts is very effective in killing the females
as they go about their daily activities.

Nuisance Vertebrate Pests

The vertebrates belong to the phylum Chordata, which have bilateral symmetry, like insects, but the nerve cord is dorsal and the circulatory system is closed. Next to the nerve cord is a stiff rod of cartilage, the notocord. The notocord in vertebrates is replaced by a series of bony segments, the vertebrae or backbones. All chordates with these vertebrae are in the subphylum Vertebrata. The vertebrates are again subdivided into the fish, superclass Pisces, and the four-legged animals, superclass Tetrapoda. The tetrapods are divided into four classes: the Amphibia (frogs, toads, and salamanders), the Reptilia (lizards, snakes, and turtles), the Aves (birds), and the Mammalia (mammals). Of these groups, some birds and mammals are occasional pests of turf, usually during their search for insect or weed food.

The class Aves is a large and diversified group. About 1700 species in over 90 families are found in North America. Fortunately, the vast majority of these birds inhabit turfgrass areas only on rare occasions. Some of the pest birds, usually imported species or blackbirds, damage turf only in their search for insect food. All other birds are considered beneficial and are protected by international, federal, and state laws. The most common damaging birds are the starling (family Sturnidae) and the blackbirds (family Icteridae). The blackbirds that may damage turf are grackles, cowbirds, and red-winged blackbird. Crows, robins, and various water birds may also be nuisances on turf. The plant-feeding water fowl, especially ducks and geese, cause problems when their flocks forage on the high-quality turf of golf courses, commercial facilities, and home lawns. Their copious feces, often laden with weed seeds can become a real nuisance.

The class Mammalia is another diverse group with more than 350 species found in North America. Mammals have fur on the body, warm blood, and mammary glands for feeding the young. Orders of mammals that commonly damage turf are the Marsupialia (opossum), Insectivora (shrews and moles), Carnivora (raccoons, skunks), Rodentia (pocket gophers, voles, and ground squirrels), and Xenarthra (armadillos). The opossums are true marsupials, with a pouch for carrying the premature young for development. The insectivores are small mammals with pointed noses and tiny bead-like eyes. The carnivores are medium to large animals with prominent canine teeth. Carnivores are general meat eaters but they may feed on fruits, nuts, and berries. The rodents are small to medium animals with two pairs of prominent front teeth that continually grow and must be worn down through chewing. The armadillo is an unusual mammal with degenerated teeth and a protective armor of hairy plates.

Common Grackle

Species: *Quiscalus quiscula* (Linnaeus) is sometimes confused with the boattailed grackle, *Quiscalus mexicanus* (Gmelin), and the great-tailed grackle, *Quiscalus major* (Phylum Chordata, Class Aves, Family Icteridae).

Distribution: The common grackle can be found in the summer east of the Rocky Mountains from southern Canada to the Gulf states. They usually migrate south for the winter and are most common in the Mississippi valley states and coastal states.

Damage symptoms: These species commonly congregate with other blackbird species during the fall and winter. Large flocks are a nuisance and health hazard in urban areas. These birds rarely attack turf but will actively forage for various insects. Flocks may scrape away grub-damaged turf in late summer and fall. Grackles are generally considered beneficial when small groups feed on insects.

Description of stages: An iridescent blackbird that is about 12 inches (30 cm) long, the common grackle has a shorter tail than its two relatives. Grackles generally have bright yellow eyes and long tails. These noisy birds make a series of clucks and rising screeches.

Life cycle and habits: These common birds are numerous in the urban habitat where they prefer to forage in lawns, parks, and open fields. They are often gregarious and prefer to breed in tall conifers. However, the large stick nest lined with fine grass may also be found in lower shrubs. Usually, five pale blue eggs with irregular black markings are laid in the nest. Usually, a single generation per year is reared and the birds form large migratory flocks with other blackbird species in the fall. This bird has an extremely diverse diet and will eat almost any animal that is small enough to swallow. They also occasionally raid other birds' nests. They will feed on grains, wild fruits, and cultivated small fruits. Urban grackles readily take food at parks and can be a nuisance around trash baskets.

Control options: See Starling. This species is generally not a pest until the gregarious migratory flocks are formed in late summer and fall.

Starling

Species: *Sturnus vulgaris* Linnaeus (Phylum Chordata, Class Aves, Family Sturnidae).

Distribution: The starling is an introduced pest from Eurasia. Since 1890, it has spread across North America from Quebec to southern Alaska down to northern Mexico.

Damage symptoms: This anthropophilic (prefers to live around humans) species is well adapted to urban life and often roosts in large numbers on buildings and in trees. Large migratory flocks are a noise nuisance and a health hazard from their feces. Flocks in the fall may attack grain crops and can damage turf that is infested with grubs. The birds insert their beak into the thatch, spread it, and make a hole. This is the common searching activity used when looking for caterpillars or other surface- and thatch-dwelling insects. Grubs are located by scraping away the turf, like chickens scratching. This species is considered beneficial when they nondestructively feed on insects.

Description of stages: The adults are slightly smaller than a robin, about 7.5–8.5 inches (19–22 cm) long. They have a short tail, generally black iridescent plumage, and are often flecked with white spots. They have a long pointed bill that is yellow-white in the summer and brownish in the winter. They have a large repertoire of calls consisting of clicks, squeaks, and whistles. They sometimes mimic the calls of other birds.

Life cycle and habits: Starlings prefer urban and suburban habitats. They build a rough nest consisting of twigs, trash, and grass. The inner cavity is lined with fine grasses and feathers. The nests are usually built in spaces on buildings, especially small confined cavities, and in larger trees. Nesting occurs at the first hint of spring and sometimes a second brood is reared in the summer. Large roosts may be located on buildings after the nesting season and thousands of birds may be present. Sometimes, starling roosts are mixed with other fall gregarious black birds. These roosts, once established, are very difficult to move or drive away. Starlings are generally more aggressive than most native birds and outcompete them for food and nesting sites.

Control options: See Birds: Introduction. These highly urban-adapted birds are best controlled by reducing nesting sites and food, or by moving roosts.

Option 1: Cultural control—restrict roosting and nesting sites: Buildings with fancy facades with many cavities should be covered with fine mesh or sticky repellents to discourage roosting.

Option 2: Cultural control—move roosts: This is a very difficult task, especially if a roost has been established for several nights. Generally, specialists armed with loud speakers to play tapes of bird distress calls and exploding cartridges are needed to relocate a roost.

Option 3: Chemical control—spray wetting agents: This chemical technique relies on a soap or other wetting agent and cold weather. Roosts are sprayed by plane or helicopter with a wetting agent just before a cold rainstorm in the fall. The birds become wet from the rain and the cold causes an outbreak of bird pneumonia. This technique is used by cities with major bird roosts that are creating health hazards.

Moles

Species: The eastern mole, *Seaopus aquaticus* (Linnaeus), and star-nosed mole, *Condylura cristata* (Linnaeus), are common turf pests in the eastern half of the United States. Several species of Seaopus occur in turf in the West Coast states (Phylum Chordata, Class Mammalia, Family Talpidae) (Figure 3.109).

Distribution: The eastern mole is common from Massachusetts west to Lower Michigan, into Nebraska and south to all the Gulf states. The star-nosed mole is usually found in damp areas in the northeast from Virginia to Minnesota and north into Canada.

Damage symptoms: Moles tunnel under the soil surface in search of insects and earthworms. This tunneling throws up ridges in turf. Deeper tunnels are often marked with cone-shaped earthen mounds. Turf often dies along the surface tunnels, especially in dry weather. Moles may become especially active in grub-infested turf or areas where other soil invertebrates are numerous.

Description of stages: Moles are plump, elongated, smooth-furred mammals with elongate enlarged skulls that end in a pointed snout. There are no external ears, the ear holes being hidden by thick fur, and the tiny eyes apparently can only detect light and dark. The front feet are greatly enlarged, clawed, and face outward. The eastern mole has a slightly pink, pointed nose with a short tail, while the star-nosed mole has a pink fleshy nose armed with 22 fleshy projections and a long hairy tail. Adult moles are 3.5–8.75 inches (9–22 cm) long.

FIGURE 3.109
All moles, like this common eastern mole, are well equipped to dig through the soil with their spade-like front legs.

Life cycle and habits: Moles are true insectivores like their smaller cousins, the shrews. The eastern mole prefers to forage in well-drained, loose soils of fields, lawns, and gardens but it may nest and set up base burrows along the edge of wooded areas. The star-nosed mole is semiaquatic and is commonly found in wet woodlands, fields, or swamps. It also digs in turf located next to streams, rivers, and ponds. Both species use earthworms as their major food item but they quickly forage for white grubs when available. Moles make longitudinal tunnels at the soil surface, which are their major thoroughfares. Deeper tunnels are constructed for hiding and resting. These deeper tunnels are usually marked by soil thrown out of a hole, the cone-shaped "molehills." Most moles are active at dusk and dawn but tunneling may be observed on overcast days. A mole can travel up to a foot per minute in loose soils. Moles are generally solitary, though considerable numbers can be found in an area. Males seek out females in the early spring and the females produce a single litter of two to six young. The young are kept in an underground nest until old enough to begin to forage on their own, usually within a month. The starnosed mole often builds its nest in a raised area that does not get swamped. Moles are considered beneficial because of their insect-feeding behavior and ability to aerate the soil. However, they are a disaster in well-maintained turf.

Control options: Moles are generally not a problem in soils with few earthworms and no insect larvae. Since moles do not feed on plant material like pocket gophers, elimination of invertebrate food will effectively control mole activity. However, this type of management, where all the invertebrates are eliminated, is not recommended for ecological reasons. Flooding the tunnels through a water hose, using automobile exhaust, and placing chewing gum in the tunnels are not appropriate or effective management options.

Option 1: Chemical control—fumigation: Fumigation with toxic gases is moderately effective if currently active mole tunnels can be located. If considerable activity is noticed in an area, roll down the tunnels with a lawn roller and watch for fresh activity. In home lawns, moles do not respect property lines and moles may have to be controlled in several adjacent lawns. Fumigation often fails if the mole detects the gas before toxic levels are reached. The moles simply plug the tunnel with soil when the gas is detected.

Option 2: Cultural control—trapping: Several types of traps (e.g., harpoon, strangle, or clasping) can be effective to eliminate individual moles. Again, currently active tunnels must be located. Select one of the major long tunnels, not the short branching side tunnels, for trap placement. Harpoon traps work by pressing down a tunnel under the trap and setting the trap trigger over the spot. When the mole reopens the tunnel, it must press against the harpoon trigger. Strangle and clasping traps usually require that the burrow be opened carefully and the trap placed in the runway. Care must be taken to not overly disturb the tunnel or to leave behind human scents. Since moles patrol currently active tunnels several times a day, if a trap has not captured the pest within 2 days, relocate the trap.

Option 3: Chemical control—poison baits: Baits marketed for gophers are not effective against moles. Moles usually will not eat the bran- or peanut-based baits. As with trap placement, inserting bait into mole tunnels takes care. Locate the straightest, long tunnels for bait insertion and use gloved hands to keep human odors from contaminating the bait.

Pocket Gophers

Species: Several species of the genus *Thomomys* are found from the Rocky Mountains to the West Coast. East of the Rocky Mountains, most pocket gophers are species of *Geomys*. The plains pocket gopher, *Geomys bursarius*, and the southeastern pocket gopher, *Geomys pinetis*, are common turf pests.

Distribution: The plains pocket gopher is common in the prairie areas from Wisconsin to eastern North Dakota south to eastern New Mexico and eastern Texas. The southeastern pocket gopher is common in southern South Carolina, Georgia, and Alabama into the north half of Florida.

Damage symptoms: Pocket gophers tunnel near the soil surface in search of plant roots and tubers. This tunneling often throws up ridges in turf areas. Periodically, pocket gophers dig deeper tunnels for food storage and shelter. These deeper tunnels are often accompanied with a mound of soil deposited at the soil surface. Since gophers are especially fond of tubers, such as spring flowering bulbs, they may be very active around flower beds or in turf containing "naturalized" bulb plantings. Pocket gophers occasionally eat insects encountered in the soil but they do not seek out insects like moles.

Description of stages: Pocket gophers are stout-bodied rodents with the front incisor teeth always exposed. This is different than the teeth found in moles. Gophers have enlarged front feet armed with long claws used for digging. They also have short fur, small eyes and ears, and a short tail. Their common name comes from the presence of large fur-lined cheek pouches that open outside the mouth. These pouches extend back to the shoulders. These pouches can be emptied by squeezing with the front legs. *Thomomys* spp. have smooth front teeth with no grooves, while *Geomys* spp. have two conspicuous grooves running down the middle of the teeth. Most gophers are 7.75–14 inches (20–36 cm) long.

Life cycle and habits: Pocket gophers are solitary animals and they usually fight over disputed territory. Males are especially fierce and may kill other male intruders. In the spring, males search for females but separate after mating. Females give birth to 2–11 young in a deep chamber. Within several weeks, the young may follow their mother in the tunnels and after 2 months, the young begin to dig their own tunnels and disperse. At 3 months, the young are sexually mature. A second litter may be produced by some individuals. A gopher working in a suitable area will dig numerous surface tunnels, collecting roots and tubers and throwing up mounds of excess soil. Periodic deeper tunnels are dug and extra food, tucked into the cheek pockets, are deposited here. Activity is reduced in the winter months and stored food is eaten when fresh roots cannot be obtained.

Control options: Pocket gophers, like moles, are extremely difficult to control because of their subterranean habits. Gophers are even more troublesome because they feed on plants rather than insects and worms. Special care should be taken to locate currently active burrows. These can be located by looking for fresh soil mounds. Attempts to drown the animals, use automobile exhaust fumes, or chewing gum are not effective for managing this group of pests.

Option 1: Cultural control—use killing traps: Several traps are available that spear or choke active gophers. Unfortunately, gophers often shy away from disturbed tunnels. They occasionally shove soil ahead of them in a tunnel and this soil is "speared" by the traps. If trapping fails the first time, repeat setting the traps in new locations.

Option 2: Chemical control—use fumigants: Several gas-producing canisters are available for gopher fumigation. These occasionally work if the gopher is trapped between the gas and a new section of the tunnel. If the gopher is in the old tunnels, it can often outrun the gas, or simply plug the tunnel with soil. Fumigants are also rarely effective in loose sandy soils.

Option 3: Chemical control—use poison baits: Gophers are rather fond of seeds and grains. Carefully drop poisoned grain into tunnels that have been cautiously opened using a sharp pointed stick. Better success is obtained if these baits are placed in more recent tunnels. Do not handle the baits with bare hands, as the human scent will often repel the animals.

Skunks and Civet Cats

Species: The striped skunk, *Mephitis mephitis* (Schreber), and spotted skunk or civet cat, *Spilogale putorius* (Linnaeus), are the two North

American pests found foraging in turf (Phylum Chordata, Class Mammalia, Family Mustelidae).

Description: The striped skunks have a head and body length of 13–18 inches (33–45 cm) with a 7–10 inch (17–25 cm) tail. They may be 6–14 lb (2.7–6.5 kg). The spotted skunks are smaller with a head and body length of 9–13 inches (23–33 cm) and tail 5–9 inches (13–23 cm). They weigh 1–2.5 lb (0.45–1.1 kg). The striped skunks are usually solid, shiny black with a broad, white stripe running from the head to midbody where it divides into two lines. The spotted skunk has several irregular broken white stripes along the side and top of the body. The spotted skunk usually has a white tip to the tail.

Description of stages: Both these mammals prefer open country or sparse woods and brushland. They commonly inhabit suburban areas where food and cover abound. The young are born in May–June and five to six are usually in the litter. The young are kept in the burrow for several weeks before they emerge with their mother to forage. Striped skunk young often follow their mother in a single file. By July and August, the young have learned to feed and care for themselves. Burrows are built in the ground, under old buildings or in rock and wood piles. These omnivores feed on fruits, vegetables, and nuts as well as small rodents, birds, and insects. Most activity begins shortly after sundown and ceases at sunrise. Apparently, skunks and civets are active most of the year but several females may den together in the winter. Because they may feed on rodents and insects, skunks are considered beneficial. However, they commonly catch rabies, which quickly spreads to most of the individuals in an area. In the fall, when white grubs reach their third instar stage, skunks often forage as single individuals or in groups. Once a suitable location has been found, the skunks will return night after night until the free meal is depleted.

Control options: Skunks and civets are most easily disposed of by shooting at night. They are difficult to handle when trapped because of the scent glands. Unfortunately, many states have fur-bearing animal laws that protect skunks and civets. Other municipalities do not allow for the discharging of firearms within city limits.

Option 1: Cultural control—reduce den habitat and food: Skunks and civets commonly nest in disorganized wood piles, hollow trees, or logs, under crawl spaces of porches or houses, or in partially closed culvert drains. Survey the area and attempt to eliminate these nesting sites. Keep wood piles neatly stacked with space between the rows, and cut down or remove hollow trees and logs. Thoroughly close in access to crawl spaces and keep drain culverts open.

Option 2: Cultural control—manage food sources: Skunks and civets usually become pests where garbage cans are not covered, dog or cat food is left outside and uncovered, or white grub populations are allowed to reach large numbers. Where trash is kept outside, make sure that lids or doors seal tightly and cannot be pried open. Remove open dog or cat food containers at dusk. Monitor for grub populations and control them before the third instar stage is reached.

Option 3: Cultural control—trapping: Various live traps are made that can capture skunks and civets. Unfortunately, captured animals usually release their scent and handling can be difficult. If a live trap is used, release the animals in a location recommended by the local animal control agency. Many municipal agencies will remove trapped animals, either for free or for a small fee. Remember that skunks and civets often carry rabies and special care should be used to avoid scratches or bites.

Option 4: Cultural control—discourage digging: Where white grub populations are in the three or more grubs per square foot range, skunks and raccoons will forage for a meal. In these situations, the grubs need to be controlled with rescue treatment insecticides. These insecticides often take 3–5 days to cause the grubs to be unpalatable. During this time, the human waste fertilizer, Milorganite™, can be spread over the treated area. The odor of this product appears to discourage animal digging for 5–7 days.

Raccoon

Species: Procyon lotor (Linnaeus) (Phylum Chordata, Class Mammalia, Family Pracyonidae).

Distribution: Common in North America south of the Canadian border. Not common in the higher areas of the Rocky Mountains.

Damage symptoms: Raccoons usually dig up irregular patches of turf that is infested with grubs. This damage occurs at night and several raccoons may be active in an area. The raccoons often return night after night if grubs are easy to find. These pests are common carriers of rabies.

Description of stages: This well-known "masked bandit" usually has gray and black body hair, a length of 18–28 inches (46–71 cm), and a ringed tail, 8–12 inches (20–30 cm) long. They may weigh 12–35 lb (5.4–15.9 kg). The head has a black-colored mask around the eyes.

Life cycle and habits: Raccoons do not hibernate but live in dens located in hollow logs, rock crevices, or ground burrows. Apparently, urban raccoons will establish residence in storm sewers and under porches or house crawl spaces. Females give birth to two to seven young, usually four, in April or May. The young are nursed in the den and open their eyes in about 3 weeks. After 2 months, the young may leave the den to forage with their mother. The mother teaches the young how to catch insects, frogs, crayfish, and other small animals. They will also feed on fruits and nuts. They often attack small grain crops such as corn. By the fall, the young are weaned and they have to locate their own territory. Raccoons are nocturnal but occasionally forage during the day.

Control strategies: See Skunks and Civet Cats. Raccoons are considered game and fur-bearing animals in most states. Be careful in handling live animals, as they may carry rabies.

Ninebanded Armadillo

Species: *Dasypus novemcinctus* Bailey (Phylum Chordata, Class Mammalia, Family Dasypodidae).

Distribution: The armadillo is currently found from mid-Georgia across to southwestern Missouri and south across Oklahoma, Texas, and Florida. This interesting animal used to be rare in Oklahoma and Arkansas, but it has been steadily expanding its range northward.

Damage symptoms: The armadillo eats any animal food that it can catch during its foraging in the soil. They prefer worms, insects, spiders, millipedes, and centipedes. However, they have been known to eat crayfish, amphibians, and bird eggs. They dig and root around in soils much like raccoons or skunks. The damage is often mistaken for damage by other animals.

Description of stages: The armadillo is the only mammal found in the United States with heavy, leathery, ring-like armored plates over the body. Adults are 24–31 inches (61–79 cm) long with a 9.5–14.5 inch (24–37 cm) long tail. The head is covered with small scaly plates and has a pointed snout. The shoulders and pelvis are covered with wide scaly plates and these are joined together by nine narrow, jointed plates. This lets the armadillo roll into a ball if sufficiently disturbed. The tail consists of a series of ringed plates.

Life cycle and habits: Armadillos are truly unique and unusual animals. They generally hide in the daytime in burrows dug into the ground, often along the banks of streams or rivers. After dark, armadillos emerge to forage for food by rooting through plants and digging in the soil. They will eat almost any small arthropod or animal that can be picked up. They often travel in bands of several individuals that grunt and snort constantly. Mating occurs in the fall and each female gives birth to exactly four identical quadruplets. Apparently, the fertilized egg divides twice before being implanted in the uterine wall. The young are born in a breeding burrow lined with leaves and grass. The young are able to walk within hours after birth, and they have their eyes open. This litter follows their mother around, often in a single file in search of food. Armadillos can run quite rapidly when disturbed but are often startled or dazed by bright lights at night. They may swim by gulping air or they may cross small streams by walking across the bottom. These animals cannot survive freezing temperatures over extended periods and are active year-round.

Control options: Armadillos digging in turf are generally an indication that insects or worms are present in abundance. They will feed for several nights in an area if ants, grubs, or worms are easy to find. They usually move on after a night of feeding, unlike skunks or raccoons. See Skunks and Civet Cats.

Further Reading

Baker, J.R., Ed. 1982. *Insects and Other Pests Associated with Turf, Some Important, Common, and Potential Pests in the Southeastern United States.* North Carolina Agr. Ext. Ser. Ag. 268: 108pp.

Brandenburg, R.L. and M.B. Villani, Eds. 1995. *Handbook of Turfgrass Insect Pests.* Entomological Society of America, Lanham, MD. 140pp.

Leslie, A.R. and R.L. Metcalf, Eds. 1989. *Integrated Pest Management for Turfgrass and Ornamentals.* U.S. EPA, Office of Pesticide Programs, Washington, DC: U.S. Government Printing Office: 1989-625-030. 337pp.

Niemczyk, H.D. and B.G. Joyner, Eds. 1982. *Advances in Turfgrass Entomology.* ChemLawn Corporation, Columbus, OH. 149pp.

Niemczyk, H.D. and D.J. Shetlar. 2000. *Destructive Turf Insects,* 2nd edition. HDN Books, Wooster, OH. 148pp.

Potter, D.A. 1998. *Destructive Turfgrass Insects, Biology, Diagnosis, and Control.* Ann Arbor Press, Inc., Chelsea, MI. 344pp.

Shartleff, M., R. Randell, and T. Fermanian. 1987. *Controlling Turfgrass Pests.* Prentice-Hall, Inc., New York, NY. 305pp.

Shetlar, D.J., P.R. Heller, and P.D. Irish. 1990. *Turfgrass Insect and Mite Manual,* 3rd edition. The Pennsylvania Turfgrass Council, Bellefonte, PA. 67pp.

Streu, H.T. and R.T. Bangs, Eds. 1969. *Proceedings of Scotts Turfgrass Research Conference. Vol. 1—Entomology. May 19–20, 1969.* O.M. Scotts and Sons Co., Marysville, OH. 89pp.

Vittum, P.J., M.G. Villani, and H. Tashiro. 1999. *Turfgrass Insects of the United States and Canada,* 2nd edition. Cornell University Press, Ithaca, NY. 422pp.

Index